FOOL ME TWICE

FOOL ME TWICE

FIGHTING THE ASSAULT ON
SCIENCE IN AMERICA

SHAWN LAWRENCE OTTO

RODALE.

© 2011 by Shawn Lawrence Otto

Book design by Christopher Rhoads

Library of Congress Cataloging-in-Publication Data is on file with the publisher.

ISBN 978–1–60529–217–5

Distributed to the trade by Macmillan

2   4   6   8   10   9   7   5   3   1   hardcover

We inspire and enable people to improve their lives and the world around them.
www.rodalebooks.com

To Matthew, Sheril, Chris, Lawrence, Austin, Derek, and Darlene

And to children everywhere, who will live with our choices

CONTENTS

PART I

AMERICA'S SCIENCE PROBLEM

CHAPTER 1

LET'S HAVE A
SCIENCE DEBATE

Now we are met on the great battlefield of a new civil war, and the greatest part of that battlefield is the global warming battle. Now I know that in American-speak you have a word for global warming. Can someone tell me what it is?

[Crowd: "Bullshit!"]

Now look here. Barack Hussein Obama has just flown over in Marine One and landed on the White House lawn. He is now hiding behind the drapes in the Oval Office. He cannot hear you. Global warming is?

[Crowd: "Bullshit!"]

That's better. I think he heard that one.

—LORD CHRISTOPHER MONCKTON, 2010[1]

HOUSTON, WE HAVE A PROBLEM

"Whenever the people are well informed," Thomas Jefferson wrote, "they can be trusted with their own government."[2] This sentiment lies at the heart of the American-style democracy that has for more than two centuries inspired the world. Much like the "invisible hand"[3] that guides Adam Smith's economic marketplace,* so too does the invisible hand of

---

* See page 247 for Smith's definition of the "invisible hand."

3

the people's will guide the affairs of men and women through the democratic process.

But today the invisible hand seems confused and indecisive. Congress seems paralyzed, unable to act on many key issues that increasingly threaten the economic and environmental vitality of the nation and the planet. Ideology and rhetoric increasingly guide policy discussion, often bearing little relationship to factual reality. And the America we once knew seems divided and angry, defiantly embracing unreason.

At the same time, science is exploding all around us. There is a phase change going on in the scientific revolution: a shifting from one state to another, as from a solid to a liquid. There is a sudden, *quantitative* expansion of the number of scientists and engineers around the globe, coupled with a sudden *qualitative* expansion of their ability to collaborate with each other over the Internet. These two changes are dramatically speeding up the process of discovery and the convergence of knowledge across once-separate fields, a process Harvard entomologist Edward O. Wilson calls consilience.[4] We now have fields of inquiry where economics merges with environmental science, electrical engineering with neuroscience and physics, computer science with biology and genetics, and many more. This consilience is shedding new light on long-held assumptions about economics, the world we live in, and the nature of life itself.

Over the course of the next forty years, science is poised to create more knowledge than humans have created in all of recorded history. How that knowledge will impact life, and whether our society and our form of government will be able to withstand the rush, depends upon how we answer political questions we are currently struggling with. There is unfortunately no similar phase change going on in our politics, and therein lies the rub. Can we manage the new science revolution to our best advantage, or will we be its unwilling victims?

At the same time that we are being overwhelmed by progress, we are facing a host of legacy challenges from the science of the last century that are now being pushed to the forefront by global development. They include climate change and energy; ocean acidification and overfishing; biodiversity loss and habitat fragmentation; pandemics and biosecurity management; freshwater resource management; transportation; waste management; chemical, biological, nanotechnological, and genetic pollution; population control; national security; science education and economic competitiveness; and economic growth on a finite planet without degrading it.

Thanks to early science, we have prospered, but it has come at a cost. We now have a population that we cannot support without destroying our environment—and the developing world is just coming online using the same model of unsustainable development. We are 100 percent dependent on science to find ways to preserve our environment and support our population while maintaining health, wealth, freedom, and opportunity.

Between these two areas—the wild future that is rapidly emerging and the unresolved past we can no longer ignore—science is poised to whipsaw us in the coming decades like never before. This has the potential to produce even more intense social upheaval and political gridlock at the very time we can least afford them. We have within our grasp the potential to become the shining city upon a hill: vibrant, creative, and pristine, a beacon to all who love freedom. But we may also slip into decline, sliding day by day, almost without notice, into an environmental and economic morass, resentful and angry with each other over what we are losing, not realizing it is because of our own actions. For health and prosperity to continue, science can no longer be separated from policy making, religion, and economics. In this new age of connectivity, these four great houses of power must learn to work more closely together.

But can they?

THE SILENCE OF THE INVISIBLE HAND

Science provides us with increasingly clear pictures of how to solve our great challenges, but policy makers are increasingly unwilling to pursue many of the remedies science presents. Instead, they take one of two routes:

Deny the science, or pretend the problems don't exist. In fact, political and religious institutions the world over are experiencing a reactionary pull-back from science and reason that is threatening planetary stability and long-term viability at the very time we need science the most, and nowhere in the world is this pullback more pronounced than in the United States.

Can it be that science has simply advanced too far, or that our world has simply gotten too complex for democracy? In a world dominated by science that requires extensive education to practice or even fully grasp, can democracy still prosper, or will the invisible hand finally fall idle? Are Americans still well informed enough to be trusted with their own government?

Judging by Congress, the answer increasingly seems to be no. In an age when most major public policy challenges revolve around science, less than 2 percent of congresspersons have professional backgrounds in it. The membership of the 112th Congress, which ran from January 2011 to January 2013, included one physicist, one chemist, six engineers, and one microbiologist.[5]

In contrast, how many representatives and senators do you suppose have law degrees—and whom many suspect avoided college science classes like the plague? Two hundred twenty-two. It's little wonder we have more rhetoric than fact in our national policy making. Lawyers are trained to create a compelling narrative to win an argument, but as any trial lawyer will tell you, that argument uses facts selectively and only for the purposes of winning the argument, not for establishing the truth.

FEAR AND LOATHING ON THE CAMPAIGN TRAIL

The problem is even more pronounced in presidential politics. Consider climate change, arguably the greatest policy debate facing the planet. In late 2007, the League of Conservation Voters analyzed the questions asked of the then-candidates for president by five top prime-time TV journalists—CNN's Wolf Blitzer, ABC's George Stephanopoulos, MSNBC's

Tim Russert, Fox News's Chris Wallace, and CBS's Bob Schieffer. By January 25, 2008, these journalists had conducted 171 interviews with the candidates. Of the 2,975 questions they asked, how many might one suppose mentioned the words "climate change" or "global warming"? Six. To put that in perspective, three questions mentioned UFOs.[6] The same could be said of any one of several major policy topics surrounding science. Not a single candidate for president was talking about them. It was like they didn't even exist.[7] But in a world increasingly dominated by complex science, these questions—and not who's wearing what lapel pin—are what will determine our future.

In the fall of 2007, this strange avoidance of science in our national dialogue was also noticed by Charles Darwin's great-great-grandson Matthew Chapman, who wondered what could be going on.

A film director and the screenwriter for such films as 2003's *Runaway Jury*, Chapman picked up the phone and began calling friends to see if they'd noticed this, as well. He reached physicist Lawrence Krauss, science journalist Chris Mooney, marine biologist and science blogger Sheril Kirshenbaum, science philosopher Austin Dacey, and this author, and we all agreed that the silence on science issues was astounding. As a group, we founded what ultimately grew into the largest political initiative in the history of science, Science Debate 2008, an effort to get the candidates for president to debate the major science policy issues.[8]

We put up a Web site, placed op-ed pieces in national publications, and reached out to contacts and leading science bloggers. One of those bloggers, Darlene Cavalier of ScienceCheerleader.com, connected us with the US National Academies and became part of our core team. Within weeks, thirty-nine thousand people from across the political spectrum had signed on, including several Nobel laureates; prominent scientists; the presidents of most major American universities; the CEOs of several major corporations; and political movers ranging from John Podesta, President Bill Clinton's former chief of staff, on the left to former House Speaker Newt Gingrich on the right. We had obviously touched a nerve. Feeling affirmed, we then reached out to the campaigns.

They ignored us. This is, of course, a classic campaign tactic. You never, ever give energy to anything that you wish would go away. You simply do not engage, because the moment you do there is a story, the thing gets legs, and if you don't have your message already developed, you can lose control of your narrative. Many a campaign has been sunk by violating this cardinal rule. The question was why they wouldn't want to engage.

We called Ira Flatow and went on NPR's *Talk of the Nation: Science Friday*. The American Association for the Advancement of Science (AAAS), the US National Academies, and the nonprofit Council on Competitiveness signed on to our group as cosponsors. Our steering committee was cochaired by two congressmen, one from each party, and included prominent Democrats, Republicans, and even the leader of the Academy of Evangelical Scientists and Ethicists. Soon we represented more than 125 million people through our signatory organizations.

Presidential Candidates Talk Religion, Not Science

Still the candidates refused to even return phone calls and e-mails. So we decided to organize a presidential debate and turned to the national media outlets for help. This being science, we brought on as broadcast partners PBS's flagship science series *Nova* and the news program *Now on PBS*. David Brancaccio, *Now*'s host, would moderate. We set a date shortly before the all-important Pennsylvania primaries and teamed up with the venerable Franklin Institute in Center City Philadelphia to host. But despite the urging of advisers like EMILY's List founder Ellen Malcolm, who was involved with Senator Hillary Clinton's (D-NY) campaign, and Nobel laureate Harold Varmus, who was supporting Senator Barack Obama (D-IL), both of those candidates refused invitations to a debate that would center on the US economy and science and technology issues; Senator John McCain (R-AZ) ignored the invitation entirely. Instead, Clinton and Obama chose to debate religion at Messiah College's Harrisburg, Pennsylvania, campus—where, ironically, they answered questions about science.

How, it's reasonable to ask, has American political culture come to a point where science can be discussed only in a forum on religion? What little news coverage of this stunning development there was didn't seem to affect the campaigns at all.[9, 10] The candidates continued their policies of nonengagement.

By then it wasn't just scientists who thought this was odd. Science Debate and the nonprofit Research!America, which works to make medical research a higher national priority, commissioned a national poll and found that 85 percent of the American public thought that the candidates should debate the major science issues. Support was virtually identical among Democrats and Republicans. Religious people clearly were not put off by the idea. It seemed that the candidates alone were reticent. Something in American political culture had made science taboo.

With the window closing for a debate before the endorsing conventions, after which point the Democratic and Republican Party–controlled Commission on Presidential Debates would take over, we recruited marine scientist and former AAAS president Jane Lubchenco to help organize a debate in Oregon in August. Obama and McCain refused this one too, opting instead to hold yet another faith forum, this time at Saddleback Church in Lake Forest, California.

The scientific, academic, and high-tech communities were stunned. They saw it as a rebuke, and a travesty of American politics. Science has been responsible for roughly half of all US economic growth since World War II,[11] and it lies at the core of most major unsolved policy challenges. How could people who wanted to lead America avoid talking about science? Intel chairman Craig Barrett reached out to former Hewlett-Packard CEO Carly Fiorina on the McCain side to encourage his participation, and Varmus redoubled his efforts to convince Obama.

Meanwhile, our supporters had submitted more than 3,400 questions that they wanted us to ask the candidates. Working with several leading science organizations, we culled them into "The 14 Top Science Questions Facing America" and released them publicly (see page 321 for the original list). The candidates still ignored us.

ARE THERE REALLY TWO SIDES
TO EVERY STORY?

To be fair, the problem wasn't limited to the candidates. They were aided and abetted in their nonengagement by American journalism. Here was virtually the entire US science enterprise, the most powerful engine for economic growth and human advancement in world history, calling for the candidates to debate key science topics like climate change, health care, and energy, the candidates were dodging the questions, and almost no media outlets would cover it. Clearly, America's fourth estate had lost its moorings.

Some of this reticence may have been because many reporters and their editors (who often direct reporters' lines of questioning), like many politicians, were not required to take science classes in college, and few seem to understand science's importance to democracy. In a time dominated by science, this poses some serious concerns.

Another part of the problem may be that journalists, scientists, and politicians approach questions of fact with differing perspectives.

> **Journalism:** *There are always two sides to every story.*
>
> *Bob says 2 + 2 = 4. Mary says it is 6. The controversy rages.*
>
> **Science:** *Most times, one side is simply wrong.*
>
> *I can demonstrate using these apples that Bob is right.*
>
> **Politics:** *How about a compromise?*
>
> *New law: 2 + 2 = 5*

The problem with the modern journalistic approach is that it falls short of full reason. Journalistic techniques used to be employed as a means of fact-checking somewhat akin to scientific research. For example, reporters would get multiple sources to corroborate a story (which is an account of events in reality), or they wouldn't run the story. But today, some journalists don't even attempt to establish the reality or truth of a story. Instead, they go out of their way to present "both sides," as if this were admirable. The first casualty of this approach is journalism's own credibility, and its ability to speak truth to power. If one "side's" account is untrue and corroboration to determine which story is correct is not pursued, journalism becomes not just a meaningless relayer of information without regard to its reliability, but also weighted toward extreme views. And this inevitably fuels the extreme partisanship we see in public dialogue today.

It should be noted that many journalists argue that their job is not to attempt to establish truth, but simply to relay information. This laissez-faire view has come to dominate mainstream national political journalism. NBC News's chief White House correspondent during the George W. Bush administration, David Gregory, put it quite clearly in his defense of the White House press corps for not pushing President Bush on the lack of credible evidence of Saddam Hussein's weapons of mass destruction and the inconsistencies in his rationale for invasion before the United States entered Iraq: "I think there are a lot of critics who think that . . . if we did not stand up and say this is bogus, and you're a liar, and why are you doing this, that we didn't do our job," said Gregory. "I respectfully disagree. It's not our role."[12]

The fact that much of the mainstream press doesn't view it as their role to record the objective story of the nation puts science in the especially precarious position of being just one of many warring opinions, and it erodes our ability to make sound judgments. Similarly, the tendency of politicians to look for compromises on disputed questions of fact instead of basing decisions on an objective standard of knowledge (which now, according to some, is no longer the role of the press) is eroding the country's ability to solve its problems, leaving it mired in ongoing rhetorical battles. And by allowing the teaching of "alternative theories" to evolution or climate change in science classes, those same politicians damage our children's ability to learn the importance of facts and critical thinking and to compete in a science-driven global economy and a world increasingly impacted by climate issues. This dumbing down of the nation for ideological reasons is of course nothing new. It is precisely what happened in China during the Cultural Revolution. But it's never proven helpful to national prosperity.

THE SCIENCE GHETTO

In trying to understand why mainstream journalists weren't covering the important story of the presidential candidates' dodging of science debates, we began probing editors and news directors, and we learned something remarkable. There is a long-standing tradition in American newsrooms for editors and news directors to forbid political reporters from covering science issues and to rarely place science stories on publications' politics pages. Science has been relegated to its own specialized section.

This segregation may have worked well in the large, well-staffed newsrooms of the mid-twentieth century, when multiple reporters could cover the same story from different angles, but the policy has since become a problem. Science is now central to most of our major policy challenges. Segregating it has, in effect, taken a large portion of the political discussion off the table. No other major human endeavor is so ghettoized. The religion and ethics beat has long since crossed over into the politics pages, as has the business and economics beat. And of course military and foreign policy have been there all along. This segregation is also unique among Western democracies. In Europe it was the journalists themselves, joined together as the European Union of Science Journalists' Associations under the leadership of Hanns-J. Neubert and Wolfgang Goede, who organized and promoted a 2009 German parliamentary science debate patterned loosely on the US effort, an act since repeated by science journalists in other countries.

It wasn't that the candidates felt inhibited about opining on issues outside their expertise. They were waxing on about foreign policy and military affairs even though none were generals or diplomats and some had no military or foreign policy experience whatsoever. They offered economic plans even though they had little knowledge of economics. They talked about morality and religion even though they were not pastors or priests. But they refused to debate on science—the issue that is having some of the most profound effects on Americans' lives.

THE GREAT DUMBING DOWN

Partly, this ghettoization is due to economics. Facing increased competition from cable TV and a free model for news on the Internet, American news media have been trimming costs. Among the first things to go were investigative and science reporters. A Joan Shorenstein Center on the Press, Politics and Public Policy report from 2005—early in the science news crisis—showed that from 1989 to 2005, the number of major newspapers with weekly science sections fell from ninety-five to thirty-four. By 2005, only 7 percent of the approximately 2,400 members of the National Association of Science Writers had full-time positions at media outlets that reached the general public.[13]

In May of 2008 the *Washington Post* killed its famed science section. In November, NBC Universal fired the Weather Channel's entire Forecast Earth staff—during the NBC network's Green Week promotion—ending the station's only environmental series that focused on global warming.[14] In December, CNN fired its entire science, technology, and environment news unit.[15] In March of 2009, the *Boston Globe,* located in a worldwide capital of scientific research, closed its world-renowned science and health section.[16]

As a result, Americans find themselves in an absurd and dangerous position: In a time when the majority of the world's leading country's largest challenges revolve around science, few reporters are covering them from the scientific angle.

In Europe, by contrast, just the opposite is happening: Science coverage has increased. A 2008 analysis of prime-time news on selected European TV stations showed that there were 218 science-related stories (including science and technology, environment, and health) among the 2,676 news stories aired during the same week in the years 2003 and 2004, an eleven-fold *increase* since 1989.[17] And in the developing world, science reporting is "flourishing."[18]

CONGRESSIONAL ANTISCIENCE

Alarmed by this erosion of science in America's national dialogue, the two physicists in Congress agreed to become cochairs of the Science Debate effort. Representatives Vernon Ehlers (R-MI) and Rush Holt (D-NJ) both said that ignorance of basic scientific concepts in Congress had become a serious problem—and not just on the major science policy issues like climate change. It was of equal concern in "those countless issues, and it really is countless, that have scientific and technological components but the issues are not seen as science issues," said Holt.[19] Of the 435 representatives in the House, he said, "420 don't know much about science and choose not to." Ehlers described how he sometimes found himself "rushing to the floor" to protest cutting funding for scientific endeavors whose importance members did not understand. The member who once was pushing to cut the budget for game theory, for example, didn't realize that it dealt with economics, not sports. Another colleague who wanted to cut "ATM" research funding argued that it should be the responsibility of the banking industry, and Ehlers had to point out that the acronym stood for "asynchronous transfer mode," a fiber-optic data transfer protocol.*

The ignorance in Congress also extends to the willful variety, with powerful members vocally and publicly rejecting science almost as a way of proving their bona fides. In April of 2002, then–House majority whip Tom DeLay (R-TX) quoted the evangelical Christian authors of a 1999 book when he told a Texas church group, "Only Christianity offers a comprehensive worldview that covers all areas of life and thought, every aspect of creation."†[20] DeLay, who would soon become House majority leader, said he wanted to promote "a biblical worldview" in American politics.[21] "Our entire system is built on the Judeo-Christian ethic, but it

---

* Holt still holds office, but Ehlers retired in 2011. The 111th Congress's third physicist, Bill Foster (D-IL), was defeated in a 2010 reelection bid by the Republican backed by the Tea Party and former Alaska governor Sarah Palin, Randy Hultgren, who denies the existence of man-made global warming.

† The authors of *How Now Shall We Live?* are Charles Colson and Nancy Pearcey. Colson, called the "evil genius" by colleagues in the Nixon administration, pleaded guilty to obstruction of justice for his role in that administration's attempt to retaliate against Daniel Ellsberg, the leaker of the Pentagon Papers (The Watergate Story: Key Players: Charles Colson. WashingtonPost.com, n.d. www.washingtonpost.com/wp-srv/onpolitics/watergate/charles.html). Pearcey is an evangelical author, editor-at-large for her husband's evangelical mouthpiece the Pearcey Report, and chapter contributor to the creationist high school–level biology book *Of Pandas and People* (Pearcey Report. About page. n.d. www.pearceyreport.com/about.php).

fell apart when we started denying God," he had said in 2001.[22] After the 1999 Columbine school shootings, he gave a speech on the House floor in which he sarcastically suggested the tragedy "couldn't have been because our school systems teach our children that they are nothing but glorified apes who have evolutionized out of some primordial soup of mud, by teaching evolution as fact."[23] Ironically, DeLay's bachelor of science degree, from the University of Houston, is in biology.

DeLay's views ran throughout his caucus, particularly among the increasingly powerful baby boomers in the House. In March of 2002, current House Speaker John Boehner (R-OH) wrote to the Ohio State Board of Education to urge that the state's science curriculum content standards require teaching creationism, saying that "It's important that the implementation of these science standards not be used to censor debate on controversial issues in science, including Darwin's theory of evolution. . . . Students should be allowed to hear the scientific arguments on more than one side of a controversial topic. Censorship of opposing points of view retards true scholarship and prevents students from developing their critical thinking skills."[24]

This language was coded to boost creationists, who were promoting their "scientific" argument for "intelligent design" in the latest attack on the teaching of evolution. There is no *scientific* controversy about the theory of evolution. Boehner's letter was antiscience doublespeak.

PRESIDENTIAL ANTISCIENCE

But the problems weren't limited to Congress. The Bush White House itself had fallen victim to the antiscience fever, which is a large part of the reason why so many scientists signed on to Science Debate, and why Holt and Ehlers considered it an important issue to get behind. Bush appointed ideologues to key agency posts throughout the federal government and empowered them to hold back or alter scientific reports they did not agree with. This represented a marked change from the Republican party of just ten years prior. Consider the following quote by President George H. W. Bush, W.'s father:

> Science, like any field of endeavor, relies on freedom of inquiry; and one of the hallmarks of that freedom is objectivity. Now more than ever, on issues ranging from climate change to AIDS research to genetic engineering to food additives, government relies on the impartial perspective of science for guidance.[25]

Then consider this one by his son's White House spokesman, Scott McClellan, thirteen years later:

> This administration looks at the facts, and reviews the best available science based on what's right for the American people.[26]

The first approach uses knowledge to form effective public policy. The second looks first to a predetermined political agenda ("what's right for the American people") and seeks only those facts that support it. It is antiscience.

After President Bush's 2004 reelection, scientists noticed that the problem was becoming even worse. One example was Bush's appointment of George Deutsch, a twenty-four-year-old Texas A&M University dropout and Bush campaign intern, to a key position in NASA's public relations department. Deutsch set to work muzzling NASA's top climate scientist, James Hansen, once refusing to allow Hansen to interview with National Public Radio because it was "the most liberal" media outlet in the country[27] and telling a Web site contractor that the word "theory" had to be inserted after every mention of the Big Bang on NASA Web site presentations being prepared for middle school students. The Big Bang is "not proven fact; it is opinion," Deutsch told the contractor. "It is not NASA's place, nor should it be to make a declaration such as this about the existence of the universe that discounts intelligent design by a creator. . . . This is more than a science issue, it is a religious issue. And I would hate to think that young people would only be getting one-half of this debate from NASA." Deutsch later resigned after it was revealed that he had fabricated his own academic credentials.

Other Bush public relations appointees were muzzling scientists at other agencies, or altering scientific information in official agency reports to fit a preconceived ideological agenda,[28] angering and dismaying many in the American science enterprise who saw these tactics as antithetical to what America stands for.[29]

The problem became so widespread and so broadly reported that in early 2007, the House Oversight committee held hearings investigating the alleged distortions.[30] The Centers for Disease Control and Prevention was forced to discontinue a project called Programs That Work, which identified sex education programs found to be effective in scientific studies, none of which were abstinence-based.[31] On the National Cancer Institute's Web site, breast cancer was falsely linked to abortion.[32] The so-called morning-after pill (Plan B), an emergency contraceptive that prevents ovulation after unprotected sex[33] and may in rare circumstances prevent an already-fertilized egg from attaching to the uterus, was held back from FDA approval for over-the-counter sale even though scientists and physicians had judged it to be safe and determined that it was actually likely to *reduce* the number of abortions.[34] "Faith based" initiatives like abstinence-only sex education were federally funded, even when they were contradicted or shown ineffective by science. And business-cozy FDA administrators failed to remove the arthritis drug rofecoxib (Vioxx) from the market even after it became apparent that it was causing heart attacks,[35] resulting in more than fifty thousand American deaths[36]—nearly as many as the number of American soldiers who died in Vietnam—and made calls to a government whistle-blower-protection attorney and a leading medical journal in an attempt to discredit the scientist who brought the problem to light.[37]

While antiscience thinking overcame the executive branch and many Republican members in Congress, most Democrats seemed incapable of even articulating why it was wrong. Though Democrats as a party had not taken any *political* stances against science, the political left was suffering from a cultural retreat from it just the same, ranging from much of the alternative medicine and antivaccine movements to New Age beliefs to postmodernism in academia and primary and secondary school education. It's little wonder that scientists felt abandoned, and that there was a growing hunger for someone—anyone—to articulate the problems that Science Debate helped push forward. But the question was, with America in a populist retreat from reason, would the public still listen, and would the candidates see any advantage whatsoever in engaging on the topic of science?

THE UNBEARABLE RIGHTNESS OF BEING

By 2006, it was an open question. Antiscience views had become entrenched as mainstream political planks of the heavily religious-fundamentalist-populated and energy-company-financed Republican Party. The focus was on three main areas: denying the science of reproductive medicine, denying the science of climate change, and denying the science of evolution. Republican Sarah Palin provided a good example of the latter in a 2006 Alaska gubernatorial debate when she came out in favor of teaching creationism in science class, saying that science teachers should "Teach both. You know, don't be afraid of education. Healthy debate is so important, and it's so valuable in our schools."[38]

By 2008 it was becoming increasingly doubtful whether a Republican candidate for president could get the party's endorsement without taking a stridently antiscience position. Democrats, in turn, seemed terrified of offending fundamentalist swing voters, preferring instead to either "out-conservative" the conservatives or avoid the subjects of science and technology entirely. Scientists hoped that John McCain could somehow rebuff this trend. McCain had long crafted a reputation as a "maverick" and a data-driven "straight shooter." If anyone could stem the tide, he could, they thought. But they couldn't get even Obama to engage, much less McCain.

Finally, on the eve of the Democratic National Convention, Obama relented and told the Science Debate team that while he wouldn't participate in a televised forum, he would participate in an online "debate." He formed a science advisory team headed by Varmus to help him answer the questions. Scientists were jubilant—finally, someone was listening.

Days later, McCain agreed as well, and the press, given a classic conflict frame, was finally interested. The Science Debate story, and the candidates' answers to "The 14 Top Science Questions Facing America" made nearly a billion media impressions. The public finally started seeing discussions of the candidates' positions on climate change, energy, health care, space, the environment, and the research drivers of economic competitiveness.*

Perhaps, scientists dared to hope, the dark days of American unreason had passed.

THE EMPIRE STRIKES BACK

But almost immediately there was a backlash. Within days Palin attacked science "earmarks" in her first major policy speech as McCain's running mate. "Sometimes these dollars, they go to projects having little or nothing to do with the public good," she said. "Things like fruit fly research in Paris, France. I kid you not!"[39]

Palin was promoting her signature issue of special-needs children, talking about how the money wasted on ridiculous science earmarks could fully fund the Individuals with Disabilities Education Act. Her nephew has autism and, ironically, there is a fruit-fly center for research into autism at the University of North Carolina. The earmark in question was secured by Representative Mike Thompson (D-CA) for research into the olive fruit fly, a widespread problem in Europe that is threatening the rapidly growing US olive industry—I kid you not.

Palin followed this with "going rogue" on McCain by appearing without permission on Christian conservative James Dobson's radio show, where she contradicted McCain's position on stem cell research as stated to Science Debate. Concerned that he might be losing Bible Belt conservatives, McCain swung to the right. During the second presidential

* Some of the coverage is archived at the Science Debate 2008 news page at www.sciencedebate.org/news08.html.

debate he criticized science spending in particular, singling out such congressional science earmarks as $3 million "to study the DNA of bears in Montana. I don't know if that was a criminal issue or a paternal issue, but the fact is that it was $3 million of our taxpayers' money. And it has got to be brought under control."[40]

Grizzly bears were first listed as a threatened species under the Endangered Species Act in 1975; the act requires the federal government to get an accurate count of their numbers—a dangerous, difficult, and expensive endeavor in the rugged Rocky Mountain terrain. The best method was aerial surveys until the US Geological Survey (USGS) hit upon the idea of putting up scent stations. The bears love to rub on them. The USGS collects the fur they leave, does a DNA study, and gets a quick count of the number of individual grizzly bears—a far cheaper and more accurate way of fulfilling the federal mandate, which should have pleased any conservative.[41]

McCain, who values his reputation as a straight shooter, can't have been proud of that moment, or of the subsequent one, when he attacked Obama for voting for another "wasteful earmark" of "$3 million for an overhead projector at a planetarium in Chicago, Illinois. My friends, do we need to spend that kind of money?"

The ridiculously expensive "overhead projector" was actually *the planetarium projector*—the machine that displays the night sky on the theater dome—at the Adler Planetarium in Chicago, the first planetarium in the Western Hemisphere. The projector, which weighed more than 5,000 pounds, hadn't been replaced in 40 years (during which time our knowledge of the heavens has expanded considerably), and parts were no longer available.[42] "Science literacy is an urgent issue in the United States," the planetarium said in a statement the next day. "To remain competitive and ensure national security, it is vital that we educate and inspire the next generation of explorers to pursue careers in science, technology, engineering and math."[43]

The subsequent election of Barack Obama signaled that perhaps the tide was turning after all. Finally, a president was going into office armed with a well-formed science policy and a sense of how it fit strategically with the rest of his agenda. He featured science in his inaugural address and prioritized it in his budgets. He chose early Science Debate supporters John Holdren, Steven Chu, Jane Lubchenco, Harold Varmus, and Marcia McNutt to lead major agencies and departments. Seeing this new focus, scientists thought they had finally won. But they had overlooked a key fact, and unfortunately things were about to get far, far worse.

CHAPTER 2

IS SCIENCE POLITICAL?

[T]here is nothing which can better deserve your patronage than the promotion of Science and Literature. Knowledge is in every country the surest basis of public happiness. In one in which the measures of Government receive their impression so immediately from the sense of the Community as in ours it is proportionably essential.

—GEORGE WASHINGTON[1]

HOW TO RUFFLE A SCIENTIST'S FEATHERS

When speaking about science to scientists, there is one thing that can be said that will almost always raise their indignation, and that is that science is inherently political and that the practice of science is a political act. Science, they will respond, has *nothing* to do with politics. But is that true?

Let's consider the relationship between knowledge and power. "Knowledge and power go hand in hand," said Francis Bacon, "so that the way to increase in power is to increase in knowledge."[2]

At its core, science is a reliable method for creating knowledge, and thus power. Because science pushes the boundaries of knowledge, it pushes us to constantly refine our ethics and morality, and that is always political. But beyond that, science constantly disrupts hierarchical power structures and vested interests in a long drive to give knowledge, and thus power, to the individual, and that process is also political.

The politics of science is nothing new. Galileo, for example, committed a political act in 1610 when he simply wrote about his observations through a telescope. Jupiter had moons and Venus had phases, he wrote, which proved that Copernicus had been right in 1543: Earth revolved around the sun, not the other way around, as contemporary opinion—and the Roman Catholic Church—held. These were simple observations, there for anyone who wanted to look through Galileo's telescope to see.

But the simple statement of an observable fact is a political act that either supports or challenges the current power structure. Every time a scientist makes a factual assertion—Earth goes around the sun, there is such a thing as evolution, humans are causing climate change—it either supports or challenges somebody's vested interests.

Consider Galileo's 1633 indictment by the Roman Catholic Church, which was at the time the seat of world political power:

1. The proposition that the sun is in the center of the world and immovable from its place is absurd, philosophically false, and formally heretical; because it is expressly contrary to Holy Scriptures.

2. The proposition that the earth is not the center of the world, nor immovable, but that it moves, and also with a diurnal action, is also absurd, philosophically false, and, theologically considered, at least erroneous in faith.

Therefore . . . , invoking the most holy name of our Lord Jesus Christ and of His Most Glorious Mother Mary, We pronounce this Our final sentence: We pronounce, judge, and declare, that you, the said Galileo . . . have rendered yourself vehemently suspected by this Holy Office of heresy, that is, of having believed and held the doctrine (which is false and contrary to the Holy and Divine Scriptures) that the sun is the center of the world, and that it does not move from east to west, and that the earth does move, and is not the center of the world; also, that an opinion can be held and supported as probable, after it has been declared and finally decreed contrary to the Holy Scripture.[3]

Why did the church go to such absurd lengths to deal with Galileo? For the same reasons we fight political battles over issues like climate change today: Because facts and observations are inherently powerful, and that power means they are political.

Failing to acknowledge this leaves both science and America vulnerable to attack by antiscience thinking—thinking that has come to dominate American politics and much of its news media coverage and educational curricula in the early twenty-first century. Thinking that has steered American politics off course and away from the vision held by the country's founders.

Wishing to sidestep the painful moral and ethical parsing that their discoveries sometimes compel, many scientists today see their role to be the creation of knowledge and believe they should leave the moral, ethical, and political implications to others to sort out. But the practice of science itself cannot possibly be apolitical, because it takes nothing on faith. The very essence of the scientific process is to question long-held assumptions about the nature of the universe, to dream up experiments that test those questions, and, based on the observations, to incrementally build knowledge that is independent of our beliefs and assumptions. A scientifically testable claim is utterly transparent and can be shown to be either most probably true or false, whether the claim is made by a king or a president, a pope, a congressperson, or a common citizen. Because of this, science is inherently antiauthoritarian, and a great equalizer of political power.

WHO DEFINES YOUR REALITY?

The scientific revolution has proven to be more beneficial to humanity than anything previously developed. By building knowledge we have been able to double our life spans* and boost the productivity of our farms by thirty-five times.† With careful observation and recording we have been able to give children to those who were "barren" and the fertile the freedom to decide when—and whether—to reproduce. Our science now influences every aspect of life, and it has freed us from a life that was, according to Thomas Hobbes, "a war . . . of every man against every man . . . solitary, poor, nasty, brutish and short."[4]

In such remnants of thinking from the Middle Ages, economics was a zero-sum game: "Without a common power to keep them all in awe," men in Hobbes's time fell into war. There was finite wealth and opportunity, and to get ahead one had to take some away from another. In its capacity to create knowledge, science had the tools to break that zero-sum economic model and generate wealth, health, freedom, nobility, and power beyond Hobbes's wildest dreams. It has given us tremendous insights into our place in the cosmos, into the inner workings of our own bodies, and into our capacity as human beings to exercise our highest aspirations of love, hope, creativity, discovery, compassion, courage, wonder, and charity.

Each step forward has come at the price of a political battle. As we continue to refine our knowledge of the way nature really is, independent of our beliefs, perceptions, and wishes for it, we must also refine our ethics and morality and assume more responsibility as humans. Inevitably this is uncomfortable, because the process compels us to give up, alter, or somehow intellectually sequester many comforting notions, notions that are often profoundly powerful because they are our most deeply rooted and awestruck explanations about the wonders of creation, the specialness of our identities, our history, and the possibility that our spirits may somehow live on after death.

---

* Average life expectancy for a white American born in 1850 was 39.4 years. For those born in 2006, it is 77.7 years (Infoplease.com. Life Expectancy by Age, 1850–2004. n.d. www.infoplease.com/ipa/A0005140.html and Xu, J., et al. Deaths: Final Data for 2007. *National Vital Statistics Reports* 58(19), May 20, 2010. www.cdc.gov/NCHS/data/nvsr58/nvsr58_19.pdf).

† In 1820, 70 percent of the US population lived on farms. By 1990, that number had fallen to 2 percent, meaning that productivity is thirty-five times greater (USDA Economic Research Service. Farming's Role in the Rural Economy. *Agricultural Outlook,* June–July 2000, pp. 19–22).

THE SCIENTIFIC METHOD

How do we create knowledge? There is no one "scientific method"; rather, there is a collection of strategies that have proven effective in answering our questions about how things in nature really work. How do plants grow? What is stuff made of? How do viruses work? Why are montane voles promiscuous while prairie voles are loyal lifelong mates? The process usually begins with a question about something, and that suggests a strategy for making and recording observations and measurements. If we want to learn how plants grow, for example, we begin by looking at plants, not rocks. These initial recorded observations suggest a hypothesis—a possible explanation for the observations that partially or fully answers the initial question. The hypothesis must make a risky prediction, one that, if true, will confirm our conclusion or, if false, will destroy it. If there's no possible way to prove the hypothesis is false, then we aren't really doing science. Saying plants grow because God wills it is a statement of faith rather than a statement of science because (a) it's not limited to the natural world, and (b) it cannot be disproved.

After we set out our hypothesis, we design and conduct experiments that test the hypothesis and try to disprove it. If we can't disprove it, we conclude that it seems to be true and write a paper detailing what we did and concluded. We send it to a professional journal, which sends it to others with knowledge of the field (peer reviewers) to see if they can tear any holes in it. Was our method sound? Did we make any mistakes? Did we control all the possible influences on the outcome? Are there other explanations we didn't think of? Was our math right?

If these critical reviewers feel our work is sound enough to stake their reputations on, they recommend our paper for publication. Others who read it may then set out to disprove it. If they can, their stars rise and ours falls a bit. But if they confirm what we found, the conclusion becomes a little more reliable.

There are lots of entry points into this process and other strategies besides these, but the main goal is to describe something in the real world that we did not understand before in such a way that it is possible to disprove it, but even so nobody has yet been able to. In this way we achieve a high degree of reliability that it is a true statement about reality. In other words, that it is knowledge. The method is fallible, since our senses and our logical processes are easily influenced by our assumptions and wishes and so they often mislead us, but over time the method tends to catch those errors and correct them. Thus, bit by careful, painstaking bit, we build upon what we know, as distinct from our beliefs and our opinions.

HOW OLD IS EARTH?

One example of knowledge, as opposed to belief or opinion, is the age of Earth itself. Geological measurements show, over and over, no matter who does the measuring, that Earth is about 4.54 billion years old.[5] This is something you can learn to measure for yourself. It's called radiometric dating. Radioactive uranium isotopes decay at known rates into stable (not radioactive) lead isotopes, and radioactive potassium isotopes decay at known rates into stable argon isotopes. By using a mass spectrometer you can buy on eBay for about $1,000, you can count how many atoms of a particular uranium isotope are left in a rock and how many atoms of its "daughter" isotope of lead there are, or how many atoms of potassium are left and how many have decayed into argon. Doing some junior high school math lets you then figure out how old that rock is. Because the whole solar system is thought to have formed at the same time, we can also look at the ages of meteorites and of rocks astronauts brought back from the Moon to get a pretty complete picture of Earth's history, and its age consistently comes up to be about 4.54 billion years old.[6] This isn't something we believe, it's something we measure, like the distance between Minneapolis and Dallas.

But these measurements conflict with translations of ancient statements in the Bible that, if taken literally, date Earth at between six thousand and seven thousand years old. These statements were made before we knew how to measure such things. Who made these statements? Fallible man, or all-knowing God? How do we account for that conflict? If we choose the careful, repeatable science of observation and measurement over the emotional comfort and reassurance of the creation stories of old, must we reject the rest of religion? Or does it still have value to guide and comfort a moral life?

WHEN DOES LIFE *REALLY* BEGIN?

Careful and reproducible observations and measurements in the biosciences have similarly forced us to repeatedly refine our traditional ideas about what life itself is and when it begins. Is a human being first a life when it emerges from the birth canal? Does it have any legal rights as a person before then? Or is it a life at the stage of development where it is able to survive independently outside of the womb even if it is removed from there early, as can happen naturally with premature birth or with a Caesarean section? But wait! Perhaps it is really a life when a fertilized egg first implants in the uterine lining, which, based on careful observations, is the medical definition of when a pregnancy begins. A woman cannot be said to be medically pregnant until her body begins the chemical and biological changes that accompany a symbiotic hosting of the embryo, can she? If it does not, the egg, even if fertilized, is simply flushed. Now here we are getting into a tricky area, because many religious conservatives say, "No, it is a life when egg and sperm meet," whether or not the fertilized egg ever implants. But then, a scientist would ask, is it *still* a life at that moment, even if you know from careful observation that one-third to one-half of all fertilized eggs never implant?[7] And of course that brings up a secondary point: What are fertilized eggs that never implant? How do we define them? As miscarriages? Abortions? Nonpregnancies? Something else? What implications might that definition have for the use of birth control pills that inhibit implantation? Is that abortion, murder, or pregnancy prevention?

As our careful observations of life continue, so does our power both to assist and prevent pregnancy. But as our skills improve, new, more troubling questions form. How about if we remove the uterus from the process entirely? Is it a life when sperm and egg are joined in a test tube at a fertility clinic and allowed to divide into a group of, say, sixteen cells that are then frozen for future implantation in a woman desperate to have children? Can the woman be said to be "pregnant" as long as this microscopic clump of frozen cells exists? What, if any, rights should these frozen cells possess? And is a child conceived in this way, a "test-tube baby," as we once called them, without a soul, as was suggested by some religious conservatives in the 1970s? Once born, are the joy they bring and the contributions they make less valuable? If we make a special exception for them, by agreeing that in vitro fertilization is not interfering with God's plan or by acknowledging that they do appear to have souls, why?

While we're pondering these quandaries, our observations lead us to yet another new understanding. We don't need sperm to fertilize an egg; we can do it with the nucleus of another cell from the same being. We try this, and sure enough, we can create many identical genetic copies of a sheep or mouse. We call them clones. But then we have to ask: Is it a life if it is just an ovum that has had its nucleus removed and replaced by the nucleus of another cell and then been chemically or electrically shocked to induce the natural process of cell division—without fertilization by sperm? If egg and sperm have never met, is it a life? Or is that creature—possibly, one day, a human—damned or soulless like "test tube babies" were once said to be?

Observations tell us that these beings produced in nontraditional ways seem to be the same as any other creatures. We have to ask, then, is *every* one that remains of the roughly 1½ million eggs a woman has in her ovaries at birth a life with rights? When does life begin? Is it true, as the comedy troupe Monty Python sang in *The Meaning of Life*, that "Every sperm is sacred"?

Is it a life if we transform adult skin cells into stem cells and those into sperm and egg and then fertilize one with the other? And is that a clone or something else? Is it a life if we design its genome on a computer, as scientists at the J. Craig Venter Institute have done,[8] buy a DNA synthesizer on eBay for $8,000 or so, use it to make fragments of the genome we designed, chemically stitch them together, inject the complete genome into a cell with an empty nucleus, and shock it into replicating? Here, we have made a living, reproducing thing starting with a computer design and a few common chemicals. What does that mean for our ideas about life? Is it wrong to be doing this? To be asking these questions? Applying these observations? Gaining these powers?

What is life? Is life an unbroken chain of genetic code, running down through the generations, endlessly recombining in new forms? Is it *software*? When does it become an individual with rights? Where do we draw the legal line? The moral line? Can we draw a line at all? Is that the right way to be thinking about it?

In each of the above cases, new knowledge was gained by applying the scientific method of making careful observations and recording the data, then testing and drawing conclusions based on the results instead of on assumptions or beliefs, and then publishing those for others to review and attempt to disprove if they can. The knowledge gained through this incredible process gives us new power over the physical world, the power to assist or prevent pregnancy, but it also forces us to reevaluate intuitive assumptions and to refine and in some cases redefine the meanings of words and values we thought we understood.

This power and these new definitions have moral, ethical, and legal implications for how we conduct our lives. As our knowledge becomes more refined and precise, so too must our social contract, and this process is inherently and continuously disruptive to moral, ethical, financial, and political authority. Science itself is inherently political, and inherently antiauthoritarian.

AUTHORITARIAN POLITICS

Because this is the case, it's reasonable to ask how science fits into political thought. Science writer Timothy Ferris reminds us that in politics there are not just two forces, the progressive left (encouraging change) and the conservative right (seeking constancy). In fact, there are four.[9] Imagined on a vertical axis, there are also the authoritarian (totalitarian, closed, and controlling, at the bottom of the axis) and the antiauthoritarian (liberal, open, and freedom loving, at the top), which one can argue have actually played much more fundamental roles in human history. Politics, then, can be more accurately thought of as a box with four quadrants rather than as a linear continuum from left to right.

Any one of the infinite gradations of political thought can be placed on the plane around these axes.

When looked at in historical perspective, it's clear that while science and republican democracy are antiauthoritarian systems of knowledge and of governance, respectively, they are neither progressive nor conservative. Both communism on the left and fascism on the right are authoritarian and opposed to the freedom of inquiry and expression that characterize science and democracy, just as fundamentalist and authoritarian religions do.

Alternatively, left-leaning progressives and right-leaning conservatives in America can find common cause in the antiauthoritarian principles of freedom of inquiry and expression, universal education, and individual human rights that go hand in hand with the liberal (meaning "free") thinking that informs science and democracy. The life of conservative writer David Horowitz, who was a part of the radical "new left" movement in the late 1960s but is now on the radical right, provides an example of how one can move 180 degrees ideologically from left to right, but still maintain the same general level of liberal antiauthoritarianism vertically.

TOP-WING POLITICS

The present-day Republican Party provides an example of how politics actually breaks down into these four quadrants, rather than in a simple left–right continuum. Its constituents range from antigovernment anarchists on the top to totalitarian fundamentalists on the bottom—but both groups are right wing. There is currently a power struggle going on between the antiauthoritarian Tea Party top wing and the authoritarian big-government "family values" bottom wing.

The George W. Bush administration did a masterful job of uniting these disparate ends of the vertical spectrum by first rebranding progressives as "liberals," thus silencing liberal conservatives, and then by using profreedom, antitotalitarian, small-government rhetoric on the one hand and antiscience, profundamentalist rhetoric on the other, directing half of the argument at each wing.

In actuality, Bush became the opposite of liberal: He increased government spending, limited individual freedoms, asserted totalitarian authority by suspending habeas corpus and detaining American citizens indefinitely, and eroded the barriers that keep state and religion safe from one another. He moved from a full-right conservative position to a full-authoritarian, bottom-wing position, becoming an authoritarian who grew the size and control of government.

This move was made possible by the reaction to 9/11, which drove the country toward fascism, a lower-right position. Bush was a very effective leader in that he moved a large portion of the public to follow his direction, but his presidency became the most illiberal and antidemocratic in American history, as well as, not surprisingly, the most antiscience. This was aided by two generations of scientists having gone silent in the national dialogue and the voices of authoritarians like conservative talk show host Rush Limbaugh and his self-described "dittoheads" holding sway.

This left a void in the top right quadrant for the Tea Party to fill. The Tea Party is not as concerned with the epithet "liberal" because it is in some ways a liberal movement in that it purports to place a high value on freedom. But the Tea Party risks being co-opted by bottom-wing authoritarians unless it extends its top-wing embrace of freedom to include science, diversity, reason, and tolerance, all of which it currently seems antagonistic to in varying degrees.

In the wake of the Bush presidency, the already-clear rift between the two dominant perspectives on the right—the small-government libertarians/anarchists and the theocratic fundamentalists—began to grow even wider. Far more than the conservative or liberal philosophy, it is who wins the argument between authoritarians, who value top-down control and conformity, and antiauthoritarians, who value bottom-up freedom and tolerance, that will drive the success or failure of the United States on the major issues of the twenty-first century. This argument has little to do with current party politics, and everything to do with science.

Like it or not, the world is now fully dependent on science. Without it, we could not sustain our population or our environment. Science is driving the entire conversation, and the country needs a political framework that will allow it to adapt to these challenges successfully. This requires making some adjustments in our understanding of the role and purpose of nations and legal and regulatory systems. Thus, the independent lover of science and the future will tend to be fiscally conservative and socially liberal, seeking a live-and-let-live ethos in the name of freedom and natural law, using government regulation to optimize freedom and level playing fields; the neoconservative fundamentalist will tend to be fiscally liberal and socially conservative, seeking big-government control of personal behavior in the name of morality and security while opposing regulation that increases freedom or levels playing fields.

DEMOCRACY: AN ENDANGERED SPECIES?

The challenge to authority that science presents is one of many reasons why it has flourished in free, democratic societies. It is not a coincidence that the ongoing scientific revolution has been led in significant part by the United States and other free, democratic societies. But it is also partly why, since the late twentieth century, the political climate has increasingly hampered US policy makers in dealing with so many critical science policy issues, and why the United States may soon cede both its leadership in scientific research and development and the economic and social benefits that leadership provides.

Thomas Jefferson's fundamental notion that, *when well informed,* people can be trusted with their own government lies at the center of democracy. Without a well-informed voter, the very exercise of democracy becomes removed from the problems it is charged with solving. The more complex the world becomes, the more challenging it is for democracy to function, because it places an increased burden of education and information upon the people—and in the twenty-first century, that includes science education and science reporting. Without the mooring provided by the well-informed opinion of the people, governments may become paralyzed or, worse, corrupted by powerful interests seeking to oppress and enslave.

For this reason and others, Jefferson was a staunch advocate of free public education and freedom of the press, the primary purposes of which were to ensure an educated and well-informed people. In 1787 he wrote to James Madison:

> And say, finally, whether peace is best preserved by giving energy to the government, or information to the people. This last is the most certain, and the most legitimate engine of government. Educate and inform the whole mass of the people. Enable them to see that it is their interest to preserve peace and order, and they will preserve them. And it requires no very high degree of education to convince them of this. They are the only sure reliance for the preservation of our liberty.[10]

But what do we do when the level of complexity actually *does* require a "very high degree of education"? Can democracy still function effectively?

Will American-style democracy, run by the people, be able to compete in the complex, science-driven global economy with nations like China, which is run by highly educated engineers and scientists who have moved into government leadership roles?* It's a critically important question, both for America and for our ideas about liberty.

* Of the nine members of the 2011 Politburo Standing Committee, China's top Communist Party leadership, eight are engineers and one is an economist (Wang, T. China's Future Leaders. Forbes.com, May 17, 2009. www.forbes.com/2009/05/17/china-leaders-stars-leadership-rising-stars.html).

PART II

YESTERDAY'S SCIENCE POLITICS

CHAPTER 3

RELIGION, MEET SCIENCE

The value of science to a republican people, the security it gives to liberty by enlightening the minds of its citizens, the protection it affords against foreign power, the virtue it inculcates, the just emulation of the distinction it confers on nations foremost in it; in short, its identification with power, morals, order and happiness (which merits to it premiums of encouragement rather than repressive taxes), are topics, which your petitioners do not permit themselves to urge on the wisdom of Congress, before whose minds these considerations are always present, and bearing with their just weight.

—THOMAS JEFFERSON[1]

IN THE BEGINNING

To understand the absurd situation America finds itself in—its major challenges largely revolving around science and yet few in Congress understanding these issues and few media outlets reporting on them—we have to understand what makes the relationship what it is. Has America always been a nation of science? Was it founded as a Christian nation? Has there always been a conflict between religion and science? What, exactly, are the relationships among science, politics, freedom, and religion in America? Why did science get so advanced here? And most important, why do we keep having conflicts over it? Can democracy—and Earth's ecosystem—survive them?

Contrary to what many fundamentalist politicians and televangelists have claimed, America was not founded as a Christian nation. The land was initially settled by Puritans, true, but the country was founded on the principles of science. In fact, 150 years after the Pilgrims' arrival, the founding fathers took great pains to expunge religious thinking from the writings that laid the legal and philosophical foundations for the country they wished to form, beginning with the Declaration of Independence.

They carefully carved out a new, secular form of government based on limited powers for the authorities and reservation of most freedoms for the people, including freedom of inquiry and expression and freedom of and from religion. The founding documents guaranteed protection of these freedoms and of the people's right to experiment with and modify their government—by using an eighteenth-century version of crowdsourcing—and this was something entirely new. It was not coincidental with the scientific revolution, but was rather, as Ferris and others have indicated, the natural outgrowth of it.[2] The liberties these founding principles afforded have in turn produced the highest standard of living, the greatest scientific and technological advances, and the greatest power in the history of the planet.

GOD'S NATURAL LAW IS REASON

Among the first to seize upon the discovery of the New World were the English merchants of Jamestown, Virginia. When their party of 104 men and boys landed in 1607, they were looking for gold, but instead they wound up growing and selling something even more profitable: tobacco.

They were followed some thirteen years later by those seeking not fortune, but freedom of religion. The Puritans began to settle Massachusetts as early as 1620, forming fundamentalist enclaves in niches carved out of the "savage" new continent. They disliked the authority of the Church of England and believed in progress and innovation, that the Bible was God's true law, and that it provided a plan for living.[3]

Puritanism wasn't just a theology, it was a whole set of ideas that included taking an antiauthoritarian, experimental, empirical approach to discovering the *natural laws* by which God's creation abided. In exercising his will, God did not contradict reason. Rather, he revealed himself to humans through two books: the Book of Revelation, made accessible by faith, and the Book of Nature, made accessible by observation and reason. Science was the "handmaiden" to theology,[4] assisting in the study of "the vast library of creation"[5] as a vehicle to religious understanding.

This idea that God does not contradict reason lies at the foundation of English common law, as first set forth in Christopher St. Germain's 1518 treatise *The Doctor and Student*, which relates a hypothetical conversation between a doctor of divinity and a student of the laws of England and established common law's moral basis. St. Germain was a Protestant polemicist during the reign of King Henry VIII, a time when a great battle was raging between the Catholic Church, which was the highest authority in all matters, and the antiauthoritarian, questioning Protestants, who promoted do-it-yourself study of the Bible and nature. In fact, *The Doctor and Student* was published just a year after Martin Luther posted his Ninety-Five Theses on a church door. As the reform movement swept through Europe, monks were thrown out of monasteries and told to marry nuns, as Luther did in 1525 when he married an ex-nun. Adherents to Luther's philosophy destroyed and looted the Catholic churches of the bones of their saints and other relics and jewels, condemning the objects as false idols. The truth was to be found in the Bible and in direct experience, they believed, not in the pronouncements of the pope in Rome.

This thinking required a reexamination of the world and the devising of a new order based not upon the authority of the church, but upon reason. The question at hand for St. Germain amid this upheaval was "what be the very grounds of the law of England."[6] He offered first that the law of God underlies reason:

> The law of God is a certain law given by revelation to a reasonable creature, shewing him the will of God, willing that creatures reasonable be bound to do a thing, or not to do it, for obtaining of the felicity eternal.

He then declared that reason and natural law are synonymous:

> As when any thing is grounded upon the law of nature, they say,
> that reason will that such a thing be done; and if it be prohibited by
> the law of nature, they say it is against reason, or that reason will
> not suffer that to be done.

Therefore, nature was knowable, and God's will could be understood by studying nature to discern its laws. This was a powerful idea that put man into an immediate relationship with God and nature, without an intermediary authority figure. To the Puritans, then, there were no conflicts in the ideas of religion, law, reason, and science. All were varying examinations of natural law.

The idea of natural law developed over the next ninety years as Protestantism flourished in England. In 1608, the great English jurist Edward Coke sought to more clearly articulate it in *Calvin's Case*. Coke was a Puritan sympathizer[7] who spent his career working to protect individual liberty and make sure the monarchy's arbitrary authority was circumscribed by the rule of law, an idea the Puritans very much favored.

His report, *Calvin's Case,* became a foundational document in English law. In it, Coke wrote,

> The law of nature is that which God at the time of creation of the nature
> of man infused into his heart, for his preservation and direction.[8]

And in an often-quoted section of his 1628 *Institutes of the Lawes of England,* Coke broadened this idea in an important way that for the first time turned to the crowdsourcing model the American founders would eventually adopt. Seeking to limit the caprice of the "Royal Prerogative" by which the king claimed the authority to do whatever he wanted because he was king, Coke argued that natural law motivated individual men, but "an infinite number of grave and learned men" working over "successions of ages" could refine and perfect the laws derived from this initial natural moral basis. Coke called this aggregation "artificiall [*sic*] reason," which he defined as perfect reason, which commands those things that are proper and necessary and which prohibits contrary things.[9]

Thus, law had a basis in physical reality through the hard-wired biological instincts of humans that God had infused into their hearts at the time of their creation, but its full force and power came from socially aggregating those insights. No one man's authority, even the king's, stood above it.

This science-friendly Protestant perspective that one could establish law and understand God's will by studying nature and then over time aggregating and refining a body of knowledge that bound even the king stood in stark contrast to the position taken by the Roman Catholic Church a few years later, in 1633, when church authorities denied the validity of astronomical science and indicted Galileo for heresy for simply observing nature.

The poet John Milton, author of *Paradise Lost*, is said to have visited Galileo at Arcetri, the hilly area to the South of Florence where he was under house arrest. Milton said that he had been counted by the "learned men" he met there "happy to be born in such a place of philosophic freedom as they supposed England was." The "inquisition tyrannies" of the church's crackdown in response to Galileo had "dampened the glory of Italian wits; that nothing had been there written, now these many years, but flattery and fustian."[10] By the end of the seventeenth century, as Anglican clergy in London were preaching Newton's science, Italian scientists were standing trial in Naples for stating "that there had been men before Adam composed of atoms equal to those of other animals."[11]

The DNA of Western Thought

Each arm of this double helix of Western Christianity—Roman Catholicism and the emerging Protestantism—embodied the two distinctly different worldviews of the authoritarian and the antiauthoritarian: that rules and methods were either proscribed from on high or built up by individuals in consensus.

These two views had always been present, but they were greatly amplified in 1517, when Martin Luther posted his Ninety-Five Theses* challenging church authorities to debate principles that seemed defensible only by virtue of the church's authority over its subjects. In Luther's view, the church had become corrupt, telling people they could buy their way into heaven by purchasing "indulgences," the proceeds of which the church used to finance building St. Peter's Basilica in Rome. "Why does the pope, whose wealth today is greater than the wealth of the richest Crassus [a legendarily greedy and unethical first-century Roman businessman]," Luther asked, "build the basilica of St. Peter with the money of poor believers rather than with his own money?"[12] Luther's theses split the church between those who clung to authority and tradition and those who believed in man's individual connection to God. Protestantism, with its streak of populist antiauthoritarianism, was born.

Luther's grand movement, and the very idea that knowledge could be accessible by individuals without resorting to a conduit provided by an intervening authority, was made possible by a 1451 technological invention: the printing press. For the first time, books could be mass produced, permitting knowledge and its attendant power to be spread widely. Luther used this new tool to spread power among the people with his 1534 translation of the Bible from Greek, Hebrew, and Latin into common German. More than one hundred thousand copies of the Luther Bible were sold within forty years of its publication (an unfathomable number for the time), and millions heard its message. People could suddenly study the Bible and come to their own conclusions without the intercession of pope or priest. The printing press, by recording and widely distributing knowledge, laid the intellectual foundation for the scientific revolution that was to come.

This marked an important moment in human history, when Western thought was split into twin, complementary paths that in many ways made the advances of the subsequent five hundred years possible. The other three major sources of human power—government, economics,

---

* This is the common name for the document, officially titled *Disputation of Doctor Martin Luther on the Power and Efficacy of Indulgences.*

and science—developed similar authoritarian, top-down and antiauthoritarian, bottom-up strains of thought over the ensuing centuries. In government, as in religion, there are authoritarian, totalitarian models such as monarchy, dictatorship, and fascism on the one hand and antiauthoritarian models like democracy and anarchy on the other. In economics, communism and capitalism are the opposing theories, as are (less extremely) the ideas of John Maynard Keynes about the need for government stewardship of the economy on the one hand and Milton Friedman's laissez-faire, free-market focus on the other. And in science, the split fell between the two complementary and competing paths of knowledge that were first proposed by the Catholic René Descartes and the Protestant Francis Bacon. The two men's thinking has been combined to underlie what we now call the scientific method.

DESCARTES VERSUS BACON

Descartes stressed the importance of deduction, a method of reasoning from the top down using "first principles," beginning with his famous line "I think, therefore I am." In his 1637 *Discourse on the Method of Rightly Conducting One's Reason and of Seeking Truth in the Sciences,* [13] he began by embracing skepticism and approaching the entire world with doubt, free of preconceived notions. There was nothing, he said, he could count on as real if all things were subject to skepticism. Ah, but wait—*he* was here, thinking these thoughts. Therefore, he must be real.

Beginning with his mind as the only reliable foundation, Descartes regarded the senses as unreliable and the sources of untruth and illusion. He concluded that reliable truth about reality could be determined only by a mind that was separate and distinct from the physical body—thereby originating the concept of the mind–body split, or Cartesian (from "Descartes") dualism. To Descartes, a conclusion is valid if *and only if* it follows logically from the premise, as did the syllogisms in Aristotle's classic book on logic, *Organon*:

All men are mortal. Socrates is a man. Therefore, Socrates is mortal.[14]

Bacon, in contrast, thought Aristotle had gotten it all wrong, and he stressed nearly the opposite approach. Bacon was a lawyer who worked under the attorney general, Edward Coke, a position he would eventually assume himself. Toward the end of his legal career he turned more of his attention to science and published what would become a foundational volume, *Novum Organum,* or "New Organon," a devastating attack on Aristotle's book and the logic of the Greeks with its emphasis on top-down reasoning and disdain for experimentation.[15] In it he argued instead for using the *inductive* method of reasoning, which underlies much of the scientific method we use today. Inductive reasoning proceeds from the bottom up by observing with the senses and then building in logical steps to reach a general conclusion about reality. An example would be:

> All observed swans are white; therefore, all swans are white.

This method clearly has a limitation: Its conclusions are provisional and always subject to disproof. All it takes is the discovery of a single nonwhite swan to invalidate the statement. This is why you hear scientists talking about the "theory" of evolution. It is not an observed fact; rather, it is a conclusion that is *supported by all the facts observed so far,* but because of the provisional nature of inductive reasoning, scientists hold out the possibility, no matter how small, that it could be invalidated. Science thus demands intellectual honesty, and a scientific conclusion will always contain a provisional statement:

> All observed swans are white; therefore, all swans are *probably* white.

In practice, Bacon's method doesn't bother scientists, or most reasonable people, because the chances of being wrong, while present, are usually very small. It is, for example, theoretically *possible* that chemical processes taking place in your body could cause you to spontaneously combust, but we don't live our lives worrying about it because the *probability* is extremely small. But since we have a limited view of the universe and cannot see everything at once, we cannot make absolute conclusions. That is why math and statistics have become such important parts of science: They quantify the relative probability that a conclusion is true or false.

PURITAN SCIENCE

Puritans, rooted as Protestantism was in a *protest* of Catholic authority, did not take kindly to the indictment of Galileo or to the idea that opinions that were supported as probable by observation of nature, and thus evidence of God's law, could be decreed contrary to holy scripture. In fact, the growing conflict between the Puritans and the Church of England—established in 1534 because the Roman Catholic Church would not annul the marriage of King Henry VIII to Catherine of Aragon, preventing him from marrying again—arose because the Puritans thought the Church of England was not anti-Catholic enough. Thus, their name: Puritans.

Their frustration led King James in 1604 to authorize a new translation of the Bible, the King James version, to address their concerns. Nonetheless, many Puritans viewed having a monarch as the spiritual leader of the church (as is the case with the Church of England) as an irreconcilable compromise, a substitution of king for pope that had been accomplished solely for the matrimonial benefit of Henry VIII. The monarchy's Royal Prerogative seemed to them yet another hypocritical corruption of authority, one akin to the Catholic Church's "indulgences."

Puritans became even more upset when James's successor, King Charles I, married the French Catholic princess Henrietta Maria within two months of his coronation, before Parliament could meet to forbid it. Soon after, he began appointing Catholic Lords to his court. He appointed William Laud the new Archbishop of Canterbury in 1633. Laud replaced the wooden communion tables with stone altars, installed railings around them, and ordered the use of candles, causing Puritans to complain that he was altering the Anglican churches to be more Catholic. Laud responded by closing Puritan churches and firing nonconformist clergy.

Separatist Congregationalist Puritans began emigrating to America again, fearing a return of Catholic absolutism and partisan political reprisals. Other Puritans organized as dissenters to King Charles's use of arcane laws to levy personal taxes and his aggressive authoritarian power grab. Edward Coke, now in Parliament, sought to limit the bottom-wing king's powers, for example by serving as chief author of the Petition of Right, passed in 1628, which laid out several basic rights that the United States would later adopt. Among these were that taxes could be levied only by Parliament, not by the king; that martial law could not be imposed in peacetime; that prisoners had to be able to challenge the legitimacy of their detentions through a writ of habeas corpus; and that soldiers could not be billeted in private residences. But in 1629, Charles dissolved Parliament and asserted personal rule by extended Royal Prerogative. This state of affairs lasted for eleven years until the English civil wars, from 1642 to 1646, in 1648, and from 1650 to 1651, which pitted Royalists (authoritarians) against the largely Puritan Parliamentarians (antiauthoritarians) and eventually led to both Laud's and Charles's beheadings.

After the civil wars ended and the Church of England was restored, many Puritans broke away. At first these "nonconformists" were again persecuted, but by the 1660s, they were tolerated. Because of their emphasis on individual liberty, their adherents included some of the greatest minds of the age, including Isaac Newton.

Newton provides an example of how the idea of "science" had not yet fully emerged as something separate from religion in early Enlightenment thinking. In fact, during the seventeenth century, the word "scientist" was not commonly used to describe experimenters at all; they were called "natural philosophers"[16] in an extension of the Puritan idea of the study of the Book of Nature. Science had also not fully emerged as a separate concept, but was sometimes thought of as a method or style of study rather than a discretely defined set of disciplines. This was true even into Thomas Jefferson's day. Jefferson himself usually used the word to mean what today we call the hard sciences, but sometimes he used it to refer simply to the rigorous study of other fields, such as the "sciences" of language, mathematics, and philosophy.[17]

By 1663, a time when Puritans were in a decided minority in England, 62 percent[18] of the natural philosophers of the famed Royal Society of London were Puritans,[19] including Newton, who wrote far more on religion and alchemy than he did on science.[20] Newton believed in the inerrancy of scripture, biblical prophecy, and that the apocalypse would come in 2060.[21] He was "not the first of the age of reason. He was the last of the magicians," said economist John Maynard Keynes,[22] who purchased a collection of Newton's papers in 1936 and was astounded to find more than one million words on alchemy and four million on theology, dwarfing his scientific work.[23]

Newton went on to create calculus and to publish *Philosophiae Naturalis Principia Mathematica,* or *Mathematical Principles of Natural Philosophy,*[24] upon which modern physics was founded.

Eighty-nine years later, *Principia* was one of the main sources Thomas Jefferson drew upon for inspiration as he sat in the two second-story rooms he had rented from Jacob Graff in Philadelphia, writing the Declaration of Independence.

THE SCIENTIST-POLITICIAN

Like Bacon, who famously died of pneumonia after conducting an experiment on preserving meat with snow, Thomas Jefferson was both an accomplished attorney and a scientist. On July 4, 1776, the day the Continental Congress adopted the Declaration of Independence, Jefferson took the time to record the local temperature on four separate occasions as a part of a broader research project he was conducting. His measurements typically also included barometric pressure and wind speed.[25] His goal was to improve meteorological science to refine farmer's almanacs and improve weather forecasting throughout the colonies, both of which were of personal importance to Jefferson as a farmer.

Jefferson also had knowledge of physics, mechanics, anatomy, architecture, botany, archeology, paleontology, and civil engineering.[26] He was an avid astronomer. He carried a small telescope with him wherever he went and recorded the eclipse of 1778 with great precision, although he was frustrated by the cloudy conditions.[27] As president, he commissioned the Lewis and Clark expedition. He sold it to Congress as an economic initiative, but he sent his presidential secretary, Meriwether Lewis, for training with the top scientists of the day and instructed him to conduct it as a scientific expedition.[28]

Jefferson's love of science is so well known among students of science policy that many can quote him verbatim. "Science is my passion, politics my duty," he said.[29] In writing to a friend just prior to retiring from the presidency of the United States, he said:

> Never did a prisoner, released from his chains, feel such relief as I shall on shaking off the shackles of power. Nature intended me for the tranquil pursuits of science, by rendering them my supreme delight, but the enormity of the times in which I have lived, has forced me to take part in resisting them, and to commit myself to the boisterous ocean of political passions.[30]

Jefferson was also very familiar with Coke, whose *Institutes* he had studied as a law student.[31] This heady mix of science, law, and politics would lead Jefferson to carve out a founding document for the United States that was based not on religion or God, but on knowledge and reason. Whereas religious authority and proximity to God could be endlessly argued between different faiths or countries, Jefferson reasoned that a country that was based on the more narrowly defined rule of men—a democracy—was removed from this religious argument, freeing both religion and the government. He needed to convince the world's nations that American independence should be respected as the only rational and correct state of the world as dictated by the laws of nature, so he had to build the most unassailable argument possible. As his friend and adviser Benjamin Franklin later noted dryly after signing the declaration Jefferson would craft, "We must all hang together or most assuredly we will all hang separately."

Racked by the threat of war and with its political power resting on uncertain ground, in June of 1776 the Continental Congress appointed Jefferson, along with Franklin, John Adams, Roger Sherman, and Robert

Livingston, to secretly draft the document. The committee delegated the writing of the first draft to Jefferson.

How Do We *Know* Things?

Holed up in his rented rooms, faced with this awesome responsibility, the thirty-three-year-old took up his quill pen. He considered Francis Bacon, Isaac Newton, and John Locke, whom he had studied at the College of William and Mary,[32] to be the three most important thinkers of all time. He called them "my trinity of the three greatest men the world had ever produced."[33] Writing on a portable "lap desk" of his own design, he labored to create a document that reflected the clear, axiomatic logic of John Locke, who instructed that "in all sorts of reasoning, every single Argument should be managed as a mathematical demonstration; where the connexion of ideas must be followed till the mind is brought to the source on which it bottoms."[34]

Like Newton and Bacon, Locke was an Englishman and a Protestant, and he is credited with creating the philosophy of empiricism, on which much of science is based. He divided human thought into two categories: knowledge and belief. Locke was aware of the many divisions within Christianity, with each faith arguing that it was the one true religion. This was true of not only the great divide between Protestantism and Roman Catholicism, but also of lesser divides between German Lutheranism and English Protestantism and between the Church of England and the dissenters—the Puritans, and within them the sects of Presbyterians, separatist Congregationalists (from whose congregations came the American Pilgrims), and Baptists. Each could not be the one true religion, so some method of ascertaining truth or falsehood had to be developed, or the conflicting claims were likely to go on forever. This led him to ask a fundamental question:

How do we know something to be true? What is the basis of knowledge?

Locke's *An Essay Concerning Human Understanding*,[35] published in 1689, just two years after Newton's *Principia*, strove to lay out what can be known empirically, how it is that we know it, and the inherent limits of knowledge. He began, building on Bacon, with observation. He then divided knowledge into three types: intuitive, demonstrative, and sensitive.

Intuitive knowledge is "self-evident" to anyone looking at it, and it carries the least doubt of the three types of knowledge. Three is more than two, black is not white, and the presence or absence of a thing are examples of intuitive knowledge.

Demonstrative knowledge, the second type of knowledge, is slightly less certain than intuitive knowledge. Agreement or disagreement is not immediately clear, but instead depends on the use of reason to demonstrate "by necessary consequences, as incontestable as those in mathematics," that something is so. Each step in a reasoning process—which, as St. Germain had described, was the process of discovering the natural law of things—must in and of itself be intuitively evident. For example:

> A feather falls more slowly in air than a penny does. When we remove the air with a vacuum pump, the feather and the penny fall at the same rate. Therefore, air changes how gravity acts on different objects.

These intermediate steps in reasoning are called "proofs." Each conclusion is reliable because it is ultimately traceable, step by demonstrable step, back to the foundation of the natural world.

Locke called the third kind of knowledge "sensitive knowledge," meaning that we get it directly from our senses. For example, we become aware of a rose by its scent, then look for its presence.

But, in a nod to Descartes, he acknowledged that our senses are often wrong. Sometimes it's not a rose we smell, but perfume; sometimes we see not a pond, but a mirage. Locke argued that sensitive knowledge is thus much less certain than intuitive or demonstrative knowledge.

Finally, he said,

> whatever comes short of one of these, with what assurance soever embraced, is but faith, or opinion, but not knowledge, at least in all general truths.[36]

THE PROGENY

This was critical to Jefferson because it laid the foundational argument for democracy, which was implicit in a different form in Coke's argument for the primacy of English common law: If we discover the truth by using reason, which according to St. Germain and Coke is driven by God's will, then *anyone* can discover the truth, and therefore *no one* is naturally better able or more entitled to discover the truth than anyone else. Because of this, political leaders and others in positions of authority do not have the right to impose their beliefs on other people. The people themselves by natural law retain this inalienable right. Based on Locke's ideas of knowledge, and Coke's ideas of law, this antiauthoritarian equality of all men in their ability to use reason to discern the truth for themselves is logically self-evident. It is intuitive knowledge. And that's the heart of America.

Jefferson worked diligently for days to craft a document that was grand and yet achieved the unassailable quality of a logical proof. The axiomatic beauty of the argument he was reaching for would indelibly tie science, knowledge, freedom, and democracy together in a single common cause of human advancement and proclaim the inalienable right of the people to reject authoritarian tyranny as illegitimate.

But in spite of his best intentions, in his rough draft, Jefferson foundered on the shoals of authoritarian religious assumptions left over from Hobbes's bleak and brutal post–Middle Ages—a fact that illustrates how deeply rooted these assumptions are, even for a scientist like Jefferson, and how slow and careful a process is required to tease out what is knowledge from what, to quote Locke, is "but faith, or opinion." The misstep occurred in the opening of the second paragraph, when Jefferson wrote:

> We hold these truths to be sacred and undeniable; that all men are created equal.

When Jefferson showed his draft to Franklin, Franklin made several firm, bold backslashes striking the words "sacred and undeniable" and, drawing on Locke, whom he too admired,[37] he replaced Jefferson's reference to divine authority with the antiauthoritarian words "self-evident"[38] that Locke had used in *Essay*:

> The idea of a Supreme Being, infinite in power, goodness, and wisdom, whose workmanship we are, and on whom we depend; and the idea of ourselves, as understanding rational beings, being such as are clear in us, would, I suppose, if duly considered and purified, afford such foundations of our duty and rules of action, as might place morality amongst the science capable of demonstration: wherein I doubt not, but from self-evident propositions, by necessary consequences, as incontestable as those in mathematics, the measures of right and wrong might be made out, to anyone that will apply himself with the same indifference and attention to the one, as he does to the other of the sciences.[39]

Franklin's edit, it may be argued, helped to make the United States into the scientific and technological powerhouse it became. At the time America's most renowned scientist, Franklin was also an admirer of Newton's *Principia* and a friend of the Scottish economist David Hume. Hume had written extensively on natural law and liberty, which Jefferson had drawn on in the sentence, and he defined liberty as freedom of choice:

> By liberty, then, we can only mean a power of acting or not acting, according to the determinations of the will; this is, if we choose to remain at rest, we may; if we choose to move, we also may.[40]

Even though Newton did not see a conflict between science and religion, neither did he insist upon applying religious thinking to the realm of science, which is the realm of "understanding," as he put it, or knowledge versus "but faith, or opinion." "A man may imagine things that are false," Newton said, "but he can only understand things that are true."[41]

Newton and Hume both instead rested their arguments on empiricism, "bottoming them out" in the natural world with evidence, and so it had to be with the argument for liberty. Hume argued that

> whatever definition we may give of liberty, we should be careful to observe two requisite circumstances; first, that it be consistent with

plain matter of fact; secondly, that it be consistent with itself. If we observe these circumstances, and render our definition intelligible, I am persuaded that all mankind will be found of one opinion with regard to it.[42]

This is the persuasive power that Jefferson was reaching for by tying his arguments back to the plain matter of fact laid bare by his venerated "trinity" and the aggregated authority of grave and learned men in English common law as established by Coke.

Franklin understood that Jefferson's words had inadvertently confused the realms of knowledge and faith, resting the principle being argued— that all men are created equal and are endowed by their creator with certain inalienable rights—on an authoritarian, religious assertion, which, as Locke himself had shown, could be argued indefinitely. It was therefore weak as a political argument, a matter of mere *belief,* rather than knowledge or evidence based.

Franklin knew Jefferson was reaching for something more powerful, something "that all mankind will be found of one opinion with regard to it," and he knew how to take it there. He instead rested the principle on reason and Locke's *intuitive knowledge,* moving the founding argument for the United States firmly out of the realm of an assertion of religious authority (as in "sacred and undeniable," i.e., "God is on our side," or "God save the monarchy") and into the realm of man, reason, and the laws of nature that flowed from empiricism, antiauthoritarianism, and nature itself.

It was *self-evident.*

In the process they created something entirely new: a nation that respected and tolerated religion in every sense, but did not base its authority on religion. A nation whose governance was instead based on the underlying principles of liberty, reason, and science.

CHAPTER 4

SCIENCE, MEET FREEDOM

Instead of being derided as geeks or nerds, scientists should be seen as courageous realists and the last great heroic explorers of the unknown. They should get more money, more publicity, better clothes, more sex, and free rehab when the fame goes to their heads.

—MATTHEW CHAPMAN*

THE BIOLOGY OF DEMOCRACY

The implications of empiricism for government were profound. Suddenly kings and popes logically, *empirically* had no greater claim to authority than anyone else. This was a self-evident truth that proceeded from careful observation of nature through our knowledge-building process—the same process that we use in science. And if it was the case that we were equal and free, then the only form of government that made sense was government of, by, and for the people. The only question was how best to implement it.

It can be argued that the method the framers hit upon, democracy, is based to some extent on biology, via its roots in natural law. This argument is supported by recent research in the field of opinion dynamics by Princeton University biologist Simon Levin. His observations of animal herds show that, like human social groups, they follow certain innate rules of organization. A vote is an expression of opinion, and

* Chapman is cofounder of Science Debate 2008 and the great-great-grandson of Charles Darwin.

herd animals, quite literally, vote with their feet when determining the overall herd's grazing patterns. But the voting is not entirely egalitarian. There are opinion leaders in these herds, just as there are in human social groups. The idea of having equality of opportunity—in practical terms, having the right to vote—exists, but just as in human society, all individuals do not have equal influence. In America, the framers of the Declaration of Independence had more influence in shaping our democracy than the odd Virginia tobacco farmer did. In herds, a very few individual opinion leaders make decisions that influence the entire herd's grazing patterns. "Individuals in reality are continually gaining new information, and hence becoming informed," says Levin. "So for sure, any individual could become a leader (just look at elections in the US). That does not mean that all men are created equal, but in terms of the ability to lead, they all have equal opportunity."[1]

The idea that natural law and hard-wired biological instincts are somewhat synonymous forces lies at the very foundation of Western law. Remember *Calvin's Case*: "The law of nature is that which God at the time of creation of the nature of man infused into his heart, for his preservation and direction."[2] In today's language, this is another way of saying "biological nature," or, more commonly, "human nature," what drives us naturally, as reasonable creatures. What democracy did was to structure and channel those natural opinion dynamics—what we call natural law—for use in organizing society. Democracy is rooted in our biology.

FECUNDITY REQUIRES FERTILE SOIL

Because of this close relationship between innate individual potential and the opportunity afforded by democracy, science has made the greatest advances in liberal (as in free and open) democratic societies that spread freedom and opportunity broadly, like fertilizer, through tolerance, diversity, intellectual and religious freedom, individual privacy, equal individual rights, free public education, freedom of speech, limited government authority, consideration of minority views, and public support for research.

These societies acknowledge the role of opinion leaders, but tie their actions closely to the feedback of the herd. In *The Science of Liberty*,[3] science journalist Timothy Ferris demonstrates how powerfully many of these qualities have promoted the wealth and progress of the nations that have embraced them, and how their restriction by regressive, fundamentalist, theocratic, and totalitarian governments—in other words, *authoritarian* governments—have led to comparatively few scientific advances and resulted in those nations falling further and further behind the liberal democracies both economically and technologically.

This has occurred largely for two reasons. The first is intellectual flight. Historically, the brightest minds have migrated to open societies, and once there have made discoveries and created works that enriched and advanced those societies. A classic example is the intellectual flight from fascist Europe in the years leading up to World War II. Persecution, particularly of Jews and homosexuals, spurred emigration that turned America into an intellectual mecca. America offered scientists and artists freedom, tolerance, egalitarianism, opportunity, and support for their work, and it had the military strength to protect those ideals. In return, the new immigrants gave America breakthroughs in chemistry, biology, and physics and an expanded Hollywood liberal narrative culture, America's chief cultural export.

But scientific leadership proceeds from not only openness but also the degree of opportunity available to citizens. By making education free and accessible to all, by stimulating cross-pollination and creativity with diversity of views and support for research and the arts, and by leveling the economic playing field to provide equal opportunity and freedom of inquiry, democratic societies have broadcast the intellectual fertilizer that helps talent develop wherever it may be.

Combined, these two factors have had a powerful effect: Even more than empowering individuals, they empower *ideas*. It is this mix of freedom, tolerance, creativity, talent, and diversity in science, in art, and in the social and intellectual interplay between art and science that has his-torically spawned the great breakthrough cultures that produce bumper crops of new ideas and fresh insights.

A NATION OF THINKERS OR TINKERERS?

Jefferson was certainly aware of some of this, and he heavily promoted sci-ence during the eight years of his presidency from 1801 to 1809, frequently writing and speaking of its value and importance to the nation and spon-soring major scientific expeditions such as that of Meriwether Lewis and William Clark. Befitting the great westward expansion, in the nineteenth century it was America's pioneer spirit and can-do attitude that produced the world's great inventors and implementers, including Eli Whitney, Sam-uel Morse, Alexander Graham Bell, Thomas Edison, and Nikola Tesla, but Europe was still the home of real *science* and the scientists who made the fundamental theoretical breakthroughs, including Charles Darwin, Marie Curie, Michael Faraday, James Maxwell, Gregor Mendel, Louis Pasteur, Max Planck, Alfred Nobel, and Lord Kelvin.

This focus on tinkering and engineering versus science and discovery was partly because America lacked the well-established academies of Europe, but it also seemed to have something to do with the American character itself. French political scholar Alexis de Tocqueville noted this focus on pragmatism when he toured America in 1831 and 1832. His report of what he learned, *Democracy in America,* contains a chapter titled "Why the Americans Are More Addicted to Practical Than to Theoretical Science." Tocqueville observed that free men who are equal want to judge everything for themselves, and so they have a certain "con-tempt for tradition and for forms." They are men of action rather than reflection, and hold meditation in low regard. "Nothing," he argued, "is more necessary to the culture of the higher sciences or of the more ele-vated departments of science than meditation; and nothing is less suited to meditation than the structure of democratic society. . . . A desire to utilize knowledge is one thing; the pure desire to know is another."[4] Americans were tinkerers and inventors, but the era of grand American *thinkers* like Jefferson, Franklin, and Livingston was over, if it had ever been. This American focus on practicality, this preference for utility over beauty, Tocqueville argued, might eventually be the country's downfall, and he related a striking cautionary tale that resonates powerfully today.

> When Europeans first arrived in China, three hundred years ago, they found that almost all the arts had reached a certain degree of perfection there, and they were surprised that a people which had attained this point should not have gone beyond it. At a later period they discovered traces of some higher branches of science that had been lost. The nation was absorbed in productive industry; the greater part of its scientific processes had been preserved, but science itself no longer existed there. This served to explain the strange immobility in which they found the minds of this people. The Chinese, in following the track of their forefathers, had forgotten the reasons by which the latter had been guided. They still used the formula without asking for its meaning; they retained the instrument, but they no longer possessed the art of altering or renewing it. The Chinese, then, had lost the power of change; for them improvement was impossible. They were compelled at all times and in all points to imitate their predecessors lest they should stray into utter darkness by deviating for an instant from the path already laid down for them. The source of human knowledge was all but dry; and though the stream still ran on, it could neither swell its waters nor alter its course.

In other words, China fell under an *authoritarian* intellectual fundamentalism that deeply honored tradition but lacked the substance, freedom, and capacity to create anything new.

Tocqueville concluded that the basic research that had the power to change the future, to alter the course of the stream, at will, was the product of European thinking, not American. His tale suggests the dangers posed by embracing tradition and precedent at the expense of openness and creativity, applied research at the expense of basic science, fear at the expense of wonder, utility at the expense of beauty, and an insistence on financially quantifiable projections before an investment is made, the idea of which runs contrary to the entire process of discovery and creativity. Imagine, for example, an insistence on the promise of financial return prior to Darwin's trips on the *Beagle* or Neil Armstrong's first steps on the moon. They would never have happened. And yet it is hard to quantify the great wealth that has spun off from those adventures into the unknown.

It's possible that Tocqueville's general assessment of America may have been correct, and that the United States would have coasted off its vast natural resource exploitation until its economy eventually ran out of growth, but would have never really led the world, were it not for three major developments. The first grew out of Jefferson's insistence on public education, which over time did indeed provide opportunity to undiscovered talent. The second, also heavily encouraged by Jefferson, was the burgeoning American university system, which was being built up to rival those of Europe. And the third was the American values of tolerance and freedom, which, serving as an intellectual beacon to the world, drew talent from elsewhere. By the first decades of the twentieth century, all three developments were beginning to pay major dividends for America, and particularly for Republicans.

Republican Science

By its very nature science is both progressive and conservative:

> **conservative:** retentive of knowledge and cautious about making new assertions until they are fully defensible
>
> *and*
>
> **progressive:** open to wherever observation leads, independent of belief and ideology, and focused on creating knowledge

It would thus be a mistake to characterize scientists as mostly Democrats or mostly Republicans. They are mostly for freedom, exploration, creativity, caution, and knowledge—and not intrinsically of one or another party. In the early twenty-first century, the party that most stands for freedom, openness, tolerance, caution, and science is the Democratic Party, which may explain why 55 percent of scientists polled in 2009 said they were Democrats while only 6 percent said they were Republicans, compared to 35 and 23 percent of the general public, respectively.[5]

Early in the twentieth century this situation was almost reversed. Republican Abraham Lincoln had created the National Academy of Sciences in 1863. Republican William McKinley, admired by Karl Rove, won two presidential elections, in 1896 and 1900, both times over the anti-evolution Democrat William Jennings Bryan, and supported the creation of the Bureau of Standards, which would eventually become today's National Institute of Standards and Technology. Bryan's strident campaigns of anti-evolutionism, culminating in the 1925 "Scopes Monkey Trial," helped to drive even more scientists toward the Republican Party.

"We have many people even here who hasten to condemn evolution without having the remotest conception of what it is that they are condemning, nor the slightest interest in an objective study of the evidence in the case which is all that 'the teaching of evolution' means," wrote an exasperated Republican, the Nobel physicist and CalTech head Robert A. Millikan in the leading journal *Science* in 1923, "men whose decisions have been formed, as are all decisions in the jungle, by instinct, by impulse, by inherited loves and hates, instead of by reason. Such people may be amiable and lovable, just as is any house dog, but they are a menace to democracy and to civilization because ignorance and the designing men who fatten upon it control their votes and their influence."[6]

Other prominent scientists noted the political divide. The great botanist Albert Spear Hitchcock, who would soon become principal botanist at the US Department of Agriculture, wrote the following spring in the same journal that "it is absurd for a scientist to shiver with fear if he sees a black cat cross his path or if he walks under a ladder. It is equally absurd to believe that all Germans or all democrats, or all Roman Catholics . . . are undesirables and a menace to society."[7]

Originally growing out of the Anti-Federalist Party of Thomas Jefferson, by the early twentieth century the Democratic Party had become
dominated on the national level by Southern religious conservatives and
was divided over culture-war issues including evolution, the prohibition
of alcohol, restrictive immigration laws, the Ku Klux Klan, and the
Catholic faith of Al Smith, the Democratic presidential nominee in 1928.
Republicans, by contrast, were the party of Abraham Lincoln and Theodore Roosevelt, of progressive optimism and tolerance married with
environmentalism and finance—the party of rationalism and national
parks. And by the 1930s, one of the most famous men in the world was a
Republican scientist named Edwin Hubble.

Hubble, who was born in 1889, grew up in Marshfield, Missouri, and
then, at the turn of the century, in Wheaton, Illinois, where he attended
public school and was famous for his athletic prowess.[8] At the time, science was considered a somewhat fanciful pursuit and a less-than-solid
career path, suited more for adventurers and wealthy "gentlemen scientists" than professionals.[9] Hubble's father wanted him to be the consum-
mate professional, a lawyer, and when Hubble earned one of the first
Rhodes scholarships while a star student of Millikan's at the University of
Chicago, he went to the Queen's College at Oxford University to study law.

Still, it was a time when great discoveries were being made in astronomy, which captivated Hubble's imagination. America was entering a
golden age of science, propelled in no small part by the massive philan-
thropic investments of two Republican men: steel magnate Andrew
Carnegie,[10] who funded public libraries across the nation, helped found
what is now Carnegie Mellon University, and funded basic scientific
research through the Carnegie Institution of Washington (since renamed
the Carnegie Institution for Science); and John D. Rockefeller Sr.,[11] who
endowed the University of Chicago as well as Rockefeller University and
Johns Hopkins University's school of public health. As Hubble began
secretly studying astronomy,[12] what had captured the imagination of the
American public was the growing fame of the former Swiss patent officer
with wild hair and a playful face, Albert Einstein.

THE HOAX OF RELATIVITY

Einstein's general theory of relativity had made the striking prediction that gravity could bend space and so disrupt the straight-line flow of light. On May 29, 1919, scientists set out to test the theory by carefully observing the way starlight behaved during a solar eclipse. If Einstein was right, the sun's gravity would bend the light of stars that were in line with it, making them appear to be slightly offset. The eclipse blocked enough sunlight that astronomers could see the stars and measure changes in their apparent locations. If they shifted, Einstein's theory would be proved.

The test's audacity drew the attention of scientists and journalists the world over. If Einstein was wrong, his reputation would be ruined. If he was right, he would be celebrated as a genius and change everything we thought about the universe. The results were dramatically presented at a November joint meeting of the Royal Society of London and the Royal Astronomical Society, and they confirmed Einstein's predictions spectacularly.

The American popular press loved the drama, and Einstein became a household name as a little tramp of a professor who happened to also be a bold genius with funny hair and a beloved violin, not unlike Charlie Chaplin and his cane. In contrast to the image of the snobby European intellectual, Einstein connected emotionally as an underdog, a trait that appealed to the antiauthoritarian American spirit and was cited in press accounts of his "hero's welcome" when he first visited America in 1921.

America's embrace of Einstein stood in stark contrast to the treatment he was getting at home. Right-wing relativity deniers, like modern American climate science deniers, mounted ad hominem attacks against Einstein and his theory, which they loudly branded a "hoax." They were led by an engineer named Paul Weyland, who formed a small but mysteriously well-funded group that held anti-relativity rallies around Germany, denouncing the theory's "Jewish nature," and culminating in a major event at the Berlin Philharmonic Hall on August 24, 1920. Einstein attended, only to suffer more personal attacks.[13] The political animosity grew so bad that Einstein decided to leave Berlin.[14]

"This world is a strange madhouse," he wrote a friend three weeks after the rally. "Currently every coachman and every waiter is debating whether relativity theory is correct. Belief in this matter depends on political party affiliation."[15] Even prominent physicists were getting into the denialism, largely along political lines. The winner of the 1905 Nobel physics prize, Philipp Lenard, who had previously exchanged flattering letters with Einstein, had since become bitter about the Jews and jealous of the popular publicity Einstein's theory was receiving. He now called relativity "absurd" and lent his name to Weyland's group's publicity brochures, and as a Nobel laureate he worked behind the scenes to make certain Einstein would be denied the prize.[16]

"THE GREATEST TRIUMPH OF SATANIC INTELLIGENCE IN 5,931 YEARS OF DEVILISH WARFARE"

At the same time, a surge in antiscience politics had been growing in the United States in reaction to the perceived evils of the theory of evolution. Like relativity, evolution was seen by social conservatives as undermining moral absolutes—in this case absolute biblical authority. The movement was led in part by the attractive, charismatic revivalist Aimee Semple McPherson, the founder of what may have been the first American evangelical megachurch, the 5,300-seat Angelus Temple in Echo Park, Los Angeles. The facility was equipped with radio towers to broadcast her sermons, and she filled it to capacity three times a day, seven days a week. McPherson stressed the "direct experience" approach to religion, not unlike the empirical spirituality of the Puritans, and like the Puritans, she challenged the mainstream Protestant churches on being too orthodox, too tolerant—and not authoritarian enough.

This revivalist spirit was propelled by the wave of immigration that followed World War I, which many Americans found disconcerting; the recovery from the traumatic flu pandemic of 1918, which had killed millions; and the return of millions more from the war, many of them still with untreated shell shock, the condition we now call post-traumatic stress disorder. Fueled by the new domestic activity and cheap labor, the stock market boomed. Moral restrictions were loosening and the country dearly needed to blow off some steam and cut a rug. Materialism soared during what F. Scott Fitzgerald called the Jazz Age in his 1925 novel *The Great Gatsby*.

The line between liberalism and running amok depends upon one's psychological keel, as *Gatsby* showed. For many this powerful mix of materialism, diversity, and tolerance was simply too much, and they began to lose their moral bearings—or to feel that the rest of America was losing theirs. McPherson was among the latter group. Like a strict church lady of the Wild West, she set about working to bring order to society, and her moral fierceness offered a bulwark in the storm to many.

By the mid-1920s McPherson's name had become a household word. She was made an honorary member of police and fire departments across the country,[17] and from this great platform she took up a campaign against the two most profound evils threatening America at that time: the drinking of alcohol and the teaching of evolution in public schools.

McPherson was not alone in this holy campaign. William Jennings Bryan, the former Secretary of State, had been the Democratic candidate for president for a third time in 1908 and had spoken throughout the United States in favor of Prohibition and against the teaching of evolution, which he believed had led to World War I.[18] If we were in fact "descended from a lower order of animals," he professed, then there was no God and, as a consequence, nothing underpinning society. Darwin himself had not seen it this way, writing to John Fordyce in 1879, "It seems to me absurd to doubt that a man may be an ardent Theist & an evolutionist,"[19] but several states passed laws banning the teaching of evolution after Bryan's stump speeches. The most notable of these was Tennessee's Butler Act, signed into law on March 21, 1925.[20] By April, the American Civil Liberties Union had recruited a substitute teacher named John Scopes to break the law in Tennessee, after which they would pay for his defense to challenge it.

McPherson was a strong supporter of Bryan during the resulting

"Scopes Monkey Trial." Bryan had been her guest at the Angelus Temple and had watched her preach that social Darwinism had corrupted students' morality.[21] The teaching of evolution was "the greatest triumph of satanic intelligence in 5,931 years of devilish warfare against the Hosts of Heaven. It is poisoning the minds of the children of the nation," she had said.[22] While Bryan was representing the World's Christian Fundamentals Association at the trial, McPherson sent him a telegram offering, "Ten thousand members of Angelus Temple with her millions of radio church membership send grateful appreciation of your lion hearted championship of the Bible against evolution and throw our hats in the ring with you." The confrontation at the trial between Bryan and Scopes's "sophisticated country lawyer" Clarence Darrow, also a Democrat, was the climax of one of the nation's earliest major science-politics-religion controversies, and it turned public opinion in support of teaching evolution in public schools.

The Vatican stayed out of this first major political debate over evolution partly because its healthy network of parochial schools meant it had little skin in the game—state laws concerning public school curriculum were of little concern. Even today in Georgia the joke is "If you want your kids to learn about evolution, send them to Catholic school, because they won't learn it in public school."[23] Evangelical Protestants and Roman Catholics had now nearly reversed their respective popular positions with regard to science, a reversal that astronomer Edwin Hubble would soon help to accelerate.

HENRIETTA THE HUMAN COMPUTER

It was into this fiery climate of the 1920s that the Protestant-raised Hubble, adorned with the cape, cane, and British accent he had adopted while at Oxford, returned after the war. He arrived at the Carnegie Institution of Washington–funded Mount Wilson Observatory outside Pasadena, California, insisting on being called "Major Hubble."[24] Looking through the great Hooker telescope—at one hundred and one inches in diameter and weighing more than one hundred tons it was by far the largest and most powerful scientific instrument in the world—Hubble was able to view the universe with a light-gathering capacity of more than two hundred thousand human eyes.

What he saw changed humanity's view of the universe forever—and would further roil the controversy over science's role in defining the origins of creation. A conservative Baconian observer, Hubble photographed a small blinking star in the Andromeda nebula that he identified as a Cepheid variable. This observation would become iconic in its power.

Another astronomer, Harvard College Observatory's Henrietta Leavitt, had in 1912 shown something remarkable about Cepheid variable stars: The longer their period, the brighter they appeared to be. This made sense. Stars were very faint, and to probe deeply, astronomers began to mount cameras to their telescopes and take very-long-exposure photographs on glass plates so they could capture light from stars that were too faint to see with the human eye. A blinking star that has a longer "on" period deposits more light on the plate than one with a shorter period.

Leavitt's brilliant observation was so self-evident as to be irrefutable, but no one had noticed it before. The accomplishment was all the more remarkable because Leavitt was not allowed to be part of the scientific staff; she was a "computer"—one of several women hired merely to identify and catalog stars and calculate light curves for the male scientists. She was paid a premium rate of thirty cents per hour, over the usual two bits, because of the high quality of her work.[25]

Danish astronomer Ejnar Hertzsprung seized upon Leavitt's discovery to create a means of measuring intergalactic distances. Using inductive reasoning, Hertzsprung determined in 1913 that if two Cepheid variable stars had similar periods but one was dimmer than the other, it was probably farther away. He found Cepheids close enough to Earth to measure the distance to them using statistical parallax and was able to compare this with their apparent brightness (their brightness as observed from Earth). The resulting formula allowed scientists to measure the distances to all Cepheid variables. These blinking stars became "standard candles" for measuring distance throughout the heavens.

This was an immense discovery, and in 1915 American astronomer Harlow Shapley, a liberal Democrat, used it and Mount Wilson's sixty-inch telescope to map the Milky Way in three dimensions. Shapley's measurements expanded the known size of the Milky Way severalfold and showed that the sun was not at the center of the galaxy, as had been thought, but was in fact located in a distant outer arm. This over-turned the concept of the centrality of humans yet again, and Shapley was celebrated as the greatest astronomer since Copernicus—a title he himself helped promote—for having achieved the "over-throw" of the heliocentric universe.[26]

THE GREAT DEBATE: A CAUTIONARY TALE

In what would become an important cautionary tale for American science—and, by extension, democracy and much of what ails American politics today—Shapley became blinded to reality by his belief that the Milky Way was the entire universe—or maybe his hubris in wanting to believe that he had mapped the whole shebang. He argued that the spiral nebulae seen in the heavens were simply wisps of gas and clouds of dust within the Milky Way, rather than entire "island universes"—galaxies—of their own, as fellow astronomer Heber Curtis posited. He debated this point with Curtis at a meeting of the National Academy of Sciences in April of 1920, in an event famously called the Great Debate.

The debate ended in a draw because there wasn't yet enough observational data to draw firm conclusions, but this was partly because Shapley, without realizing it, had stopped using Locke's and Bacon's inductive reasoning to build knowledge from observation and was instead trying to prove his point with a *rhetorical* argument—an a priori, top-down, Cartesian approach of first principles that had him arguing more like an attorney than a scientist. When his assistant Milton Humason showed Shapley a photographic plate that seemed to indicate the presence of a Cepheid variable in Andromeda, Shapley shook his head and said it wasn't possible. Humason had unimpressive formal credentials—he had been elevated to assistant from a lowly mule driver and had but an eighth-grade education—but it was Shapley's a priori idea that occluded his vision. He took out his handkerchief and wiped the glass plate clean of Humason's grease pencil marks before handing it back.[27] No one realized it at the time, but Shapley's career as a major scientist ended at that moment.

Soon after, Shapley moved on to run the Harvard College Observatory while his rival Edwin Hubble took over the telescope. Adopting the uncredentialed but nevertheless brilliant Humason as his assistant and conservatively adhering to a strict Baconian observational methodology, Hubble identified Cepheids in Andromeda and used them to show that the spiral "nebula," as he called the galaxy, was not part of the Milky Way at all; it was in fact nearly a million light-years distant—more than three times farther away than the diameter of Shapley's entire known universe. The Great Debate was settled, and Hubble became an overnight sensation.

A Republican Expansion

Riding some five stories up in the velvet darkness alongside the Jules Verne–like structure of the Hooker telescope under cold and intensely starry skies, Hubble followed this accomplishment up in 1929 by showing that there is a direct correlation between how far away a galaxy is and its redshift, the degree to which its light waves are shifted to the red end of the spectrum. Light waves are emitted at known frequencies. Redshift is caused by a star's light waves stretching out, apparently due to the star's rushing away from Earth, making them lower in frequency and thus shifted toward the red end of the visible light spectrum.

Scientists had already established that the redshift suggested that light waves were subject to the Doppler effect. We notice the Doppler effect in everyday life when the sound of a train's whistle or a police siren lowers in pitch as it races away from us. The scientists believed redshift could be used as a measure of the speed at which a star appears to be moving away from us. Hubble correlated that redshift with distance, and then showed by painstaking observation, performed mostly by Humason, whom Shapley had recommended for promotion to the scientific staff, that the farther away a star is, the greater the redshift. Hubble found this to be uniformly true in every portion of the sky. This suggested that the universe itself was probably expanding at an even rate.

To picture how this could be, imagine blowing up a perfectly round balloon until it is no longer flaccid, but not yet taut. Now take a marker and mark spots in a grid pattern all over the entire surface of the balloon, each spot exactly one inch distant from the next. Now finish blowing up the balloon and watch what happens. As you blow, every dot moves farther away from every other dot. The space between each pair of dots expands. In addition, the dots that are, say, five dots apart from each other move apart five times faster than the dots that are only one dot apart do because the surface is expanding uniformly. Five times the distance, five times the expansion. This is a close analogy to what Hubble saw happening in three dimensions in the universe.

This fundamental velocity-distance relationship came to be known as Hubble's Law, and it is recognized as one of the basic laws of nature.[28] It also implied something even more momentous.

A CATHOLIC PRIEST'S BIG BANG

Georges Lemaître was a pudgy, pinkish Belgian Jesuit abbé—a Catholic priest—who also happened to be a skilled astronomer. Lemaître had noticed that Einstein's general theory of relativity would have implied that the universe was expanding but for a troublesome little mathematical term called the cosmological constant that Einstein had inserted into his equations. Lemaître saw no convincing reason why the cosmological constant should be there. In fact, Einstein himself had originally calculated that the universe was expanding, but he was a theoretician, not an astronomer. When he turned to astronomers for verification of his theory, he found that almost all of them held the notion that the universe existed in a steady state and there was no motion on a grand scale. So in deference to their observational experience, Einstein adjusted his general theory calculations with a mathematical "fudge factor"—the cosmological constant—that made the universe seem to be steady.

Lemaître had independently been working off the same mathematical principles that Einstein had originally laid out, and in 1927 he wrote a dissenting paper in which he argued that the universe must be expanding, and that if it was, the redshifted light from stars was the result of this expansion. This redshift had been observed by a number of astronomers, but until then there had been no consensus on what the cause could be.

Lemaître saw Hubble's self-evident observations and clear logic and immediately realized that it confirmed his math and refuted Einstein's general theory. Furthermore, he deduced, if the universe was expanding equally in all directions, it must have initiated in a massive explosion from a single point. This meant to him that the universe is not infinitely old; it has a certain age, and that the moment of creation—which British astronomer Fred Hoyle later mockingly called the "big bang"—was analogous to God's first command at the beginning of the good abbé's most cherished book, the Bible: *Let there be light.*

Hubble's meticulously reported logic and observations convinced Einstein that he had been wrong about the cosmological constant. He made a pilgrimage from Germany to Mount Wilson Observatory outside of Pasadena, where he joined Hubble, Humason, Lemaître, and others to make a stunning public announcement. Unlike Shapley, Einstein changed his position and removed the cosmological constant from his general theory of relativity, later calling it "the biggest blunder of my life." The universe was indeed expanding.

SCIENCE ROCK STAR

This dramatic mea culpa by Einstein drew even more attention to Hubble and the striking depictions of the immense universe coming from the swashbuckling astronomers atop the 5,715-foot Mount Wilson. Breathless newspaper headlines screamed across several columns about the gargantuan distances and the millions of new worlds Hubble was discovering. He began to lecture on science to standing-room-only crowds of five thousand people[29] and became one of the most famous men in the world.[30]

Decades later, the Hubble Space Telescope would be named in his honor, but in the 1930s, Hubble's work captured the public interest like that of few scientists before him. He became one of the first great popularizers of science with his traveling and speaking, even delivering, as part of a scientific lecture series, a ten-minute national radio address heard by millions during an intermission of a New York Philharmonic broadcast. He and his wife Grace were the special guests of director Frank Capra at the March 4, 1937, Academy Awards ceremony, where Capra, the academy's president that year, won a best director Oscar for *Mr. Deeds Goes to Town*. Hubble became the toast of Hollywood, and a long line of actors and directors made the pilgrimage up Mount Wilson to peer through the lens of the telescope with him. Unlike the admiration that his strikingly clear papers produced, this activity engendered the wrath of his fellow scientists, who scorned him as an arrogant egotist and shameless self-promoter. His protégé, Allan Sandage, said that Hubble "didn't talk to other astronomers very much, but he was certainly not arrogant when he was in the company of other people."[31] He had the talent to back up his celebrity and had done a great public service of the sort that scientists should as a rule be more appreciative of.

Among the many celebrities Hubble came to know was Aimee Semple McPherson, who visited him on the mountain. According to Sandage, Milt Humason, who was a famous womanizer, claimed that in 1926, during a month-long disappearance that McPherson attributed upon resurfacing to having been kidnapped, tortured, and held for ransom in Mexico, the attractive radio evangelist had actually been up on Mount Wilson enjoying Humason's special attentions in the Kapteyn Cottage.[32] If true, it would seem an example of the not uncommon phenomenon of some strident preachers and lawmakers who attempt to impose rules and structure on society perhaps seeking to control their own overpowering urges.

Hubble maintained a tolerant but skeptical relationship toward all

religions, but when it came to politics, according to Sandage, a Democrat, Hubble was a strong Republican who enjoyed colluding with other
Republican scientists to schedule known Democrats for telescope time
on Election Day to prevent them from voting.[33]

In 1951, the Holy Father, Pope Pius XII, gave a momentous speech in
which he addressed Hubble's work and the big bang theory, stating that
the big bang proved the existence of God by showing there was a moment
of creation, so there must be a creator. It was an adoption of the old Puritan view of science being the discovery of God's laws. A friend of Hubble's read the text of the pope's speech in the *Los Angeles Times* and
wrote to him.

> I am used to seeing you earn new and even higher distinctions; but
> till I read this morning's paper I had not dreamed that the Pope
> would have to fall back on you for proof of the existence of God.
> This ought to qualify you, in due course, for sainthood.[34]

Hubble, heralded by scientists as the greatest astronomer since Galileo,
had brought the relationship between science and the Roman Catholic
Church full circle.

CHAPTER 5

GIMME SHELTER

Now we're all sons-of-bitches.

—KENNETH BAINBRIDGE
to J. Robert Oppenheimer* after the July 16, 1945, atom bomb test[1]

AN INTELLECTUAL WEAPON

Science took an important leap in America's public consciousness during World War II, when it was transformed from an exploration of nature into a means to win the war for democracy. Radar and the atomic bomb had major impacts on the war's outcome, as did sonar, synthetic rubber, the proximity fuse, and other key wartime innovations, with many of the efforts being led by immigrants from an increasingly anti-science Third Reich.

The war didn't start out this way. In fact, during the 1930s Adolf Hitler was an early adopter of the latest science and technology, which he used to great political advantage. He barnstormed twenty-one cities by airplane—a first—in his 1932 campaign for president against Paul von Hindenburg, an effort the campaign called "Hitler über Deutschland."[2] The Nazi party mounted a relative novelty, gramophones, on vehicles and used the public's attraction to them to broadcast a uniform political message. Hitler lost the presidential election, but won enough publicity to be named chancellor in

---

* Bainbridge was director of the test, the first explosion of an atomic bomb. Oppenheimer was director of the Manhattan Project.

73

1933. That year, the Third Reich introduced another weapon with which to spread Nazi ideology. They underwrote the *Volksempfänger*, or "people's receiver," and offered it to the public at low cost, with great success. It had no international shortwave bands, only domestic, which the government filled with propaganda and patriotic music. The world's first regular television broadcast was instituted in Germany beginning in March 1935, with similar goals, and the Third Reich pioneered the use of the classroom filmstrip to inculcate uniform Nazi ideas about politics and racial pseudoscience in students.

As Hitler's minister for armaments, Albert Speer, recounted at his trial in Nuremberg after the war, "Hitler's dictatorship differed in one fundamental point from all its predecessors in history. It was the first dictatorship in the present period of modern technical development, a dictatorship which made complete use of all technical means for the domination of its own country. Through technical devices like the radio and the loudspeaker, eighty million people were deprived of independent thought."[3]

Germany also made great strides in mechanized warfare and made key technological advancements to the submarine and ballistic missile. But the intolerance and uniformity of the Nazi regime and the elevation of ideology and propaganda over knowledge and science began to work against the ideals of freedom that lie at the foundation of science. Berlin was the capital of world art, culture, and science in the late 1920s and early 1930s, but the Nazis considered the city's artistic and scientific cross-pollination degenerate. As they elevated rhetoric and ideology over science and tolerance, Germany's intellectuals began to either conform to Nazi ideology or flee. Within a decade, German scientific and technological progress ground to a halt as the Third Reich lost many of its most creative minds to Britain and the United States.

Presiding over the American science war effort was Edwin Hubble's boss Vannevar Bush, an engineer and the president of the Carnegie Institution of Washington. There had been a certain lack of cooperation between the Europe-friendly science enterprise and the military during

World War I, and certain administrative barriers to the military's adoption of new technologies,[4] that Bush was anxious to prevent the United States from repeating, particularly with the vast influx of talent the country was reaping as a result of growing Nazi intolerance. Albert Einstein was the most famous of these immigrants, but there were many others—most of them Jewish. Bush was strongly of the opinion that science and technology would lead to military superiority for whichever country best exploited them. After the Germans invaded Poland in September 1939, Bush became convinced of the need to establish a federal agency that would coordinate US research efforts. He was able to schedule a hasty meeting in June 1940 with President Franklin D. Roosevelt, who approved the agency in less than ten minutes.

The National Defense Research Committee (NDRC), the forerunner to today's National Science Foundation, was established on June 27. The open society, the wartime esprit de corps, the federal dollars, and the pooling and marshaling of talent organized the American science enterprise into an intellectual weapon unlike any seen before. Under the auspices of this and a related agency, Bush initiated and oversaw the development of the atomic bomb until it was taken over by the military, as well as the development of radar, sonar, and numerous other inventions critical for the war effort and several significant medical advances, including the mass production of penicillin.

THE END OF INNOCENCE

One of the four top scientists Bush would appoint to lead the NDRC was Harvard president James B. Conant, whom he initially placed in charge of chemistry and explosives. When the NDRC took on the goal of making an atomic bomb before the Germans could, Conant recruited a former Harvard chemistry major, the charismatic and popular University of California at Berkeley theoretical physics professor J. Robert Oppenheimer, who was recommended by his friend and fellow Berkeley physicist Ernest Lawrence. It was to be physics' finest hour, and Oppenheimer,

the poetic son of German Jewish immigrants, who read the Bhagavad
Gita in Sanskrit and studied philosophy under Alfred North Whitehead,
threw himself into the problem with abandon, assembling a crack team
of the best minds in physics, including several of his own top students[5]
and several European immigrants. In September 1942 the project was
turned over to the military under the command of engineer and briga-
dier general Leslie Groves. Groves recognized Oppenheimer's brilliance
and ambition and appointed him scientific director of the now code-
named Manhattan Engineer District or, more simply, the Manhattan
Project. The work was "without doubt the most concentrated intellectual
effort in history."[6] Science was to be America's greatest defense against
tyranny in a race against time.

But then in the blink of an eye, everything changed. The project suc-
ceeded, and on August 6, 1945, the United States dropped the first of two
of its new bombs of light, code-named Little Boy, on the Japanese city of
Hiroshima. On August 9, Fat Man fell on Nagasaki. The bombs proved
the power of knowledge once and for all, and Oppenheimer, as the director of the project, was the first public spokesman for the awesome power
of science in a new era.

After the euphoria of winning the war had ebbed, however, the idea
that the United States had used science to kill an estimated 110,000 Japanese civilians without any warning, with another 230,000 dying from
radiation injuries over the next five years—a side effect the United States
at first officially denied[7]—weighed heavily on Oppenheimer's conscience, and it weighed on America's conscience too.

But in addition to moral unease, Oppenheimer, like many other leading scientists, had a mounting concern that the Soviet Union, with its
vast uranium deposits, would engage the United States in an arms race.

The Allies—and America in particular—had up to this point regarded
themselves as fighting the good fight—honorable, fair, and true, with one
hand tied behind their backs like Superman. The obliteration of two cities
of civilians avoided what would surely have been a horrible and bloody
invasion against a radicalized nation that was using suicide bombers, but
it also exposed a dark side of the power that science could unleash, and the

horrific consequences that are possible when ethics lag behind knowledge. Mainstream Americans, who had been largely proscience during the 1920s and 1930s, now became deeply ambivalent. Was it right, what we had done? Was it honorable? And could it come back to hurt us?

Science, and with it America, was growing up, and with the increased power came the dawning of a new age of responsibility. Seven weeks after the Hiroshima and Nagasaki blasts, William Laurence, the *New York Times* reporter whom the War Department had contracted to be the atomic bomb's official historian, characterized this visceral feeling.

> The Atomic Age began at exactly 5:30 Mountain War Time on the morning of July 16, 1945, on a stretch of semi-desert land about fifty airline miles from Alamogordo, NM, just a few minutes before the dawn of a new day on this earth. . . . And just at that instant there rose from the bowels of the earth a light not of this world, the light of many suns in one. [8]

There was a sense that scientists had unlocked a power whose use crossed an ethical boundary, and that this act had somehow soiled science and might even destroy humanity. Oppenheimer, the poet physicist, who thought of a verse from the Bhagavad Gita upon seeing the first atomic detonation at Trinity test site in New Mexico—"I am become death, the destroyer of worlds"[9]—spoke of his growing misgivings at the American Philosophical Society in November.

> We have made a thing, a most terrible weapon, that has altered abruptly and profoundly the nature of the world. We have made a thing that by all [the] standards of the world we grew up in is an evil thing. And by so doing, by our participation in making it possible to make these things, we have raised again the question of whether science is good for man, of whether it is good to learn about the world, to try to understand it, to try to control it, to help give to the world of men increased insight, increased power.[10]

Albert Einstein, who had played a key role in alerting President Roosevelt to the possibility of making such a bomb, shared Oppenheimer's feelings. He sent a telegram to hundreds of prominent Americans in May of 1946, asking for $200,000 to fund a national campaign "to let the people know that a new type of thinking is essential if mankind is to survive and move toward higher levels. . . . This appeal is sent to you only after long consideration of the immense crisis we face." The telegram contained what has become one of the most famous quotes in science:

> The unleashed power of the atom has changed everything save our modes of thinking and we thus drift toward unparalleled catastrophe.[11]

Everything has changed "save our modes of thinking" can speak not only of the bomb, but also of climate change, biodiversity loss and habitat fragmentation, ocean trawling, geoengineering, synthetic biology, genetic modification, chemical pollution, and a host of other science challenges we now face. Science and technology have delivered awesome power to governments and industry, but they have not granted us complete awareness of the potential consequences of this power or mechanisms for its sustained use without causing destruction. We are ever like teenagers being handed the car keys for the first time.

After the war, the feeling that our ability with science had outstripped our moral and ethical development as a society, perhaps as a species, was not limited to physicists, but rather was spread across the sciences. The Austrian Jewish biochemist Erwin Chargaff immigrated to the United States to escape the Nazis in 1935. His work would lead to James Watson and Francis Crick's discovery of the double-helix structure of DNA. Chargaff's autobiography described his changed feelings about science.

> The double horror of two Japanese city names [Hiroshima and Nagasaki] grew for me into another kind of double horror: an estranging awareness of what the United States was capable of, the country that five years before had given me its citizenship; a nauseating terror at the direction the natural sciences were going. Never far from an apocalyptic vision of the world, I saw the end of the essence of mankind—an end brought nearer, or even made possible, by the profession to which I belonged. In my view, all natural sciences were as one; and if one science could no longer plead innocence, none could.[12]

Military leaders shared a similar concern. Omar Bradley, the first chairman of the US Joint Chiefs of Staff and one of the top generals in

North Africa and Europe during World War II, gave blunt voice to the cultural angst in a 1948 Armistice Day speech.

> Our knowledge of science has clearly outstripped our capacity to control it. We have many men of science, but too few men of God. We have grasped the mystery of the atom and rejected the Sermon on the Mount. Man is stumbling blindly through a spiritual darkness while toying with the precarious secrets of life and death. The world has achieved brilliance without wisdom, power without conscience. Ours is a world of nuclear giants and ethical infants.[13]

THE ENDLESS FRONTIER: FROM WONDER TO FEAR

In November of 1944, Roosevelt had asked Bush to consider how the wartime science organization might be extended to benefit the country in peacetime, to improve national security, aid research, fight disease, and develop the scientific talent of the nation's youth.[14] After the war was won, Bush submitted his report to President Harry S. Truman. *Science, the Endless Frontier*, made the case that the creation of knowledge is boundless in its potential.[15] The report is widely credited with laying the groundwork for the second golden age of American science, during which the federal government, not wealthy philanthropists, became the principal funder of scientific research in peacetime as well as in war.

In his report, Bush argued that science was of central importance for freedom, an argument that was powerfully underscored when, in the dog days of August 1949, the Soviet Union detonated an atomic bomb of its own, as Oppenheimer had feared it would. America's sense of impending doom over the power scientists had unleashed by splitting the atom turned into a profound fear of a clear and present danger. In less than a year, a bill creating the National Science Foundation (NSF), a subject that had been languishing, was signed into law, and science began to undergo a largely unnoticed but profound change in its central relationship to American culture. The grand enterprise that had for two centuries been motivated by a sense of *wonder* on the part of noble idealists and adventurers, wealthy visionaries, and civic-minded philanthropists was now motivated by federal investments propelled by a sense of *fear*.

DUCK AND COVER

It was a fear that would impact the broader American culture for the next fifty years. In the first decade of the cold war, American defensiveness bordered on hysteria. We knew what a nuclear weapon could do: We'd done it. And now the possibility was very real that it would boomerang back on us. And the horror was that what was at risk was our most precious asset: our children. The baby boom.

The Federal Civil Defense Administration determined that the country that would win a nuclear war was the one best prepared to survive the initial attack. Achieving this required a homeland mobilization on an unprecedented scale, and our children needed to know what to do when nuclear war came. They commissioned a nine-minute film called *Duck and Cover* that showed Bert the turtle pulling into his shell to survive a nuclear explosion that burns everything else. The film exhorted millions of schoolchildren to "duck and cover" like Bert by immediately covering the backs of their heads and necks and ducking under their desks if they saw a bright flash. The film didn't mention that the gamma-ray burst, which carries most of the lethal radiation, arrives with the flash. The children were reminded regularly that because a nuclear attack could happen at any time, they, like soldiers in a combat zone, needed to maintain a high level of alertness, forever ready to duck and cover. They did drills in school, and participated in citywide mock Soviet atomic bomb attacks.[16]

RUN LIKE HELL

This was terrorism on a new scale. Imagine that al-Qaeda is in charge of a country the size of the Soviet Union and has nuclear weapons trained on the United States, and you can get a sense of the fear that was driving the nation. We knew what these weapons could do, and we knew they could be used again. Our only option was to plan for an attack on American soil. This knowledge changed American culture and its relationship to science.

For example, it has long been the prevailing opinion that American suburbs developed as a result of the increased use of the car, GI Bill–funded home construction, and white flight from desegregated schools after the 1954 Supreme Court decision in *Brown v. Board of Education of Topeka*. But in reality the trend had started several years before *Brown*.

In 1945, the *Bulletin of the Atomic Scientists* began advocating for "dispersal," or "defense through decentralization" as the only realistic defense against nuclear weapons, and the federal government realized this was an important strategic move. Most city planners agreed, and America adopted a completely new way of life, one that was different from anything that had come before, by directing all new construction "away from congested central areas to their outer fringes and suburbs in low-density continuous development," and "the prevention of the metropolitan core's further spread by directing new construction into small, widely spaced satellite towns."[17]

Nuclear safety measures drove the beginning of the abandonment of our cities. After being told that "there is no doubt about it: if you live within a few miles of where one of these bombs strike, you'll die" and "We can always hope that man will never use such a weapon but we should also adopt the Boy Scout slogan: Be prepared," getting far enough out of the "target" city so that the blast might be survivable seemed wise. Those who could afford to left. Those who remained were generally less affluent, and minorities made up a disproportionate share of the poor.

A far worse development for minorities in America came in 1954, when the federal Atomic Energy Commission realized that with the advent of the vastly more powerful hydrogen bomb, "the present national dispersion policy is inadequate in view of existing thermonuclear weapons effects."[18] But by then it was too late; the suburbs were growing, but offices were still by and large downtown. A new strategy was needed.

President Dwight D. Eisenhower instead promoted a program of rapid evacuation to rural regions. As a civil defense official who served from 1953 to 1957 explained, the focus changed "from 'Duck and Cover' to 'Run Like Hell.'"[19]

Cities across America ran nuclear attack drills, each involving tens of thousands of residents, practicing clearing hundreds of city blocks in the shortest possible time.[20] It became clear that this would require massive new transportation arteries in and out of cities. The resulting National Interstate and Defense Highways Act of 1956 was the largest public works project in history. It created a system that provided easier access from the suburbs into cities as well as a way to more rapidly evacuate cities in case of nuclear war. The new freeways had to be built in a hurry and were naturally routed through the cheapest real estate, which usually meant plowing through richly tapestried and vibrant minority communities, displacing millions. Although poverty had been concentrated in these very neighborhoods, their destruction ripped apart the social fabric of America's uprooted minority communities for years, destroying social support networks and leading to a generation of urban refugees.

These accommodations for defense brought about an immense change in the fabric of America, altering everything from transportation to land development to race relations to modern energy use and the extraordinary public sums that are spent on building and maintaining roads—creating challenges and burdens that are with us today, all because of science and the bomb.

THE PROTECTION RACKET

The fear that was changing the nation stepped up another notch with the Soviet launch of *Sputnik 1*, the first Earth-orbiting satellite, on October 4, 1957. Its diminutive size—about the size of a beach ball—made it perhaps the most profoundly influential twenty-three-inch-diameter object in history. Traveling at roughly eighteen thousand miles an hour, the shiny little orb circled the planet about once every hour and a half, emitting radio signals that were picked up and followed by amateur radio buffs the world over, but nowhere more closely than in the United States.

Sputnik shocked America in ways that even the 1949 Soviet nuclear test had not, because for the first time, the Commies were not just catching up, they were ahead—and now they had the high ground. The fear was that America stood at risk of being overrun by yet another totalitarian, closed society. This was the cultural context that *Sputnik* brought to a focal point, and so it was different from the current characterization that China is now presenting the United States with an economic "Sputnik moment."

The little orb focused the amorphous fear and placed the entire country in danger, at least psychologically. Broad debates had been swirling since the 1949 Soviet nuclear test about the need to invest more in education, particularly in science, technology, engineering, and mathematics (often referred to as simply STEM) because of their critical importance to national defense, but they foundered on the shoals of congressional indifference, much as they are today. Now those debates came into sharp focus. As historian JoAnne Brown put it:

> The struggle for federal aid may have been won in the sky, but it was fought in the basements, classrooms and auditoriums, as educators adapted schools to the national security threat of atomic warfare and claimed a proportionate federal reward for their trouble.[21]

Within a year, the National Defense Education Act of 1958 was passed, with the goals of improving education in defense-related subjects at all grade levels and bolstering Americans' ability to pursue higher education, and science would become a major issue on the presidential campaign trail in 1960. If we didn't recommit to science and technology, we might lose the cold war! Our entire way of life, perhaps our very survival, was now at stake, and it all hinged on what we could do to protect ourselves by reinvesting in science and technology to catch up to[22] and then to resoundingly beat "those damn Russkies." The NSF's budget, which had been quite low, jumped dramatically in 1957 and continued to grow.

American public opinion, which had been one of great moral ambivalence about science for twelve years since the bombings of Hiroshima
and Nagasaki, began a new relationship with it almost overnight: Scientists might be sons of bitches, but they were our sons of bitches.

GIMME SHELTER

At the same time that the federal government was promoting *Duck and
Cover* and transforming on an unprecedented scale how and where we
lived and traveled, the Office of Civil and Defense Mobilization (OCDM)
was broadly distributing public service pamphlets whose intent was to
instill in everyone a sustained alertness to danger, the better to prepare
the country to survive the first wave of a nuclear attack. These pamphlets were bundled with vinyl record albums of recorded survival
instructions such as the below, from Tops Records:

> Our best life insurance may be summed up in four words: be alert;
> stay alert. This will take some doing on your part. It will take inge-
> nuity; it will take fervor; it will take the desire to survive. . . . There is
> no doubt about it: If you live within a few miles of where one of these
> bombs strike, you'll die. It may be a slow and lingering death, but it
> would be equally as final as the death from the bomb blast itself.
> You'll die, unless you have shelter. Shelter from the intense heat, and
> the radiation that is the by-product of a nuclear explosion. . . . Let's
> assume bombs fall before you have time to prepare a shelter, or while
> you wait, in the belief atomic war will never come. We can always
> hope that man will never use such a weapon, but we should also
> adopt the Boy Scout slogan: Be prepared. . . . It may be safe for you to
> leave your house after a few hours, or it may be as long as two weeks
> or more. Two weeks with very little food or water . . . tension . . .
> unaccustomed closeness. Two weeks with sanitary facilities most
> likely not operating. No lights. No phone. Just terror.[23]

Fallout shelters were built around the country. New commercial
buildings had them. New homes had them, and residents stocked them
with water, canned goods, candles, blankets, and tranquilizers. Public

buildings had them installed. The possibility of sudden nuclear annihilation at the hands of the evil Communists became part of everyday life, and many families rehearsed and planned for living for extended periods in dark and rancid basement shelters, and for being separated from one another indefinitely if an attack came when the children were in school. Some schools distributed metal dog tags similar to those worn by soldiers so that the burned bodies of students could be identified in the days and weeks after a nuclear attack. What grief and what horrors science had brought upon us. Our most precious new baby boom of children was growing up in a climate of terror.

WHEN SCIENCE WALKED OUT ON POLITICS

By this time it was clear that science was the answer to the twin threats of the arms race and *Sputnik*. And that America was, in fact, in a *science race,* as Vannevar Bush had essentially argued in his report *Science, the Endless Frontier.* Science had become one of the primary weapons in a new kind of war. The nation that invested the most in science and engineering research and development would lead the world.

Science had in the span of two short decades attained a sacred cow status enjoyed by few other federal priorities. Gone were the days of scientists needing to reach out to wealthy benefactors to justify and explain their work. The adoption of science as a national strategic priority changed the relationship between science and the public. Over the course of a single generation, government funding allowed scientists to turn inward, away from the public and to their lab benches.

But this relationship came with the conflicting emotions of need and resentment. Though their work is by nature antiauthoritarian, scientists became figures of authority in white lab coats, bland, dry, value neutral and above the fray. This new image of science, implanted in baby boomers by hundreds of classroom filmstrips, couldn't have been less inspiring or further from the truth. By their nature, scientists are very often intensely passionate and curious, and these are the very qualities that typically motivate their interest in science—the exploration of creation—to begin with. But very little of that characteristic passion and curiosity would be communicated to the general public for the next *fifty years*. Science became *the other* culture of quietly cloistered intellectual monks.

With tax money pouring in from a vastly expanding economy and the public respect afforded the iconic authority of the white lab coat, two generations of scientists instead had to impress their own university departments and government granting agencies to keep research funds coming their way. But they no longer had to impress the public.

At the same time, science was becoming less accessible even to other scientists. As research became increasingly specialized, no one could keep up on all the latest findings. There was simply too much information. With scientists not being able to follow each other outside their own fields, reaching out to the public seemed a hopeless exercise. What mattered was not *process*, but *results*. University tenure tracks rewarded the scientists who had successful research programs and multiple professional publications, but gave no similar consideration to great science communication or public outreach. "Those who can, do," the attitude of scientists became, "and those who can't, teach." It was a horrible mistake.

Locked in a subculture of competitive, smart, and passionate people focused on their own research, on *doing*, scientists seemed to forget that they were responsible to—indeed, part of—the community of taxpayers that funded much of their work and so deserved a say in what they did, or at least *to be sold*. Scientists as a group became notoriously cheap donors of both time and money and withdrew from civic life in other ways. Giving back and participating in the greater civic dialog just wasn't part of their culture. As in any cloistered society, attitudes of superiority developed within the science community—attitudes that ran counter to the fundamental antiauthoritarian nature of scientific inquiry.

Many scientists came to view politics as something dirty and beneath them. Arguing that they did not want to risk their objectivity, they eschewed voicing opinions on political issues. So while science was entering its most dizzyingly productive and politically relevant period yet, very little of this creativity was relayed to the public. Only the results were publicized. From the public's perspective, the science community had largely withdrawn into its ivory tower and gone silent. This proved to be a disaster.

PUBLIC SENTIMENT IS EVERYTHING

American democracy is often confounded by the mistaken idea that *politics* is the lowly part of the business—what you have to put up with in order to enact *policy*—but in fact the opposite is true. This mistake is made especially often by scientists, who view politics as dirty. Abraham Lincoln eloquently illustrated this point when he debated his opponent Stephen Douglas in the 1858 Illinois campaign for the US Senate. Lincoln lost that election, but he forced Douglas to explain his position on slavery in a way that alienated the Southern Democrats. That set Lincoln up to defeat him in the race for president two years later. "Public sentiment," Lincoln said, "is everything. With public sentiment, nothing can fail; without it, nothing can succeed. Consequently he who molds public sentiment goes deeper than he who enacts statutes or pronounces decisions. He makes statutes and decisions possible or impossible to be executed."[24] Thus politics, which moves the invisible hand of democracy, is more important than policy. It reflects and shapes the will of the people. It is the foundation on which policy is based. Lincoln's thinking in this regard echoed that of Thomas Jefferson.

Scientists were certainly smart enough to realize this, but the structure that was put in place under Vannevar Bush's grand vision worked against it, for who now had to worry about shaping public sentiment? As president of the Carnegie Institution of Washington, Bush was very familiar with all the time and resources that fund-raising required, and his goal was to lift the onus of obtaining research funding off scientists and universities to propel the nation forward in a more coordinated way. But the need to sell the worth of one's work to the public and donors, to converse about new discoveries and their meaning, and to inspire and excite laypeople may be the only thing that keeps the public invested and supportive in the long term. Bush may have done the job too well. The shift to public funding changed the incentive structure in science.

This might not have been a problem if scientists had valued public outreach, but by and large they didn't. As economists are quick to point out, people often adjust their behavior to maximize the benefit to themselves in any given transaction, and the economics of the new science structure rewarded research but not public outreach or engagement. As a result, most scientists ignored it and science coasted off the largess of the taxpayers' fear of the Soviets.

THE TWO CULTURES

The growing divide between science and mainstream culture was happening in England too, and the situation in both countries was famously articulated by British physicist, novelist, and science adviser C. P. Snow, a man who straddled both worlds like the scientists of old had and who warned in a famous 1959 lecture titled "The Two Cultures and the Scientific Revolution" that the widening communication gulf between the sciences and the humanities threatened the ability of Western peoples to solve their problems.

> A good many times I have been present at gatherings of people who, by the standards of the traditional culture, are thought highly educated and who have with considerable gusto been expressing their incredulity at the illiteracy of scientists. Once or twice I have been provoked and have asked the company how many of them could describe the Second Law of Thermodynamics. The response was cold: it was also negative. Yet I was asking something which is about the scientific equivalent of: "Have you read a work of Shakespeare's?"
>
> I now believe that if I had asked an even simpler question—such as, "What do you mean by mass, or acceleration," which is the scientific equivalent of saying, "Can you read?"—not more than one in ten of

the highly educated would have felt that I was speaking the same language. So the great edifice of modern physics goes up, and the majority of the cleverest people in the western world have about as much insight into it as their neolithic ancestors would have had.[25]

Scientists didn't see this as a warning or an invitation to reach out; rather, they viewed it by and large as a criticism of the willful ignorance and snobbishness of those practicing the humanities. To a certain extent, their view was justified: Intellectuals weren't giving their work its due. The fast-growing importance of the sciences was garnering scientists considerable funding and public regard in exchange for the new powers and freedoms they were giving society, yet that same society's highbrows still refused to acknowledge their work's significance.

But life in survival mode, as the "Soviet menace" necessitated, made what could save us the priority and relegated other important things to the status of "luxuries." Suddenly we didn't have the "luxury" of indulging wonder, or the humanities, to the extent we once had. And science had adeptly proven its utility to society, as Snow argued. Although he criticized scientists who could scarcely make their way through Dickens with any understanding of its subtleties, he saved his harshest criticism for British universities, which had underfunded the sciences to the benefit of the humanities despite the former's contributions, and for the snobbishness of literary intellectuals. "If the scientists have the future in their bones," he said, "then the traditional culture responds by wishing the future did not exist." It was a statement that could just as easily describe the US Congress some fifty years later.

The lecture was printed in book form and widely debated in the United States as well as in Britain. It has been declared one of the one hundred most influential Western books of the last half of the twentieth century.[26]

Snow envisioned a third culture emerging of people schooled in both the sciences and the humanities, but that is not what took place. A great change had begun in Western universities, and humanities professors felt themselves inexorably slipping from the top spots and being supplanted by scientists, who generally seemed like they couldn't have cared less about the humanities. Why bother with all the reading and writing and talking when science was actually *doing* things? But this was equally shortsighted, and in the shift we let go of something precious: our grasp on the classics that had informed Western culture. Since scientists couldn't be bothered with civics, democracy continued to draw its elected leaders primarily from the humanities, and so we were also creating a culture war that is with us to this day, threatening the ability of Western democracies to solve their problems—just as Snow feared.

THE SCIENTIFIC-TECHNOLOGICAL ELITE

Democracy is, as we now know, rooted in science, knowledge, and the biology of natural law that is at the base of our legal framework. But most of our elected leaders have not had significant training in science, or in how the ideas rooting law and democracy relate to science. In the middle years of the twentieth century, this was beginning to pose a problem.

The twin threads of fear and resentment created a growing sense that science might be outpacing the ability of a democracy to govern itself. The situation was alarming enough that it compelled President Eisenhower to warn the American people about it. On January 17, 1961, in his farewell address to the nation, he famously warned of the dangers of the emerging "military-industrial complex." Ike blamed the growing funding of science by the federal government as the primary cause of the rising behemoth and complained that the solitary inventor was being overshadowed by teams of scientists in cloistered labs hidden from the watchful eye of the public and awash in taxpayer money. "In holding scientific research and discovery in respect, as we should," Ike warned, "we must also be alert to the equal and opposite danger that public policy could itself become the captive of a scientific-technological elite."[27]

How far we had come from the days of the first State of the Union address, when George Washington told Congress that "there is nothing which can better deserve your patronage than the promotion of Science and Literature." The very possibility of democracy had been created by a scientific-technological elite that had included Thomas Jefferson, Benjamin Franklin, George Washington, Benjamin Rush, and other founding fathers. Elitism had been something to aspire to. Now it was something to be feared.

CHAPTER 6

Science, Drugs, and Rock 'n' Roll

Be wise, strain the wine; life is short; should hope be long? While we speak, envious time has already slipped away. Seize the day, trust as little as possible in the future.

—HORACE[1]

THE GRADUATES

This warning about science by a well-liked outgoing president fed into the momentous changes afoot in the general culture. Traumatized by a dozen years of maintaining a high alert to the threat of nuclear holocaust at any moment, the public's patience was wearing thin. The enormously more powerful H-bomb had made "duck and cover" a ludicrous farce. Advances in government- and industry-funded science were viewed with increasing skepticism, and the baby boomers found their power by questioning the authority of the government and, by extension, government science. Parents didn't have any good answers. This was new territory. The social order was being upended. Scientists and engineers as a whole were suddenly seen as vaguely jingoistic and associated with the military-industrial complex, as were traditional organized religion and other sources of authority. The general culture began moving more toward nature, hedonism, and spiritualism.

And why not? Faced with what seemed like the collapse of the main-stream culture's moral authority but lacking the power to change it, many baby boomers over the next decade either railed with rage in anar-chistic riots or tuned in, turned on, and dropped out, setting the twin paths of anarchy and spiritualism that have come to define the genera-tion. World-renowned British mathematician Bertrand Russell captured the dour pessimism in a 1963 *Playboy* magazine interview.

> The human race may well become extinct before the end of the present century. Speaking as a mathematician, I should say that the odds are about three to one against survival. The risk of war by accident—an unintended war triggered by an explosive situation such as that in Cuba—remains and indeed grows greater all the time. For every day we continue to live, remain able to act, we must be profoundly grateful.[2]

With towering intellectual figures like Russell, who cowrote the iconic *Principia Mathematica*, won the Nobel Prize in Literature, and began cham-pioning logic and freedom at the beginning of the twentieth century, mak-ing these sorts of dire pronouncements, living just for today—and being profoundly grateful for the chance—seemed like an entirely rational idea.

THE DARK SIDE OF THE MOON

The next president would attempt to turn the tide of fear and restore a sense of wonder to science, not from a basis of vision as much as despera-tion. By May of 1961, John F. Kennedy had been in office for only a little more than three months and had already stumbled into deep trouble. The recession of 1958, the worst since the end of World War II, had been quickly followed by another one that began during the presidential cam-paign of 1960 and lasted into 1961. The Bay of Pigs had already occurred, resulting in an embarrassing failure for the three-month-old administra-tion. Five days before the thwarted invasion, the Soviets had sent the first human into orbit, pulling the rug out from under Kennedy's campaign rhetoric about besting them in the space race. Soviet premier Nikita

Khrushchev was testing him in every way he could, keeping him on the ropes, a pattern that would continue all year long and include the building of both the Berlin Wall and the nuclear missile sites that resulted in the Cuban Missile Crisis. Back on his heels, his credibility on the line, Kennedy looked weak and outclassed. He needed a way to turn the political boxing match around and assert his leadership. He turned to the Moon—not for science's sake, but to use science to beat Khrushchev. He later admitted as much in a November 1962 meeting with NASA administrator James Webb. "I am not that interested in space," he told Webb. The main reason he wanted Apollo was because of its importance in the cold-war rivalry with the Soviet Union.[3]

The Russians had led with *Sputnik 1* in 1957 and a month later had sent a dog into space on *Sputnik 2*. Two years after that they had crash-landed *Luna 2* on the Moon. Now they had beaten the United States yet again by sending the first human, Yury Gagarin, into space on April 12, 1961. On May 25 of that year, Kennedy addressed a joint session of Congress and laid out several "urgent national needs," among them getting America back into the space race.

> Recognizing the head start obtained by the Soviets with their large rocket engines, which gives them many months of leadtime, and recognizing the likelihood that they will exploit this lead for some time to come in still more impressive successes, we nevertheless are required to make new efforts on our own. For, while we cannot guarantee that we shall one day be first, we can guarantee that any failure to make this effort will make us last.[4]

He laid out a bold agenda, a desperate and visionary agenda, to regain the lead—militarily, ideologically, and in national prestige—and to at the same time turn around the economy by landing a man on the Moon and returning him safely to Earth. The effort would require a peacetime science mobilization on a par with the Manhattan Project, requiring the building of entire cities to support it. At its peak, the Apollo Program would employ some 400,000 people.[5]

There were just two problems: We didn't have a clue about how to do it, and we didn't have the money either, with the country in a major recession and federal tax revenues down. To make matters worse, Kennedy's inspirational ideas for domestic policy were getting shot down. He had painted a grand vision of the New Frontier and the War on Poverty, but he couldn't get Congress to pay for a major expansion of social programs, at least not yet. But if a program were tied to the cold war, he thought he could get Congress to support it like they had the National Interstate and Defense Highways Act five years before and the National Defense Education Act of 1958.[6] Apollo could be a bold new vision and a jobs program, an economic stimulus with benefits. The cultural anxiety was so high by then that the very idea of Russians crawling all over the Moon caused a visceral reaction in many Main Street Americans. Kennedy thought he had a winner.

But once the budget numbers came back, they showed that the program would cost almost $20 billion over eight years, eating up all the discretionary funds that Kennedy needed for his War on Poverty. If he wanted Apollo, he would likely have to sacrifice everything else. He began looking for a way out.

He realized that if he took away the cold-war justification, he'd lose the support of fiscal conservatives, and he could use that loss to move the deadline back indefinitely by defunding it. So he reached out to Khrushchev, his worst enemy, at the Vienna Summit and suggested over lunch that they bury the space-race hatchet and go to the Moon together as a cooperative venture. Surprised, Khrushchev at first said no. Kennedy looked at him and Khrushchev said, "All right, why not?" and then thought about what it was going to cost and changed his mind again, saying that disarmament was a prerequisite for cooperation in space.[7]

Kennedy sold the idea in a speech at the University of California, Berkeley, in March 1962, and another at the United Nations in September 1963. Fiscally conservative Democrats who had backed the program now saw Kennedy's support wavering and started jumping ship.[8] The Senate proposed cutting NASA's funding and scrapping the Apollo Program.[9] Kennedy's political escape plan was working.

In general, Americans were more concerned about domestic issues like poverty, race relations, and the economy,[10] and suspicion of science because of the many bad things it could produce—and had produced— was growing. In 1962, marine biologist Rachel Carson's book *Silent Spring* came out and made a permanent impact on the national psyche, shocking America into an awareness of chemical pollution, reaffirming Eisenhower's warnings about the scientific-technological elite, and launching the US environmental movement. That same year, only about 35 percent of Americans thought Apollo was worth the cost.

At the same time, opposition to the idea was building in the science community. By 1963, the Senate Committee on Aeronautical and Space Sciences was holding hearings and scientists were testifying that human spaceflight was inordinately expensive and risky, with its paramount objective being not research, but bringing astronauts back alive. Unmanned spaceflight with robots, they said, would cost much less and return much more information.

The nation was adrift, and the public felt hard-pressed to think of science as a good thing. Why couldn't we spend our money on putting our best scientific minds to work solving issues here on Earth? The cost of funding the newly created NASA and the Apollo Program was incredible—by 1966, it would reach as high as 4.5 percent of the federal budget, an astronomical figure compared to today, which is about one-ninth of that level.[11]

But then, in November of 1963, everything changed. Kennedy was assassinated in Dallas, and the next year the elections went overwhelmingly to the Democrats. Newly installed president Lyndon Johnson fully committed to the Apollo Program and the Great Society in honor of Kennedy's memory. Very few could stand opposed, and in a dramatic turnaround, both initiatives were funded.

The funding of Apollo stood in increasing contrast with the mainstream culture. In 1965, Ralph Lapp, the former head of the nuclear physics branch of the Office of Naval Research,[12] captured the growing fear of the nation when he published *The New Priesthood*, in which he

argued Eisenhower's thesis that the "scientific elite"—people who understood how science and technology work—were starting to supplant our
elected leadership in control of things. The book's argument reflected an
emerging idea that "democracy faces its most severe test in preserving its
traditions in an age of scientific revolution."[13]

CARPE DIEM

The live-for-today ethos was changing the way Americans approached
living, consumption, and finance. The change was particularly evident
among the emerging baby-boom generation. With the older generation
having lost its moral credibility, the boomers felt the older crowd had no
right to tell young people anything. "Today there is great concern among
my generation that an era of permissiveness has resulted in unrest among
our young people," said California governor Ronald Reagan on May 20,
1967. "But just to keep things in balance there is a widespread feeling
among our young people that no one over 30 understands them."[14]

These baby boomers, feeling powerless, needed an outlet for their anger
and moral distrust of the older generation, so they adopted the protest
songs of folk music. Singer-songwriter Bob Dylan became an overnight
sensation, the "poet to a generation."[15] Satire became a dominant cultural
art form, lampooning all kinds of authority for its hypocrisy and failure.
The Graduate, Catch-22, One Flew Over the Cuckoo's Nest, MAD magazine, *Dr. Strangelove, Saturday Night Live,* and many other satirical cultural touchstones were fueled by rage against the dominant culture. The
hilarious, humanitarian, childlike but darkly pessimistic novels of Kurt
Vonnegut Jr. became runaway hits. In 1970, Vonnegut gave a commence-
ment address at Vermont's Bennington College in which he famously
said, "Everything is going to become unimaginably worse and never get
better again."[16] He cautioned the baby boomers that "we would be a lot
safer if the Government would take its money out of science and put it
into astrology and the reading of palms. I used to think that science
would save us. But only in superstition is there hope." American kids

loved his cynical pessimism because he gave voice to their fear and anger. A year before he died in 2007, Vonnegut showed the same satirical mix of anger and regret when he penned a "Confetti print," which he sold numbered silk screens of:

> *Dear future generations:*
> *Please accept our apologies. We were roaring drunk on petroleum.*
> *Love, 2006 A.D.*

But by that time, the baby boomers were in charge, and few were still listening.

LADIES AND GENTLEMEN, STEP RIGHT UP!

Into this inferno of cynicism went NASA. Because of the nationalistic and competitive nature of its mission, NASA, more than any other US science agency, had come to understand Lincoln's dictum on the importance of public sentiment. In a way, the whole Apollo Program was a big show of American technological prowess and innovation—a show intended to influence public sentiment—that relied on science and just happened to spin off immense benefits. But the showmanship, wonder, and politics were lost on most other scientists, who saw no value—or worse, a negative value—in showing personality or telling engaging stories about their work intended to inspire the public. Hubris, after all, is what had ruined Harlow Shapley's career; it was the gateway to the rosy path of a priori principles.

From a pure science viewpoint, human spaceflight was wasteful. But from a public engagement and thus overall funding viewpoint, it was sheer genius, because it gave the public a narrative for appreciating science. It was in some senses an example of the third culture C. P. Snow had hoped for—a marriage of literary resonance with "doing the big things," as Kennedy had urged in his United Nations address.

The remainder of the 1960s and 1970s, however, brought increasing disillusionment with the authority of the mainstream culture and the military-industrial complex that science had become closely aligned with. NASA battled this growing antiscience backlash with an increasingly powerful narrative that culminated in the dramatic 1968 Apollo 8 mission, in which humans left Earth's orbit for the first time and traveled to orbit the Moon in a capsule named the *Beagle 2*, after Darwin's vessel. In a Christmas Eve television broadcast that was the most-watched TV program to that time, the astronauts read the first ten verses from the Book of Genesis. The mission returned to Earth on December 27, the day on which Darwin's HMS *Beagle* had departed England 137 years before.

This symbol-rich unifying of science with its progenitor, religion, in a common sense of wonder at the great questions of the universe, and in humanity's sense of curiosity and exploration, captivated the nation. Pop culture put astronauts everywhere. The *Peanuts* comic strip character Snoopy, a beagle, wore a spacesuit on the popular Snoopy the Astronaut pins and was the most popular cartoon character of the day—so much so that it became the symbol of the Apollo Program. The journey of Apollo 8 set up the nation for the incredible dramatic climax of Apollo 11 less than seven months later, when American astronauts would actually land and walk on the Moon, fulfilling Kennedy's goal before the decade was out.

But in the months leading up to the grandest of all human adventures, voices of concern began to be heard anew. Despite the incredible successes of the Apollo Program and the spiritual sense of wonder and moment its narrative generated, the growing distrust of the quasi-military organization and culture that NASA had adopted to get the job done recalled Eisenhower's warning of eight years before.

This sense of distrust in the American science enterprise was particularly true in the nation's debates over race and poverty, which had been fueled by the civil rights movement, the destruction of many minority communities by highway construction, and the increasing disparity in wealth between the predominantly black urban cores and the new white suburbs.

The Reverend Ralph Abernathy, who had cofounded the Southern

Christian Leadership Conference with the Reverend Dr. Martin Luther King Jr., carried on with King's new focus on economic justice for the poor of all races after King's assassination on April 4, 1968. In 1969, Abernathy led a national march against the Apollo 11 Moon launch, criticizing the incredible spending as an "inhuman priority"[17] at a time when so much suffering existed in the nation. "One fifth of the population lacks adequate food, clothing, shelter, and medical care,"[18] he argued.

> A society that can resolve to conquer space; to put man in a place where in ages past it was considered only God could reach; to appropriate vast billions; to systematically set about to discover the necessary scientific knowledge; that society deserves both acclaim and our contempt . . . acclaim for achievement and contempt for bizarre social values. For though it has the capacity to meet extraordinary challenges, it has failed to use its ability to rid itself of the scourges of racism, poverty and war, all of which were brutally scarring the nation even as it mobilized for the assault on the solar system.[19]

NASA administrator Thomas O. Paine met with Abernathy on the eve of the Apollo 11 launch and eloquently said that if he could alleviate the suffering by not pressing the launch button he would, but that sending three men to the Moon was "child's play" compared to "the tremendously difficult human problems" Abernathy and his people were concerned with.[20] Incredibly, the exchange ended agreeably, with Abernathy agreeing to pray for the lives of the astronauts.

There is a very good chance that, had Kennedy lived, the United States would not have put a man on the Moon by 1969, or even perhaps by 1979, because the costs and the politics were tilting heavily against it. The conflict between social and science spending lives on to the current day.

The Apollo Program's ultimate triumph of both technology and narrative inspired the pop culture in profound ways. The mix of science and religion in the context of awe and adventure seemed, in spite of the times, to be about the best of humanity, and it was memorialized in songs like David Bowie's "Major Tom," Deep Purple's "Space Truckin'," Elton John's

"Rocket Man," Joni Mitchell's "Willy," Pink Floyd's "Brain Damage" and "Eclipse" from *The Dark Side of the Moon,* and many more, and in films, including the ambivalent *2001: A Space Odyssey* and later the more optimistic *Star Wars.* It brought America together, briefly, in a moment of shared pride, wonder, and imagination. And it spun off scientific and technological advances that laid the groundwork for the digital revolution.

THE SEVEN STAGES OF TECHNOLOGICAL ADAPTATION

But space was the exception. In other ways, science was taking it on the chin. As the country was roiled by antiauthoritarian demonstrations against the Vietnam War, against poverty, against racism, against "the man," science came to be seen as a part of "the man"—the insensitive, cloistered, jingoistic, empowered white and largely male cultural elite Eisenhower had warned the nation about. Things seemed out of control and America's youth felt trapped, like Jack Nicholson, in a "cuckoo's nest" of insanity where it was the authorities and caretakers who were brutal and insane.

By the 1970s, compounding the problems of racial inequality, nukes, poverty, sex discrimination, and the war, the destructive environmental consequences of what could be termed "object-oriented" science and technology were becoming abundantly clear. What had been viewed as great breakthroughs in the nuclear, chemical, and agricultural sciences had been rushed into broad application in a more naive and optimistic time and were beginning to boomerang back with hopelessly broad and horrible environmental consequences. What had been a manageable impact in the smaller scale of the past was a devastation on the vastly larger scale that science and technology now made possible.

In 1971 *The Lorax,* a children's book by Dr. Seuss that would become a classic of environmental literature, was published. A parable about the environmental costs of industrialized technology, this book also presented the theme of scientific and technological power outpacing moral and ethical development, for which humanity has no similar tools of rapid advancement. Year after year, more and more environmental disasters were bearing this out as industrial chemicals and processes were found to have unanticipated and sometimes catastrophic environmental consequences, establishing a pattern that became the all-too-familiar hallmark of industrial science and muted the euphoria over the Moon shot. In the coming years, it would be the conservative business types who were proscience and the liberal environmentalists who grew antiscience—nearly the opposite of the politics of the 2010s. The process followed seven stages, over and over.

DISCOVERY

A new process or tool (for example, a chemical or, today, a nanotechnology or genetic technology technique) is discovered that vastly expands utility, power, convenience, or efficiency.

APPLICATION

Object-oriented industrial applications are developed to apply the process or tool to a real-world problem, often vastly increasing productivity and lowering costs, but the science of biocomplexity and ecology—of how the process or tool will affect and be affected by the broader context, from the body to the environment,—lags behind.

DEVELOPMENT

Industries grow up around the new application and its use intensifies.

BOOMERANG

A tipping point is reached where the application has broadly noticeable negative effects on health or the environment. Fueled by the public outcry, scientists study the degree of the systemic effect to determine what to do.

BATTLE

Regulations are proposed, but vested economic interests sense a potentially lethal blow to their heavily invested production systems and fight the proposed changes by denying the environmental effects, denying association, maligning and impeaching witnesses, and/or arguing that other factors are causing or are also involved in the mounting disaster. A battle ensues between the adherents of old science and those of new science. This stage is caustic to science's credibility, because it becomes a rhetorical tool and facts are cherry-picked to win arguments.

CRISIS

Evidence continues to accumulate until the causation becomes irrefutable, often through dramatic deaths or disasters finally tipping the politics in the direction of reform.

ADAPTATION

Regulations are passed or laws are changed to stop or modify the use and to attempt to mitigate the effects, and the mainstream perspective grudgingly shifts (because the science is complex and counterintuitive) from an object-focused tool approach to a complex systems approach that takes into account the relationships between the object and the environmental and/or physiological context . . . or not, in which case the process returns to stage 3, development.

This seven-stage pattern has been repeated since the 1960s with escalating stakes. The same drama of object-oriented science and the development of well-intentioned but simplistic and naive technological solutions has led to greed and denial of consequences over and over again, stumbling from DDT to asbestos to acid rain to Love Canal to the hole in the ozone layer to Three Mile Island to the Dalkon Shield to toxic shock syndrome to lead paint to atrazine and frog deformities to the Vioxx scandal and eventually to the granddaddy of them all, global climate change.

WONDERS UPON WONDERS

By the late 1970s, science had become known for environmental and health debacles, all playing out against the nuclear backdrop, and even the victories of NASA couldn't stem the growing antiscience sentiment. A 1979 NBC/AP poll showed that just 41 percent of Americans thought the benefits of the space program outweighed its costs.[21] Philip Handler, the president of the National Academy of Sciences, cautioned that scientists needed to find a way to reverse the trend of increasing public skepticism toward science.[22]

Seeing this national retreat from science, American astronomer Carl Sagan took up Hubble's mantle and joined a small group of science popularizers—including his friend the evolutionary biologist Stephen Jay Gould—that was trying to turn things around. The discoveries Hubble and his contemporaries had made about the universe had inspired Sagan from the moment he had first learned of them as a four- or five-year-old at the 1939 New York World's Fair. "I was a child in a time of hope," he said. "I wanted somehow to immerse myself in all that grandeur. I was gripped by the splendor of the Universe, transfixed by the prospect of understanding how things really work, of helping to uncover deep mysteries, of exploring new worlds—maybe even literally."[23]

But now it was a new era in which science had, with the exception of NASA, gone somewhat silent and become disengaged from the broader culture, when its justification for funding had shifted from that special brand of wonder Sagan had felt as a child to the darker, smaller-minded fear and pragmatism of business and national defense. Sagan saw in the growing divide between science and society the rise of a "demon-haunted world" as the prevailing national dialogue shifted away from science to the ESP, New Age mysticism, faith healing, miracles, and UFO abductions that increasingly occupied the popular press.

In his view, this threatened the fabric of America:

> I have a foreboding of an America in my children's or grandchildren's time—when the United States is a service and information economy; when nearly all the key manufacturing industries have slipped away to other countries; when awesome technological powers are in the hands of a very few, and no one representing the public interest can even grasp the issues; when the people have lost the ability to set their own agendas or knowledgeably question those in authority; when, clutching our crystals and nervously consulting our horoscopes, our critical faculties in decline, unable to distinguish between what feels good and what's true, we slide, almost without noticing, back into superstition and darkness.[24]

Sagan saw in the growing public mysticism and antiscience a hunger for the lost sense of wonder and adventure into mystery that science had once provided. By then, the public funding of science had been driven by fear for a generation. He set out to change that and created the 1980 television series *Cosmos,* "the greatest media work in popular science of all time," as Gould would call it.[25]

With *Cosmos,* Sagan sought to put an end to the fear and to inspire the kind of wonder Hubble's lectures had inspired in the 1930s and 1940s and the Moon landing had inspired in 1969. The series was enormously successful. For the first time since Hubble, a huge audience was engaged in exploring the grand questions of life, nature, the structure of the universe, mythology, and what it might all mean, how it might all fit together, the mystery of it all. It examined how our search for meaning through science and our accumulation of observations and knowledge were the grandest of all human quests.

The show was seen by an estimated five hundred million people, then about a ninth of the world's population, in sixty countries, nearly as many as the Moon landing—an enormous viewership for the time, and by far the largest for any science show ever. And yet Sagan was later denied admission to the National Academy of Sciences by his peers, who voted against his nomination on the grounds that his research work as a scientist was not strong enough to justify admission[26]—likely a stalking horse for the real reason: the animosity scientists felt toward Sagan's celebrity as a TV star and as a spokesman for their work. The scientist who nominated Sagan, origins-of-life researcher Stanley Miller, described the animosity he perceived. "I can just see them saying it: 'Here's this little punk with all this publicity and Johnny Carson. I'm a ten times better scientist than that punk!'"[27]

Following the shocking rejection, and Sagan's rejection in his bid for tenure at Harvard, scientists developed a new term—the "Sagan effect"—in which popularity with the general public was considered to be inversely proportional to the quantity and quality of one's scientific work,[28] a perception that in Sagan's case, at least, was false. He published, on average, once monthly in peer-reviewed publications over his thirty-nine-year career—a total of five hundred scientific papers.[29] More recent research suggests that all scientists who engage the public tend to be better academic performers as well.[30]

In the years following Sagan's drubbing by the very National Academy whose president had called for increased science outreach, the public's once-shining perception of science continued to erode. By 1999, less than half of all Americans—just 47 percent—said that scientific advances were one of the country's most important achievements. By 2009 that number had fallen to only 27 percent.[31]

Sagan's rejection became a poignant and symbolic example of how scientists had lost a sense of the value of their relationship to the society around them, a relationship that was so critical to their future and the future of the country—but that was slipping through their fingers even as they voted against him, like the sands of an ancient streambed, almost forgotten and now run dry.

CHAPTER 7

AMERICAN ANTISCIENCE

Enlightenment is man's emergence from his self-imposed immaturity. Immaturity is the inability to use one's understanding without guidance from another. . . . It is so easy to be immature. If I have a book to serve as my understanding, a pastor to serve as my conscience, a physician to determine my diet for me, and so on, I need not exert myself at all. I need not think, if only I can pay: others will readily undertake the irksome work for me.

—IMMANUEL KANT, 1784[1]

GOD HELP US

At the same time that the American science enterprise was turning away from the national dialogue, American religion was organizing. Unlike science, churches still depended on engaging the public for their financial support, and they were alarmed by the deep skepticism toward religion and religious authority that was being shown by many baby boomers and the broader public in the 1970s. The prevailing feeling was that science disasters and the bomb were rendering life increasingly hopeless, but organized religion seemed out of touch and unable to help parse the new moral and ethical challenges. With membership and collections falling, many emerging Protestant leaders believed the answer lay in reaching out in new ways. Suddenly, evangelism was relevant again, and this time its leading figure was Billy Graham.

Graham's overnight success had been two decades in the making. On September 25, 1949, just a month after the Soviet Union shook the world with its successful test of a nuclear bomb, the thirty-year-old Baptist preacher from North Carolina by way of Minnesota stepped onto the stage inside a giant tent on Washington Boulevard at Hill Street in Los Angeles. He billed it as the Greater Los Angeles Billy Graham Crusade at the Canvas Cathedral with the Steeple of Light. Backed by hundreds of Christian leaders from across Southern California, Graham drew an average of fifty-four hundred people every night, with thousands more standing out in the twilight, straining to hear, or listening over their car radios. The spectacle went on for sixty-five sermons over eight weeks, drawing three hundred and fifty thousand people.[2] Graham's evangelical sermons were the pitch-perfect blend of Northern progressivism and Southern conservatism that gave him a mainstream, Billy Sunday–style brand that went over well in California, and then across America. It was a formula that others would copy.

Graham quoted scripture and railed against science as he talked about his experiences as a traveling preacher in the days after World War II. "All across Europe, people know that time is running out," he told his worried audience. "Now that Russia has the atomic bomb, the world is in an armament race driving us to destruction."[3] The time to accept Jesus was now or never.

> First of all, I want you to see the need in the philosophical realm. We have just come through an era of materialism, an era of patronism, and humanism in the educational circles of this country. We have been deifying man. We have been humanizing God. And all over the religious world there is a stark unbelief in the supernatural. All through this country of ours we have denied the supernatural, outlawed the supernatural, and said that miracles are not possible now. And we have taken up with things, rather than the spirit of God. . . . Because of the goodness and grace of God I can say tonight that I am not ashamed of the Gospel of Jesus Christ, for it is the power of God and the salvation to everyone that believes. I do not believe that any man, that any man, can solve the problems of life without Jesus Christ. There are tremendous marital problems. There are physical problems. There are financial problems. There are problems of sin and habit that cannot be solved outside the person of our Lord Jesus

Christ. Have you trust in Christ Jesus as savior? Tonight, I'm glad to
tell you as we close that the Lord Jesus Christ can be received, your
sins forgiven, your burdens lifted, your problems solved by turning
your life over to him, repenting of your sins, and turning to Jesus
Christ as savior. Shall we pray?[4]

Graham was offering a message of hope and wonder in the face of fear.
It was a message that for the prior forty years had been offered by sci-
ence. But now, Graham was inviting people back into, in the eyes of
Immanuel Kant, the "cowardice . . . of lifelong immaturity."[5]

In his 1957 crusade, the year of *Sputnik 1,* Graham continued to use anxiety over the H-bomb to sound an antiscience theme,[6] saying, "When Sir
Walter Raleigh had laid his head on the executioner's block, and the officer
asked if his head lay right, Sir Walter Raleigh said 'It matters little, my
friend, how the head lies, provided the heart is right.' The heart has come to
stand for the center of the moral, intellectual, and spiritual life of a man."[7]

Scientists who would dismiss Graham's impact should take note that
his work has reached an estimated 2.2 billion people and, according to
Gallup polls, he has ranked among the ten most admired men for half a
century.[8] They would do well to understand the chord Graham strikes—
an intensely personal emotional and spiritual chord. The desire to create
knowledge that motivates science ultimately shares some of the same
drives as that of its progenitor, religion: to understand the mystery and
wonder of the world and our place in it, to find meaning and hope, and
to make life better. These are courageous aspirations in the face of fear,
which scientists would do well to trumpet—along with science's track
record of actually achieving them.

THE "CIVIL WAR OF VALUES"

By the 1970s, the evangelical movement Graham had played a large part
in reviving had grown exponentially, thanks to technology. Pioneering
televangelists Oral Roberts, Robert Schuller, Pat Robertson, Jerry Fal-
well, James Dobson, Jim and Tammy Faye Bakker, and other leading

televangelists produced gospel shows that together reached, according to Arbitron, more than twenty-two million viewers per week.[9] The common and primary aim of these televangelists was conversion, as dictated by Matthew 28:18–20 and Mark 16:16 (New American Standard Bible).

> And Jesus came up and spoke to them, saying, "All authority has been given to Me in heaven and on earth. Go therefore and make disciples of all the nations, baptizing them in the name of the Father and the Son and the Holy Spirit, teaching them to observe all that I commanded you. . . . He who has believed and has been baptized shall be saved; but he who has disbelieved shall be condemned."

This demanded the delivery of constant, strident, emotional, and inspiring sales pitches to any and all who would listen—just the opposite of what was going on with science, which, as it enjoyed the fruits of Vannevar Bush's ability to secure government funding for the sciences, was cloistered in its own abbeys, laboratories.

A logical outgrowth of making "disciples of all the nations" is to co-opt the political process, which is why evangelicals moved into the political sphere. Graham was a registered Democrat who publicly opposed intolerance and authoritarianism and said that religion should not choose political sides. But he became a minority, as religion went political in a big way. Sara Diamond, a sociologist who follows the growth of the Christian Right, describes it this way.

> It is a political movement rooted in a rich evangelical subculture, one that offers participants both the means and the motivation to try to take dominion over secular society. The means include a phenomenal number of religious broadcast stations, publishing houses, churches, and grassroots lobbies. The motivation is to preach the Gospel and to save souls, but also, with equal urgency, to remake contemporary moral culture in the image of Christian Scripture. On the front lines of our persisting battle over what kind of society we are and will become, the Christian Right wages political conflict not just through the ballot box but also through the movement's very own cultural institutions.[10]

While the voice of science, the very root and foundation of secular democratic society, had gone silent, the voice of Protestantism had grown evangelical, wild, angry, fearful, militant, antiscience, and intensely political, engaging in a "civil war of values," as James Dobson's *Focus on the Family* radio ministry put it, declaring that the 1990s would be "the civil war decade,"[11] in effect a cultural revolution with the goal of remaking America to conform to Christian scripture.

Few scientists saw any connection to the absence of science from the public debate. A survey of *Science* magazine, the leading publication of the American science enterprise, shows no mention of the phrase "religious right" until a November 1989 article about attempts to teach creationism in science classes—two months before Dobson's proclamation.[12] Prior to this date, references were chiefly to "fundamentalists," the dogged but easily dismissed foes of evolution whose periodic school board flare-ups had been chronicled in the magazine since the days of George McCready Price and William Jennings Bryan in the early 1920s.[13] The fact that they had become a national political force, and that the voice of science in the national dialogue was weakening, appears, with few exceptions, to have been largely overlooked or ignored in the professional conversation among scientists.

Dobson's point of view is a perfectly legitimate one, part of a strain of American thought going back to the Second Great Awakening in the early nineteenth century. But American democracy relies on a plurality of voices representing economic, scientific, and religious perspectives to arrive at balanced public policy. With the voice of science going silent in our public and political dialogue, America no longer had that plurality. The country's policies and politics became increasingly unbalanced, and a generation grew up regarding science as increasingly irrelevant in shaping the public dialogue, even as it was impacting their lives more and more powerfully. As policy challenges came increasingly to revolve around science issues, their proposed solutions increasingly came to revolve around faith.

POSTMODERN ANTISCIENCE

While scientists were turning inward to their lab benches and the religious right was organizing, a new form of antiscience thinking arose on the secular left. It was a reaction against the ideas of rationalism that lay at the foundation of the Enlightenment and its attempt to bottom out arguments in the physical world, as John Locke had counseled. The central idea of this "postmodernist" thinking was that both traditional religion and the Enlightenment had gotten it wrong and there is no such thing as objective truth.

Instead, these thinkers held, truth is subjective and to be found in the language context, cultural identity, and personal perspective of the observer. A middle-aged African American male will have a different experience and so a different truth than a young white American male, who will have a different truth than an older Hispanic American female. Their perspectives determine "what is true for them," and anything they say on a given subject must be qualified by their political right to make those statements.

Like fundamentalism, postmodernism began in the late nineteenth century and grew slowly at first, but then, after World War II and the atomic bomb and during the civil rights movement, it expanded exponentially into what it is now. It goes by a number of other names as well, including social constructivism, multiculturalism, deconstructionism, poststructuralism, and cultural or moral relativism. It would have an antiscience effect on American society as profound as the religious right's.

The top-wing "rationalist" view of the world that the more extreme postmodernists object to can be summed up as follows.

1. There is a world. It is real. It is filled with objects and processes that exist independently of us and our beliefs about them.

2. The goal of science is to create descriptions of reality that are independent of us and our opinions or beliefs. We call these descriptions knowledge.

3. To create this knowledge, we use the scientific method, which is a collection of several techniques, including observation, hypothesizing,

induction, experimentation, unique prediction, recording, and critical peer review. These techniques have evolved over time and will likely continue to evolve.

4. Like our senses, the scientific method is fallible and often leads us astray. But it is the best method we have come up with so far, and it has proven to be very powerful.

The religious right took issue with these claims when they conflicted with dogma or a literal reading of the Bible. Postmodernists took issue with them on principle, arguing that they are suspect at their root because they are based on unexamined assumptions of the Western white male dominant culture that created modern science.

Postmodernists also objected to the claims of absolute truth made by organized religion, which was also largely Western white male dominated. Postmodernism was thus a secular reevaluation of every claim of "truth," all of which were suspected of being linguistic means of domination. Truth was subjective. Over the course of twenty years, this thinking would come to influence all of American discourse, education, and politics and become central to America's culture wars.

THUS SPAKE THE ANTISCIENTIST

Postmodernism's roots ironically lie in the writings of German philosopher Friedrich Nietzsche, who famously proclaimed that God is dead and invented the idea of the *Übermensch,* or Superman, in *Thus Spake Zarathustra,* in which he set humanity the goal of creating a super race. "Out of you who have chosen yourselves, shall a chosen people arise: and out of it, the Superman."[14] It was a concept that later would inform Nazism.[15] Reacting against the Enlightenment (and suffering from paranoia), Nietzsche questioned the very idea that there could be objective truth, arguing instead for something he called "perspectivism," which held that truth is a matter of your perspective.

Following Nietzsche, a number of German and French philosophers, among them Martin Heidegger (who became a Nazi in 1933), Michel Foucault, Jacques Derrida, Jean-François Lyotard, Jacques Lacan, Julia Kristeva, and Bruno Latour, together with a few Americans, including Richard Rorty and Paul Feyerabend, began rejecting the idea that reality and facts existed independently of our thinking about them.

The intellectual descendants of Descartes, these philosophers pointed to examples from quantum mechanics, relativity, and cultural anthropology to illustrate their argument that truth was in the mind of the beholder. Cultural anthropology in particular became their tool as they subjected science to study as if it were a foreign culture, but the field was also a justification for their arguments in that it seeks to shed cultural bias by observing indigenous cultures within their own frames of reference. Thus we had anthropologist Carlos Castaneda taking peyote and writing about his mystical vision quests. Albert Einstein was showing us that our measurements were dependent on our frame of reference in special relativity. German physicist Max Planck was demonstrating how the observer affects the event being observed in quantum mechanics. By elevating subjectivity and reducing science's claim of objectivity to just one of many cultures, the troublesome implications of science, true or not—that our minds might simply be expressions of our anatomy, for example, or that the humanities had less to offer society's forward march than the sciences—were easily dispensed with, and suddenly everything seemed new and mysterious again.

In the American humanities, and subsequently in American politics and education, this came to mean that all systems of thought had equal merit and only had to be internally consistent. They were all just different "languages" or "constructs" for assembling our experience of reality. Thus, postmodernists thought of themselves as tolerant and nonjudgmental.

In America, this thinking merged with new political ideas about affirmative action and became widespread. Postmodernism provided a secular, progressive, inclusive interpretation of reality for those who felt that there were many worthy groups like African Americans, women, Native Americans, gays, lesbians, bisexuals, transgendered persons, and others who had been disenfranchised by the Western white male dominant culture, of which science was a powerful part.

The postmodern view fit well with the growing ambivalence toward science after the bomb and during the cold war. People reasoned that perhaps science didn't really provide an objective view of reality after all. Maybe it was just a bill of goods. Scientific "truth" had included a lot of oppressive tools and erroneous conclusions—the bomb, chemical pollution, eugenics, phrenology, not to mention the exclusion of women and minorities from its ranks. The list of offenses seemed endless. Perhaps it really was simply the cultural expression of the privileged Western white male society from whence the Enlightenment sprang. Perhaps its so-called objectivity was really a smoke screen to hide its attempt to exploit and hang on to power.

This view was enthusiastically embraced by the largely left-leaning academics in the humanities at many universities, who had found themselves being deposed by science in the battle of the two cultures. They found common cause with political activists representing feminism, environmentalism, African Americans, Native Americans, the working poor, humanism, the peace movement, gay rights, animal rights, antinuclear activists, and other disempowered groups. Science came to be seen as the province of a hawkish, probusiness political-right power structure—polluting, uncaring, greedy, mechanistic, sexist, racist, imperialistic, oppressive, and not to be trusted.

KUHNIANISM

In 1962, this broad ambivalence toward science crystallized with the publication of American philosopher Thomas Kuhn's *The Structure of Scientific Revolutions*. Stunningly, it became one of the most cited academic books of the twentieth century. It sold about a million copies, a figure unheard of for a philosophical text, and gave clarity to Americans' increasingly ambivalent attitudes toward science and the nature of reality. Science was not the gradual and painstaking accumulation of knowledge, Kuhn said, but rather a sociological and thus political phenomenon that happens in sudden paradigm shifts. These shifts were akin to religious revelations or quantum leaps in the energy states of electrons, which accumulate energy and then "leap" to higher orbits in discrete, sudden jumps.[16]

The politics the book ascribed to science resonated closely with prevailing attitudes. Scientists ("the Man") resist new (baby boomer) ideas, clinging to old (Western white male), outdated theories even as the evidence they are being willfully blind to accumulates (discrimination) like energy in an electron until it finally becomes overwhelming (the civil rights movement). Then, suddenly, in a crystallizing moment (revelation), the ruling order is displaced (comeuppance) and the intellectual understanding of the old (bigoted) paradigm (attitude) shifts to a new, wider-orbiting (more tolerant and inclusive) paradigm that incorporates (affirmative action) previously discounted outliers (disempowered groups).

Kuhn was striving to describe science not as we think it is, but as it really is, and so was likely strongly influenced by his times. He pointed to several past scientific revolutions as examples and argued that they had not been intellectual so much as sociological upheavals. As evidence he quoted Max Planck, who, along with Einstein, founded the revolutionary field of quantum mechanics. "A new scientific truth does not triumph by convincing its opponents and making them see the light," Planck said, "but rather because its opponents eventually die, and a new generation grows up that is familiar with it."[17]

This speaks to the politics of science that is the subject of this book. But it does not speak to the fundamental truth or falsity of a new theory itself. Kuhn's great error was to intertwine the politics of science and the discovery of truth and call them one. People have vested interests. Abandoning them to accept a newly or more completely revealed truth is done at some personal, emotional, and often financial and political cost, and that is hard. That is why the intellectual honesty demanded by science is both so brutal and so nourishing, so cherished and so beautiful. Thus "bad" or even "old" science—science colored by what Bacon called the "numberless" ways our assumptions, prejudices, political motivations, and "affections" color our understanding,[18] and science that is colored by the limitations of our senses or instruments—is eventually replaced by "good" or more accurate science. Eugenics and phrenology are discredited, for example. But so is Newtonian physics.

But it is usually only the progenitor of the previous paradigm that hangs on so dearly, not the entire scientific community, as Kuhn implied. If something new is found that better explains things, the scientific community is all over it because that's where the excitement and opportunity lie. Kuhn was writing as a social critic as much as a philosopher.

LONG DIVISION

Many of our ideas are revised upon closer observation, and it doesn't have anything to do with political biases per se. It can simply be because we have finally developed the tools to make close enough observations. Locke said that sensory knowledge is the least reliable, and so it is. Refinements in our tools and observations allow us to see that things which once appeared real were misinterpretations based on limited observation.

This process of refinement charts the history of science itself. Modern science began as natural philosophy and then, with Galileo's finer telescope, astronomy was carved out of that to become its own science, followed by Robert Boyle's 1661 masterpiece *The Sceptical Chymist,* which separated chemistry, and then Newton's carving out of physics in 1687 with *Principia Mathematica.* After Darwin's 1859 publication of *On the Origin of Species,* biology became firmly established as a separate field, and with the 1875 publication of Wilhelm Wundt's *Principles of Physiological Psychology,* we carved out psychology. In the twentieth century we finally separated neuroscience, the science of the nervous system and mind—the subject from which much of natural philosophy originally sprang.

Why so late? Because much of neuroscience became possible only with the development of electronics, computer technology, and the imaging systems required to study, measure, and stimulate brain function. These tools were then combined with new understandings of the interaction of software and hardware and the biology of how chemical and electrical signals give rise to one another. We are still asking the same questions we were in the days of natural philosophy, but now we can increasingly ask them in a scientific context.

CONCEPT COLLAPSE

Neurophilosopher Patricia Churchland calls this process concept collapse. As examples she offers impetus in physics and vitalism in biology. Impetus was thought of for centuries as an inner force that kept things moving, but then Newton revealed that to be an illusion. "We had something that had seemed to be observable, and it turned out to not be a real thing at all," Churchland says.

It was similar with vitalism—the life force that was thought to distinguish living things from nonliving things, what makes a rock different from a living being and a living body different from a dead one. "In 1900, we used to think it was one thing. Some vital force or spirit," says Churchland. "In fact, we now know that it is many things. With the discoveries of ATP [adenosine triphosphate] and the basic chemical building blocks of life, the understanding of the role of mitochondria, of ribosomes, of cell biology, the chemical nature of DNA, the folding and unfolding of proteins, and so on, that concept too has been revealed as illusory, about eighty years ago now."[19]

Another example is our age-old idea of fire. What is fire, really? It is the burning of wood. It's the fire of the sun. It's the fire of lightning. It's the magical fire in a firefly's tail. But when we applied closer observation, this one concept collapsed into four very different things, none of which has very much at all to do with any other. The burning of wood, we learned, is oxidation, much more akin to rusting than it is to the fission going on in the sun. And the splitting of atoms is altogether different from the incandescence of lightning, which turns out to have nothing at all to do with the phosphorescence produced by the chemicals in a firefly's tail. The only thing they have in common is that they appear bright; otherwise, they are not similar at all.

IT'S ALL RELATIVE

If, as we made closer observations, fire, vitalism, and impetus were revealed not to be as we thought they were, does it mean there is no objective truth, but just an endless regression of ideas? Kuhn suggested the answer was yes. As evidence he offered the work of Einstein.

> I do not doubt . . . that Newton's mechanics improves on Aristotle's and that Einstein's improves on Newton's as instruments for puzzle-solving. But I can see in their succession no coherent direction of ontological development. On the contrary, in some important respects, though by no means in all, Einstein's general theory of relativity is closer to Aristotle's than either of them is to Newton's.[20]

This suggested that there was no real progress in science. It was simply a grand circle, or an endless regression. But Kuhn's theory, while dramatic and captivating, was incorrect. Individual scientists, like Harlow Shapley, may fall off track and become overly invested in a priori, Cartesian first principles and thus become blind to observational evidence, but overall scientific progress is real. It has political implications because new knowledge gives new power, but it is not merely a social exercise that relies on outliers changing the minds of the curmudgeonly majority or waiting for them to die off. There is an observable reality on which empirical science is based and from which knowledge is derived. That is why it has power: It enables us to affect the physical world in ways we couldn't before. The observations and knowledge extend incrementally, by the contributions, risks, and suffering of many. They do not extend in sudden and dramatic paradigm shifts, and they didn't in Einstein's day, either.

In fact, many of the ideas Einstein developed were done collaboratively, with considerable debate, a prime example being the cosmological constant. His early papers were extensions of the work of Max Planck, the Austrian physicist Ludwig Boltzmann, and others, and his revolutionary findings on Brownian motion were independently discovered by Polish physicist Marian von Smoluchowski, who was also building on Boltzmann's work. Hubble's revolutionary discovery of the expansion of the universe also extended from ideas that were talked about for years.

The redshift was first noted by American astronomer Vesto Slipher in 1912—nearly two decades before Hubble's discovery. Galileo's revolution was an extension of Copernicus's writings of some seventy years before, which were widely discussed. The discovery of the double-helix structure of DNA was revolutionary, but it too was an extension, building on the work of biochemist Erwin Chargaff.

It is true that science does not proceed linearly; it proceeds more like a pack of dogs sniffing out a fox, but that is because of its trial-and-error, observational approach that adopts whatever new tools become available, applies metaphor, builds on the latest recorded knowledge ("the literature," as scientists call it), and makes and tests bold predictions to better see the reality of the thing instead of our prejudices or assumptions or beliefs or opinions or hopes and dreams. Science is our very best tool against prejudices and unexamined attitudes, not the cultural expression of them.

THE AGE OF EQUALITY

At the time, Kuhn's work seemed to offer a resounding refutation of the claims of the truth and power of science, and it served as a catalyst, shifting the paradigm of America's entire relationship to science. If science was simply one way of knowing about the world, other, previously discounted ways of knowing might be equally valid. This seemed obvious. The world wasn't mechanistic after all. We could throw out the past and take up whatever view of reality best suited us. We could be free. It was the dawning of the Age of Aquarius.

In politics, this thinking became entwined with the goals of the civil rights movement: examination of power structures, discovery of voices not valued by history, cultural tolerance, acceptance of diverse viewpoints, affirmative action, mindfulness of the biases of the speaker, a pullback from Western exceptionalism and white supremacy, and the celebration of the self-evident truth that lies at the foundation of the nation—that all people are created equal. If science was the voice of the Western white male culture, then it was not the voice of other, discounted cultures.

Academics, writers, politicians, and teachers, in a sort of intellectual affirmative action, took the idea to its logical conclusion: from all people being created equal to the notion that all cultures are created equal and from that to the idea that all ideas are created equal. Suddenly, truth was a matter of your perspective. There was no objective truth; there was feminist truth, male truth, indigenous truth, African American truth, Latino truth, GLBT truth, Islamic truth, working-class truth, and so on, all of which had to be equally respected—and not just their contributory aspects, but in their entireties, and without judgment. Thus if someone from a disempowered political group did something morally reprehensible, he or she still had to be given extra understanding, because it was probably partly because of disenfranchisement.

Cultural conservatives objected to this on a rationalist basis and were crowded into the bottom-right political quadrant with scientists, who didn't belong there, but suddenly rationalism and modernism seemed like old and conservative ideas, expressions of authority. The political left lost many otherwise liberal thinkers who could not stomach elevating a political goal over the ideas of reason and the Enlightenment. Their opposition was often judged not on its intellectual merits but through a political lens, and considered to be ignorant, racist, sexist, supremacist, ill read, or right wing—vast, sweeping indictments of everyone who did not embrace the new politics, whether they actually were racists or simply rationalists. In fact, the postmodernists argued, rationalism itself was part of the whole problem. What rationalists were unwilling to admit, they argued, was that irrational processes, not rational ones, lay at the core of Kuhn's revolutions in science. Rationalism itself was simply a thinly veneered tool of domination.

WOLVES IN SHEEP'S CLOTHING

Building on the work of Kuhn, whose manuscript of *The Structure of Scientific Revolutions* he had critiqued in 1960, postmodernist Paul Feyerabend succinctly summed up the thinking.

> The world, including the world of science, is a complex and scat-
> tered entity that cannot be captured by theories and simple rules. . . .
> There is not one common sense, there are many. . . . Nor is there one
> way of knowing, science; there are many such ways, and before they
> were ruined by Western civilization they were effective in the sense
> that they kept people alive and made their existence comprehensi-
> ble. . . . The material benefits of science are not at all obvious.[21]

Feyerabend has been widely quoted by antiscience postmodernists on
the left, but also by antiscience religious authoritarians on the right. In
1990 the archconservative German cardinal Joseph Ratzinger was in
charge of Roman Catholic doctrine. Ratzinger was a strict authoritarian
who would eventually become Pope Benedict XVI, steering the church
back into the political bottom wing. He gave a major speech in which he
condoned the 1633 trial and conviction of Galileo for heresy, using a
quote from Feyerabend to make his argument.

> "The church at the time of Galileo was much more faithful to reason
> than Galileo himself, and also took into consideration the ethical
> and social consequences of Galileo's doctrine. Its verdict against
> Galileo was rational and just, and revisionism can be legitimized
> solely for motives of political opportunism."[22]

THE END OF OBJECTIVITY

"While such views remained contained within relatively limited intellectual and political groups, little attention was paid to them by the
mainstream scientific community," reflected the editors of the science
journal *Nature* in a 1997 opinion piece. "But, over the past few years,
their influence has appeared to flourish not only in the academic
world—including school-teaching—but also in the wider community."[23]

The entire movement was tinkering with the foundations of democracy in ways few understood. By making objectivity supremacist, the
subjectivism that America's founders had sought to free it from (but had
partially failed by excepting slaves and women from those created
"equal") was restored to the throne, and much of American education

and thought after the 1970s lost its grip on reality and became embroiled in "but faith, or opinion, but not knowledge," in the words of John Locke.

It was an unfair and vastly oversimplified criticism of science to say that because it was a field predominantly populated by white men, it was simply another subculture that was trying to retain its seat of power at the top of the heap. Many professional endeavors at that time were the field of white men. The thinking mistakenly focused on scientists as a group of some particular background rather than on science as a process of ideas, something people do, a cultural expression that, like art, cuts across all groups and is ultimately a method that works against prejudices to capture truth.

In fact, it was the thinking of science itself that laid the very foundation for the values of tolerance and diversity used to sell postmodernism by originally advancing the idea that, based on our equivalent opportunities to observe reality and derive knowledge, all men and women are created equal. That this was not immediately achieved is to make the perfect the enemy of the good. None of it would have happened at all without the conceptual foundation, which rests on the early Enlightenment thinking of science.

As a foundation of democracy, science itself is a font of tolerance, and it is also the greatest beneficiary of diversity because it thrives on challenges from differing viewpoints to make its conclusions stronger. In the sense that it is now international, with scientists from around the world collaborating on research projects over the Internet, sharing a common language of science, the global science enterprise is in many ways the most diverse and yet universal undertaking in human history.

When viewed from this perspective, it is clear that if taken to an extreme, postmodernism can embroil society in the same sort of culture wars that roiled Europe at the beginning of the Enlightenment. These conflicts motivated Locke to seek a way to define what is universal and separate from denominational identity—knowledge—and inspired Thomas Jefferson, Benjamin Franklin, and others to found the United States on that bedrock.

The culture wars the United States is currently experiencing are really one three-front antiscience war—a fundamentalist backlash against science, a propaganda war being waged by vested business interests, and an assault from postmodern identity politics that are based not on religious denomination, but on gender, sexual orientation, race, etc., and are as sacrosanct as religions, with their authority undebatable and any questioning akin to blasphemy. Similar to the denominational battles between Protestants and Catholics and among the various sects of Protestantism, today's secular denominations claim the authority of a truth no member of another denomination can know.

Science Class without Objective Truth

The 1960s and 1970s were a time of momentous social change, particularly related to civil rights. For the first time, the United States was making a serious effort to educate African American children to the same standard as white students. One of the primary methods employed was school desegregation. This posed complex challenges for teachers, who were tasked with educating more diverse classrooms as black students whose communities had been uprooted first by highway construction and then by busing found themselves thrown into the mix with more advantaged white students. It seemed unrealistic to demand equal performance from students who did not have the same level of socioeconomic support or shared cultural references.

Similar conclusions were coming from science educators involved in the efforts to transfer scientific knowledge and educational methods to developing countries. "Why should we suppose that a program of instruction in botany, say, which is well designed for British children, familiar with an English countryside and English ways of thinking and writing, will prove equally effective for boys and girls in a Malayan village?" they asked. "It is not merely that the plants and their ecology are different in Malaya; more important is the fact that the children and their ecology are also different."[24]

Beginning in the 1970s, science educators began to sense "a growing awareness that, for science education to be effective, it must take much more explicit account of the cultural context of the society which provides its setting, and whose needs it exists to serve."[25] While white teachers once taught white students using white cultural references, now all teachers had to develop strategies to reach diverse classrooms. These new strategies were built on the assumptions of social constructivism, wherein learning is regarded as a social process. "Educators have long viewed science as either a culture in its own right or as transcending culture. More recently many educators have come to see science as one of several aspects of culture," wrote science education professor William Cobern, referring to the idea that science is the cultural expression of Western white men.[26] But if that is true, isn't teaching science a form of cultural genocide? The only answer to this logical contradiction lay in redefining what we meant by "truth."

Social constructivist thinking became the mainstream paradigm in Western teacher education in the 1970s and 1980s, going on to influence the educations of tens of millions of American students. Parts of this transformation have been very positive for science—the emphases on hands-on learning and on process over product. But constructivist thinking in education also came to hold the well-intentioned but incorrect belief that "there is no representation of reality that is privileged, or 'correct.'"[27] One education professor described it this way.

> Because reality is in part culture dependent, it changes over time, as cultures do, and varies from community to community. Knowledge is neither eternal nor universal. . . . We must think increasingly in terms of "teachers and students learning together," rather than the one telling the other how to live in a "top-down" manner. This is necessary both so that the values and interests of students are taken into account, and so that the wealth of their everyday experience is made available to fellow students and to the teacher.[28]

This confusion of reality with culture devalues knowledge and presumes that students have some "wealth of everyday experience" that is of equal value to the lesson at hand and the teacher does not have access to despite extensive training. What could that be? The student's experience as a member of a political identity group—race, gender, sexual orientation, disability, age, etc.—that is different from the teacher's. What does this teach the student? That there is no real knowledge "out there" that we all strive to attain. Instead, we each construct our own reality, and the perspective already held is of equal value to anything to be learned. This emotional and political goal may make the student and thus the teacher feel good, but science has proven through its profound fecundity that it is simply not true. Scientists argued that the purpose of education had shifted from teaching knowledge and skills to providing a learning environment in which students constructed their own knowledge.

While inclusiveness is important in closing the education gap, it may be argued that high expectations are also important, and that overemphasizing political identity or "culture" undermines the knowledge teachers are trying to impart, shortchanging students. There is a difference between being inclusive and elevating all ideas to the same level, and there is a difference between not forcing assimilation and cheating students of truth. Knowledge knows no color; it is of higher value because it is tied to physical reality. What defines it as knowledge is its separability from the individual; when it is not separable, it is opinion. This reasoning goes back to Locke and Bacon.

The confusion this causes makes some scientists simply want to throw up their hands in despair. Consider the following introduction to a 1998 science education paper presented at a national conference.

> As Richard Dawkins likes to put it, there are no epistemological relativists at 30,000 feet. But today some will say, "Not so fast!" Dawkins offers a brute definition of universality completely devoid of any nuance of understanding and equally devoid of relevance to the question at hand. No one disputes that without an airplane of fairly conventional description, a person at 30,000 feet is in serious trouble. The question of universality does not arise over the phenomena of falling. The question of universality arises over the fashion of the propositions given to account for the phenomena of falling, the fashion of the discourse through which we communicate our thoughts about the phenomena, and the values we attach to the phenomena itself and the various ways we have of understanding and accounting for the phenomena—including the account offered

by a standard scientific description. In today's schools there are often competing accounts of natural phenomena especially where schools are located in multicultural communities. There are also competing claims about what counts as science.[29]

The teaching that there is no objective reality, but rather many subjective realities, or in this case, that the subjective realities are on an equal par with the objective reality (you're dead!) in turn influences students' views of the primacy of knowledge. To critics, history is no longer the search for what really happened, but rather the victor's interpretation as seen through the lens of power and oppression, and it bears a cultural and political focus. Literature is no longer a study of what the author meant, but of the feelings it arouses in the reader because of his or her cultural perspective. Reading the classics is no longer required because they're sexist and racist and not germane to today's political realities. Truth must be evaluated in the context of the speaker's socioeconomic frame of reference, and teachers, when they lack the political authority of having the same cultural identity as their students, cannot presume to teach them, but can only be "guides at the side."

By the late 1980s, classics professor Allan Bloom lamented the effects of postmodernist education on the thinking of the students coming into his classroom at the University of Chicago. His concern spawned a nationwide discussion when he wrote about it in *The Closing of the American Mind.*

> The relativity of truth is not a theoretical insight but a moral postulate, the condition of a free society, or so [the students] see it. They have all been equipped with this framework early on, and it is the modern replacement for the inalienable natural rights that used to be the traditional American grounds for a free society. That it is a moral issue for students is revealed by the character of their response when challenged—a combination of disbelief and indignation: "Are you an absolutist?," the only alternative they know, uttered in the same tone as "Are you a monarchist?" or "Do you really believe in witches?"[30]

Absolutism is considered morally objectionable because it leads to intolerance, but that is only true when it is applied to a matter of "faith, or opinion, but not knowledge." In the case of science, precisely the opposite has proven to be true. By acknowledging that there is an objective reality, and that we can form knowledge about that reality by using science and observation, we remove questions of fact from the authoritative argument. This is the great insight that the United States was founded upon. The importance is in the knowledge, not the speaker, just the opposite of the postmodern perspective. Bloom was criticized as a conservative, a sexist, and a racist, all of which he denied.

> Openness—and the relativism that makes it the only plausible stance in the face of various claims to truth and various ways of life and kinds of human beings—is the great insight of our times. The true believer is the real danger. The study of history and of culture teaches that all the world was mad in the past; men always thought they were right, and that led to wars, persecutions, slavery, xenophobia, racism and chauvinism. The point is not to correct the mistakes and really be right; rather it is not to think you are right at all.
>
> The students, of course, cannot defend their opinion. It is something with which they have been indoctrinated. The best they can do is point out all the opinions and cultures there are and have been. What right, they ask, do I or anyone else have to say one is better than the others? If I pose the routine questions designed to confute them and make them think, such as "If you had been a British administrator in India, would you have let the natives under your governance burn the widow at the funeral of a man who had died?," they either remain silent or reply that the British should never have been there in the first place. It is not that they know very much about other nations, or about their own. The purpose of their education is not to make them scholars but to provide them with a moral virtue—openness.[31]

COLLAPSING HERMENEUTIC
ISOLATION: THE QUANTUM STUDIES
APPROACH TO POLITICS

By the 1990s, a full generation's worth of university departments had grown up around "cultural studies"—women's studies; gay, lesbian, bisexual, and transgendered studies; science studies, etc.—all purporting to be among the social sciences, but arguing the a priori political notion misappropriated from quantum mechanics that objective reality is subject to the observer. Academics would often borrow the language of science in this way, though only when it supported their views, and use jargon to make sometimes outlandish political statements sound highbrow.

In 1994, Rutgers University mathematician Norman Levitt and University of Virginia biologist Paul Gross published a polemic attacking this appropriation of scientific terminology called *Higher Superstition: The Academic Left and Its Quarrel with Science*.[32] Like Kuhn's *The Structure of Scientific Revolutions* and Bloom's *The Closing of the American Mind*, *Higher Superstition* charted a new course in the culture wars and became a bestseller. Gross and Levitt characterized the conflict as a clash between the academic left and the scientific right, and academics called the book right wing. But there were plenty of scientists on the left and others who were ideologically unaffiliated who were tiring of the raging argument over the culture-centric nature of reality that by then was being called "the science wars." Among the most notable was eminent Harvard University entomologist E. O. Wilson, who declared in a New Orleans speech that "multiculturalism equals relativism equals no supercollider equals communism."[33]

Then, in 1996, humanities scholar Andrew Ross, the editor of the leading postmodernist journal, *Social Text,* made a fateful decision. He decided to devote an issue to discussion of the science wars and accepted for publication a paper by Alan Sokal, a New York University physicist and self-described leftist who had been inspired by *Higher Superstition* to submit a paper titled "Transgressing the Boundaries: Toward a Transformative Hermeneutics of Quantum Gravity." The only problem was that the paper was a hoax—a parody mash-up of the most ridiculous postmodernist writing Sokal could find—that appropriated the language of science to argue that there was no reality. It was the kind of politically correct, heavily jargonized, but intellectually vapid nonsense Woody Allen's Alvy Singer had famously called "mental masturbation" in the 1977 film *Annie Hall*. It was tailor-made to please its intended audience, complete with the popularly employed quotation marks around certain words to imply an inferior, "so-called" status:

Deep conceptual shifts within twentieth-century science have undermined this Cartesian-Newtonian metaphysics; revisionist studies in
the history and philosophy of science have cast further doubt on its
credibility; and, most recently, feminist and poststructuralist critiques have demystified the substantive content of mainstream Western scientific practice, revealing the ideology of domination concealed
behind the façade of "objectivity." It has thus become increasingly
apparent that physical "reality," no less than social "reality," is at bottom a social and linguistic construct; that scientific "knowledge," far
from being objective, reflects and encodes the dominant ideologies
and power relations of the culture that produced it; that the truth
claims of science are inherently theory-laden and self-referential; and
consequently, that the discourse of the scientific community, for all its
undeniable value, cannot assert a privileged epistemological status
with respect to counter-hegemonic narratives emanating from dissident or marginalized communities.[34]

Shortly after that, Sokal published an article in *Lingua Franca,* a sort
of *People* magazine of the academic world, that described the hoax and
how it showed postmodernists were incapable of distinguishing between
a real argument and nonsense. "What concerns me," he wrote, "is the
proliferation, not just of nonsense and sloppy thinking per se, but of a
particular kind of nonsense and sloppy thinking: one that denies the
existence of objective realities."[35] The worldwide media loved the story
because of its quality of "The Emperor's New Clothes." Pompous college
professors being exposed as vacuous dupes is a narrative not without
appeal. The right, in particular, which had been railing for years against
the political correctness that was coming out of these identity politics,
saw it as a long-awaited skewering of the "effete, elitist academic who's
'above it all,'" as commentator Rush Limbaugh said of the postmodernist
Stanley Fish, the publisher of *Social Text*[36]—a skewering that dismayed
progressives.

"He [Sokal] says we're epistemic relativists," complained Stanley Aronow-
itz, cofounder of *Social Text.* "We're not. He got it wrong. One of the reasons
he got it wrong is he's ill-read and half-educated."[37] But this snobbish non-
response only bolstered Sokal's—and cultural conservatives'—criticisms.

"Conservatives have argued that there is truth, or at least an approach to truth, and that scholars have a responsibility to pursue it," wrote Janny Scott in a *New York Times* article about the hoax, news of which spread worldwide. "They have accused the academic left of debasing scholarship for political ends."

But it wasn't just conservatives. In academic circles, the hoax also reinforced the view of traditionalists on the left, including several prominent feminists, that "women's studies" and "science studies" were so much nonsense. Prominent gender historian Ruth Rosen wrote in the *Los Angeles Times*:

> It took a New York University physicist named Alan Sokal to expose the unearned prestige that the Academic Emperors have heaped upon themselves. A self-described progressive and feminist (to which I can attest; I helped with his exposé), Sokal became fed up with certain trendy academic theorists who have created a mystique around the (hardly new) idea that truth is subjective and that objective reality is fundamentally unknowable. To Sokal, the denial of known reality seemed destructive of progressive goals.[38]

Michael Bérubé, a Pennsylvania State University literature professor and one of Sokal's most vocal critics, wrote an apologia and mea culpa in 2011 describing how the hoax gave ammunition to the right, but also to a previously silent group on the left "who believed that class oppression was the most important game in town, and that all this faddish talk of gender and race and sexuality was a distraction from the real struggle, which had to do with capital and labor."[39]

Most important, the paper that "punk'd," as Bérubé called it, *Social Text* exposed the erosion of the standing of reason in American schools and universities, the educators of the next generation of leaders. This diminution has since proved problematic for democracy in a time when it is more important than ever to ground arguments in facts instead of fact-simile—"bullshit" that has the appearance of fact. What resulted was "a pseudo-politics, in which everything is claimed in the name of revolution and democracy and equality and anti-authoritarianism, and nothing is risked," wrote feminist essayist Katha Pollitt in *The Nation*.[40]

POSTMODERNISM AND THE ANTISCIENCE RIGHT

Most of today's journalists and policy makers did not come up through the sciences and were exposed in high school and college to the political correctness of postmodernists who populated humanities, education, and political science departments. Conservative students and those who would eventually become conservative did not forget the arguments they learned there—that science was just another "way of knowing" and that people in authority could get away with passing off bullshit as truth as long as their arguments sounded credible and included cherry-picked bits and pieces of science.

This cynical doublespeak seemed to confirm the idea that was at the very heart of what would become the neoconservative movement: The winners write the history books. These guys were the living proof. They could even, as these professors had for close to a generation, successfully attack the credibility of science and reason itself, on which the secular world order was based, by casting it as just another worldview.

It is perhaps no coincidence that as these very students have assumed power as senior journalists in the media and policy makers in Congress and other elected offices, democratic society has been swept into an era of endless rhetorical debate over faith and opinion and of national discussions dominated by identity politics and partisanship, an intellectual morass justified as the "marketplace of ideas."

Thus Wall Street can build a multitrillion-dollar edifice on worthless paper and the opinions of the economic seers at Harvard who pronounce it good with a few formulas and a heaping helping of intimidating and often nonsensical jargon. The major media outlets can give equal platforms to scientific outliers and celebrities on important science issues ranging from climate change to vaccines and autism. It's not their role to establish the objective truth of the matter, some leading members of the press argue, any more than it was to question the lack of a substantive rationale for going into Iraq. The times, we're told, are new. The old rules no longer apply. We must be open to new things, to hearing new voices. We must not pass judgment. The teachers must learn from the students.

Science has proven by its results that this is wrong. There is objective truth to be learned by observation, and the knowledge gained gives power that other "ways of knowing" have not. But the more dangerous problem with postmodernist thinking is its a priori nature. Not truth, but a political goal has to be served—in this particular case the goal of openness, or tolerance without judgment. But without acknowledging objective truth, all arguments become rhetorical and therefore can go on forever—and we are either paralyzed by it or we must resort to authority instead of objectivity to make decisions, which collapses us all the way back to Thomas Hobbes's war of every man against every man: predemocracy, pre-Enlightenment.

Thus, when taken to an extreme, postmodernist thinking inevitably leads, through its dependence on authority, to the brutality it sought to avoid, a brutality that was becoming increasingly evident in political arguments as the later baby boomers took charge in the culture. As the first children taught under this philosophy, they were generally unable to articulate positions based on accumulated knowledge or data because they hadn't been taught about them or taught to value them, and so were left with "but faith, or opinion" and a belief that what matters most is finding the nuggets that match your argument—and in never, ever compromising, because the winner writes the history books.

Physicist Lawrence Krauss, a cofounder of the Science Debate effort, wrote presciently of the political ramifications for the country in a 1996 *New York Times* opinion piece in which he described how presidential candidate Pat Buchanan had recently professed that he was a creationist. Although journalists questioned other Buchanan campaign planks like trade protectionism and limits on immigration, Krauss said, there were no major articles or editorials declaring the candidate's views on evolution to be nonsense.

> Why is this the case? Could it be that the fallacies inherent in a strict creationist viewpoint are so self-evident that they were deemed not

to deserve comment? I think not. Indeed, when a serious candidate for the highest office of the most powerful nation on earth holds such views you would think that this commentary would automatically become "newsworthy." Rather, what seems to have taken hold is a growing hesitancy among both journalists and scholars to state openly that some viewpoints are not subject to debate: they are simply wrong. They might point out flaws, but journalists also feel great pressure to report on both sides of a "debate."[41]

This erosion of our capacity to simply tell the truth began to have profound effects that aided and abetted the rise of the antiscience right. These effects would only become clear to postmodernists later, when the antiscience right repurposed postmodernist arguments to make the case that climate science is not objective, but rather subjective and politically suspect "junk science," the product of a "culture of corruption" by "an environmental priesthood" of scientists intent on world domination.

Today, serious candidates for Congress and the presidency can openly state views that run counter to all known science and history, and many journalists don't feel it is their role to point out that the emperor has no clothes.

The dissociation from history and the hard-won knowledge of science and the classical humanities has thus led to a generation of leaders that are at once arrogant and ignorant, and may be unable to lead the nation out of the morass because we have nothing solid to build upon. We have little foundation outside of the immediate concerns of politics and pragmatism and our own feelings. We embrace the forms of tradition but not the substance, focused only on winning, unable to discern between what feels good and what is true. It is a condition that threatens to leave the country permanently damaged.

NEW AGE ANTISCIENCE

Emerging from the ideas of postmodernism was its cousin the New Age, a pop culture spiritual movement built on postmodernist ideas.

Traditional religions rejected this notion, and so in many ways the New Age became the religious aspect of the secular postmodern movement. Much of its early formation can be traced to the writings of novelist and poet Jane Roberts, who, in 1963, the year after Kuhn's *Structure of Scientific Revolutions* became a massive bestseller, began "channeling" the words of a disembodied spirit named Seth. In 1970, Roberts published *The Seth Material,* closely followed by *Seth Speaks: The Eternal Validity of the Soul* in 1972. The books became megabestsellers, and are in some ways similar to Nietzsche's *Thus Spake Zarathustra* in that Seth argues there is a higher self, a sort of *Übermensch* driver of the vehicle that is your mind and body, and to the postmodernist view that we create our own reality, which became the foundation concept of the New Age.

Over the course of the 1970s, the New Age exploded as a spiritual movement gentrifying a collection of ideas and practices culled from non-Western and indigenous "previously discounted" religions. New Age retreats, which at the time were attended almost exclusively by white, middle-class people, featured psychic readings by clairvoyants, spiritual lectures by trance mediums like Roberts, aura readings, laying on of hands, homeopathy, extrasensory perception, Reiki healing, *A Course in Miracles,* hypnosis and past-life regression, transcendental meditation, Uri Geller and spoon bending, telekinesis, psychokinesis, remote viewing, astral projection, sweat lodges, vision quests, psychic channeling, the *I Ching,* the *Kama Sutra,* the *Tao Te Ching,* Bach flower remedies, Jungian dream analysis, reincarnation, astrology, the Edgar Cayce recordings, tarot readings, the existence of Atlantis, chakra adjustment, crystal healing, UFOs, automatic writing, spirit guides, Ouija boards, biofeedback, consciousness raising, yoga, and chanting the word "om," which is said to contain all the sounds in the universe.

In short order the movement became big business. The retreats became like carnivals. They were held around the country and included workshops put on by New Age teachers and featured big-name keynote speakers who were revered like holy gurus. A flood of bestsellers followed, creating the New Age genre: *Jonathan Livingston Seagull; Illusions;*

Remember, Be Here Now; The Dancing Wu Li Masters; The Teachings of Don Juan; Handbook to Higher Consciousness; The Hundredth Monkey; Love Is Letting Go of Fear; The Crack in the Cosmic Egg; A Course in Miracles; Your Erroneous Zones; and *The Aquarian Conspiracy,* each purporting to peel away the gauze and reveal the true, underlying, and magical nature of a reality that you could create simply by believing it was so.

If all "ways of knowing" are equal, then homeopathy, in which a substance that would create the patient's symptoms in a healthy person is successively diluted until none of its molecules remain (the more dilute, the greater the supposed potency) is administered as a cure, should be as efficacious as conventional pharmaceutical treatment by anyone who values "open-mindedness." Anyone who challenges that idea is just a skeptic (the New Age version of a spiritual bigot) who is trying to oppress "holistic," "higher" thinking because of his or her paradigm of Western white supremacy.

Philosophy professor Theodore Schick Jr. and Lewis Vaughn, once the managing editor of *Prevention* magazine, have written extensively and critically about this spiritual version of postmodern antiscience. They summarized how New Age thinking plays out through the popular culture in their book *How to Think About Weird Things: Critical Thinking for a New Age.*

> There's no such thing as objective truth. We make our own truth. There's no such thing as objective reality. We make our own reality. There are spiritual, mystical, or inner ways of knowing that are superior to our ordinary ways of knowing. If an experience seems real, it is real. If an idea feels right to you, it is right. We are incapable of acquiring knowledge of the true nature of reality. Science itself is irrational or mystical. It's just another faith or belief system or myth, with no more justification than any other. It doesn't matter whether beliefs are true or not, as long as they're meaningful to you.[42]

Thus, Reiki healing may offer as much hope as chemotherapy, and if it fails, it is because the patient is "blocked," unable to accept the healing energy, unable to let go of anger from a past life, or what have you, and so the fatal illness progresses. It's his or her own fault. The reasons for failure are emotional and spiritual and have nothing to do with the practitioner or the method or the physical world.

But like other a priori approaches, such as religion and postmodernism, the New Age lacks a method of establishing knowledge independent of the authority of the practitioner. It thus requires faith in the guru, faith in the invisible, faith in the energy, trust the Force, Luke. This road is appealing because it offers ready access to mystery and hope—exactly the fare that science once sold to the public. It is nonjudgmental and welcoming precisely because all judgment must be suspended. But in the end, without objective standards, it must fall back on authority as the arbiter of what is true, and so it too ultimately leads back to brutality.

The brutality of the blame-the-victim aspect of New Age thinking is on ready display when the patient fails to be healed. This is a common scenario in New Age healing classes, which often draw desperate people with illnesses for which medical science has not yet been able to effect cures. For admission, they pay hundreds or thousands of dollars, hoping to be healed. The healer, who claims to be tuned into the energy of the universe and his or her higher self and a very powerful and open channeler, lays on hands, and may get other students in the group to channel their higher energy as well. Only the poor patient is "blocked." The rest, these very special students, can see it clearly by clairvoyantly looking at the blackness in his aura and feeling the coldness in the parts of the aura affected by the disease. This block is likely the fault of some anger or other negativity he is holding on to. Because we manifest our own reality, physical maladies are usually the symptoms of deeper spiritual causes. It may even be something he is not aware of, from a past life. But it is why he is sick. If only he could see the truth of this and face himself, he could stop refusing the healers' loving energy and be healed. But alas, he cannot. We can all see it. Why can't he?

It's not that the coarseness of this thinking is intentional; many New Agers have the best of intentions. After all, they embrace tolerance and openness. But when confuted by reality, for example by the deterioration and death of a well-adjusted and otherwise spiritually "innocent" patient, they have nothing to fall back on because none of this thinking is based in reality, so they are forced to either reject New Age principles or become intellectually dishonest. Rejecting both a belief system and a warm and embracing but very political community is a hard thing to do—just ask Thomas Kuhn. It is what he argued scientists are unable to easily do. A paradigm change like this comes at great cost to one's vested interests, be they social, emotional, financial, or political.

But in refusing to accept objective truth and the possible pain of intellectual honesty, one is left with the heartlessness of hubris, false hope, and blame. If we are unwilling to let go of our a priori principles and muster the courage to look at the situation as it is—that good people sometimes become sick and die through no fault of their own—rather than as we wish it would be, if we insist on harboring "sciences as one would," as Bacon called it, colored by our feelings,[43] then it must be the other that is wrong. Thus Shapley wipes Humason's photographic plate clean, disease victims are at fault and die due to their own bottled-up anger or other spiritual shortcomings, and the emperor's tailor is celebrated and paid a fortune from the people's hard-earned stores for his most beautiful weave, as many New Age gurus are. It is a retreat into superstition and darkness with heartbreaking human consequences— and even more heartbreaking political ones—rendered under the auspices of openness, tolerance, and love.

CHAPTER 8

THE DESCENT OF THOUGHT

The human understanding is no dry light, but receives an infusion from the will and affections; whence proceed sciences which may be called "sciences as one would." For what a man had rather were true he more readily believes. Therefore he rejects difficult things from impatience of research; sober things, because they narrow hope; the deeper things of nature, from superstition; the light of experience, from arrogance and pride; . . . things not commonly believed, out of deference to the opinion of the vulgar. Numberless, in short, are the ways, and sometimes imperceptible, in which the affections color and infect the understanding.

—FRANCIS BACON, 1620[1]

POISON

Throughout the 1970s and 1980s, ambivalence toward science was expanding; fundamentalism was organizing; educators, politicians, and pop culture were embracing subjectivity; and we were becoming increasingly aware of the unanticipated effects of chemicals on the environment and in the human body.

As all of these things were happening, the use of synthetic chemicals in everyday products was exploding, "oppressing" (as postmodernists would argue, rightly in this case) those who did not have the training to understand the language. Take, for example, butylated hydroxytoluene, a common food additive. It takes knowledge of organic chemistry and biology to understand how a chemical that is commonly used as an additive in jet fuel, rubber, electrical transformer oil, and embalming fluid can possibly be good for you to eat.

In food, it serves as an antioxidant, keeping baked goods made with oil and butter from going rancid. But do we really understand all of its effects in the human body? Now multiply that question by 3,325, the number of additives currently approved for use in foods, which are listed in the FDA's "Everything Added to Food in the United States" list, more commonly known as the EAFUS list. Suddenly, science seemed alien, out of step with mainstream culture, and it seemed we might be unwittingly poisoning ourselves in the same way we had unwittingly weakened the shells of bald eagles' eggs with the "miracle" pesticide DDT. People who did not speak the language of organic chemistry, and even many who did, began to worry about unanticipated biocomplex effects—those arising from the complex biological, chemical, physical, and behavioral interrelationships between living organisms and their environment— spawning the organic food movement.

Others grew worried about high levels of exposure to electromagnetic fields as people were increasingly bombarded by electromagnetic waves from electronic devices. Like understanding chemical additives, understanding the risks of electromagnetic fields requires an understanding of science—in this case, physics and biology. While the growing field of environmental science offered new insights that began to transform the chemical and food industries, with electromagnetics, science couldn't corroborate any concerns, and public fear, which had begun to lose confidence in science, went off the rails of reason.

Beginning in 1989 the *New Yorker* magazine ran a series of stories about electromagnetic fields by staff writer Paul Brodeur.[2] They eventually formed the basis for his popular 1993 book *The Great Power-Line Cover-Up*.[3] By the time the book was published, there was a widespread conviction that electromagnetic fields from power lines and household appliances like microwave ovens were linked to childhood leukemia and other cancers, despite there being little or no evidence of this in broad epidemiological studies.[4]

Helping to bring this emerging national fear to a sharp focus was David Reynard's January 1993 appearance on the television show *Larry King Live*. Reynard was suing the cell phone industry, insisting that his late wife's brain cancer had been caused by her cell phone. "She held it against her head, and she talked on it all the time," he said.[5] The interview fed into the public's distrust of science and had an electrifying effect. The worry was immediately adopted into the mainstream consciousness. Reynard's suit seemed to corroborate what Brodeur had been writing about, and the new fear continued to spread even after more broad epidemiological studies showed no links at all between cell phone use and an increased incidence of brain cancer.[6]

Physics explains why no links were found. The microwaves used in cell phone transmissions do not have enough energy to break the chemical bonds of DNA, which is how cell mutations occur and cause cancer. How do we know this? Light and other parts of the electromagnetic spectrum, including microwaves, radio waves, infrared waves, and ultraviolet light waves, are all forms of radiation. A single unit of radiation is called a photon. A photon can be thought of either as a particle or as a wave. A century ago, Albert Einstein showed that the energy (E) of a photon can be calculated as Planck's constant (h) times the frequency (v) of its wave form, or $E = hv$. This formula was set out in one of his early papers, among his most famous, which was published in 1905 and is one of the reasons for which he won the 1921 Nobel Prize in Physics. Photons with low frequencies are at the red end of the spectrum. They include radio photons, whose waves can be as long as a football field and thus fly past us with low frequency and low energy. Microwaves are slightly stronger, followed by infrared radiation and then visible light waves. Waves at the high end of the spectrum fly past us at much faster frequencies (and thus have more energy) and appear more blue. This part of the spectrum includes ultraviolet light, followed by the even more energetic X-rays, followed by gamma rays, whose waves can be shorter than the diameter of the nucleus of an atom and thus very high frequency and very highly energetic.

Microwaves are slightly more energetic than radio waves, but far less energetic than even the infrared radiation that our skin gives off, which is how we can be seen by someone wearing infrared night vision goggles. Both visible light and microwaves can be used to cook food and heat up your coffee by concentrating them in very large amounts, such as in a solar oven or a microwave oven. The concentrated waves excite the molecules in a way that increases their vibration, and the friction that produces increases their temperature, but they still don't have anywhere near enough energy to break chemical bonds. If they did, the food would turn into goop.

Microwaves are much weaker than visible light, though, so it takes a lot more of them to make an oven work than it does to make a solar oven work, which is why microwave ovens are such electricity hogs. It's not until you travel into the red end of the visible light photons, then on to the yellow emitted by incandescent bulbs, then to the blue that illuminates many fluorescent bulbs, and then on past the visible spectrum into ultraviolet light that you get photons that have a high enough frequency and thus enough energy to break DNA bonds. These photons have about a million times more energy than microwave photons, enough that they can act like cue balls and knock electrons out of atoms, ionizing the atoms and changing their chemical nature. This is what Einstein's 1905 paper showed. Just imagine the force of a wave the length of a football field being concentrated into a wave the length of a molecule and then flashing past you over and over and over, and you can get a sense of the vast difference in the power of the two. That is what can break the bond between two carbon atoms, damaging DNA and causing cancer. But these ultraviolet photons still don't have enough energy to penetrate us very deeply, so they can only give us skin cancer.

Electromagnetic radiation above this level grows increasingly dangerous. X-rays are sometimes stopped by our skin, but if we are bombarded by enough of them they have enough energy to penetrate through us, a few of them being absorbed by skin and muscle, many more by our bones, which are denser, and many shooting that cue ball of a photon clear through us, which is why we can use them to make images of the inside of our bodies. X-rays can and do cause cancer, but our bodies can almost always stop them if the exposure is low enough. Gamma rays are so energetic that they can penetrate us, kill cells, and cause cancer very easily, and in high exposures they cause radiation poisoning, which kills much more quickly by damaging our bone marrow and gastrointestinal tract.

To the general public, the word "radiation" came to mean danger, never mind that our bodies give it off. Physicists like Bob Park tried to point this out and show that, like water, not all radiation is bad for us. Some is even necessary for life. Tsunamis kill, but streams nourish—the difference is in the amount and energy of the water. But science's credibility had by then been damaged by a constellation of silence, postmodernism, failure to anticipate biocomplex consequences, environmental mismanagement, and rhetoric. When that confidence has been shattered, all that is left to rely on is superstition, magic, religion, opinion, partisan pundits, television celebrities, and "prudent avoidance"—all "ways of knowing," but none of them *knowledge*. A new form of antiscience skepticism had taken root.

DEAF, DUMB, AND BLIND

The intellectual erosion of the 1970s, 1980s, and 1990s, as science sat silently on the sidelines and antiscience rose to rule on both the left and the right, was greatly worsened in August of 1987 when, during the administration of President Ronald Reagan, the Federal Communications Commission (FCC) abolished what was called the "fairness doctrine" in an historic 4–0 vote, severing one of the last ties to a common public foundation of knowledge and its cousin, the carefully researched public record that journalists had worked for sixty years to build.

The doctrine had its roots in the Radio Act of 1927 and had been a formal policy of the FCC since 1949, when television went mainstream. It required those who held licenses to broadcast over the public airwaves to present programs on controversial issues of public importance

and to present them in a way that was (in the FCC's view) honest, equitable, and balanced.

After the policy was abolished, Congress recognized that there was
danger here and tried to codify the doctrine into law, but President Reagan vetoed the legislation. As a result, broadcasters were unburdened
from the requirement to present balanced news coverage, and the age of
yellow journalism was reborn. Chief among the early gainers were angry
and opinionated baby-boomer talk jocks like Rush Limbaugh, who
began engaging in political rants that charged up listeners' amygdalae
with outrage in a sort of pro wrestling of politics, attacking examples
and perpetrators of the pet peeves of cultural conservatives, driving
audience numbers sky-high.

It was as if the *National Enquirer* and *Star* magazine had bought up the
nation's broadcast media. The problem for science is that the nation lost
its common sense of reality. The balanced programming broadcast on
public airwaves had long been held as a public standard and the voice of
objectivity. What was said on the air was taken to be generally true and
impartial. This is why Orson Welles's 1938 radio play about an alien
invasion, *The War of the Worlds,* could cause a nationwide panic.

To better understand what is happening, consider a bit of educational psychology and the cognitive styles of field-dependent and field-
independent personalities. Field-dependent personalities are more
socially oriented than others and require externally defined goals and
reinforcements.[7] Field-independent personalities are more analytical
and tend to have self-defined goals and reinforcements. Educators have
developed strategies for teaching people with each learning style. Those
field-dependent citizens in the population who valued authority and
looked to it for guidance needed the protection of the fairness doctrine
to give them a perspective on reality. In its absence, many of today's
field-dependent people have drifted under the influence of the conservative dominance in the commercial media, creating one of the most
divided political climates in American history.

It made sense that this would happen. Field-dependent personalities need more externally supplied goals and structure than field-independent types, and talk radio provided them. By freeing broadcasters from the requirement of fairness and thus the need to ground their messages in established knowledge and fact, the repeal plummeted the country into a partisan public dialogue where one-sided rhetorical arguments backed by outrage and sheer wattage held sway over facts and reason. It was a further undoing of Locke's ideas about knowledge, upon which the country's founding principles are based. By removing objectivity from the debate, it set America up for endless arguments between warring pundits.

Proponents of the repeal, who have since opposed congressional attempts at correction, argued that market forces will unleash freedom and competition in the marketplace of ideas that will stimulate more broadly ranging and higher-quality discussion, but in fact just the opposite has occurred, with discussion becoming less diverse and more polarized. Without a common "objective" standard of knowledge, listeners turn into opinion "dittoheads," as Limbaugh's followers happily describe themselves. Diversity of thought has been quashed in favor of an uncritical, authoritarian, and vehemently partisan groupthink, a new conservative identity politics directing an "In your face!" response to the feminist studies, African studies, diversity studies, and science studies identity politics of the postmodern era. The talk jocks act as chorus masters, conducting the audience members' political opinions in us-versus-them sports narratives that maximize audience share and direct the anger and political contributions of millions without any accountability to the facts.

This is the opposite of the protection of diverse minority viewpoints that was intended by the founding fathers, and it has inevitably run up against science on several occasions, with the most current and prominent example being the complete rejection of climate science and scientists as a manufactured political project of former vice president Al Gore, or, as Limbaugh lampooned him with the same degree of sarcasm he once directed at political correctness, "Algor"—the word for "a sensation of coldness."

WHAT MARKETPLACE?

But the argument for doing away with the doctrine was based on a faulty assumption: that ideas exist within a social dialogue that is akin to a marketplace in which journalistic truth has market value—that people want to hear it enough to pay extra for it, and media with a variety of viewpoints will compete to deliver the highest-quality journalism for the best price. But in fact, ideas do not exist in anything akin to a marketplace, and journalism's job is as often as not to tell people what they *don't* want to hear and would often rather avoid, but is important to know anyway. Its job is to report the *news,* meaning the facts of recent events as well as they can be determined, in order to provide the public with a common ground for debate and discussion.

The incorrect, unexamined assumption that there exists some sort of marketplace of ideas is destroying mainstream broadcast news and, with it, the balanced, moderate political weltanschauung of the country, dividing the media outlets into opposing groups of partisan extremists whose primary mantra is to seek contrast and wedge it wider by applying outrage to produce a sort of instant drama that is the triumph of opinion and applies the emotional charge necessary to compete in a ratings battle. This approach applies to the news the principles of building drama instead of those of journalistic reporting on facts, and so it is authoritarian and ultimately leads to the tyranny of might makes right, evidence of which we have seen accelerating throughout our media in the recent decades.

Hollywood producers will immediately tell you why this idea could never have worked: There is no marketplace of ideas; it's a marketplace of emotions. Given the choice, the majority of people want drama, sex, violence, and comedy, the four horsemen of entertainment. These four elements have driven plays, paintings, and stories for all of history. But they are not *news.* They are "but faith, or opinion, but not knowledge." News is knowledge.

The only way you can have a market-driven model for news is to apply a reward structure other than finance. You could do this by rewarding quality with a peer-review system like that used in scientific publishing, but review by outside peers is difficult in the extremely short time frame of competitive news gathering. Or you could get broadcasters to perform an abbreviated form of the peer-review process internally, which is what the fairness doctrine did. By repealing the policy, the FCC forced news programs into financial competition with entertainment—the market-place of emotions. Broadcasters were forced to turn to the tools that deliver audience share, and those tools are the four horsemen. It was the success of highly partisan, emotion-driven commentators like Limbaugh that took market share away from broadcasters who presented the news more evenhandedly. Without enforced standards or peer review, news was cut loose from knowledge and the emotions of outrage and comedy were increasingly relied upon to sell the news.

Imagine what would happen in the scientific community if peer review were abolished and scientists could, in the name of freedom of speech and market economics, publish anything they wished without regard for the truth. The major journals would compete for market share and give a portion of the proceeds to the scientists. A free-for-all would eventually ensue, with scientists competing for the most dramatic papers. Some would reflect good science, but many would not. Some scientists would stridently argue for the value of truth while others would quickly see the rewards of emotion. Scandals would even be manufactured in order to get more market share—say, a study manipulated to link the measles, mumps, and rubella (MMR) vaccine to autism—as occasionally slip through now despite the safeguards of peer review. Satirists would arise, pointing out how ridiculous it all is. Serious scientists would soon have to dress their findings in the enticing clothing of sexiness or drama, much as some of them try to do now in an effort to be relevant to the public. Thus we saw dramatic tales of an entire shadow biome because a NASA scientist was able to train bacteria to tolerate high levels of arsenic—tales that collapsed into controversy upon closer scrutiny.[8] Many would object to the new situation, arguing that by substituting market economics for standards of truth and quality we were losing something precious: a common ground upon which to build knowledge, which is essential to the free functioning of the scientific community. But others would argue that this was a healthy development because it opens things up to a "marketplace of ideas" and draws in a wider readership from the general public, and that's always good, right? The journals *Science* and *Nature* would likely increase their circulations, but eventually science itself would be lost.

This would not be freedom; it would be its opposite. Science would descend into darkness, and tyranny would again reign. For the sake of freedom and the public good, we must not use market economics as the sole arbiter of value in broadcasting. We need a reward system tied to truth. But to have that we need to abandon the postmodern idea that truth is subjective. The fairness doctrine was a tool of self-governance that encouraged peer review and helped ensure that.

A Congressional Lobotomy

Despite all the growing cultural ambivalence toward science, it was protected until the 1990s by the same motivation that had held sway since the Russians exploded their first atomic bomb in 1949—fear. But in 1991 the Soviet Union collapsed. With the collapse, America won the cold war and lost its major competitor in science. The American science enterprise, without realizing it, suddenly lost the rationale Vannevar Bush had used to get government funding for science.

Two generations of scientists had depended largely upon government largess, but now science itself, which had come to be seen as a means to the end of maintaining competitiveness and securing our national security and all the things that made America better than the Soviet Union, no longer had a specific mission. What's more, it was increasingly clear, thanks to the efforts of environmentalists, that science had produced a lot of messes.

Leading physicist Leo Kadanoff described the dour public mood in an October 1992 *Physics Today* piece titled "Hard Times."

> Today when the public thinks of the products of science it is likely
> to think about environmental problems, an unproductive arma-
> ment industry, careless or dishonest "scientific" reports, Livermore
> cheers for "nukes forever" and a huge amount of self-serving noise
> on every subject from global warming to "the face of God."[9]

Over the course of four decades of disconnection, specialization, and
tunnel vision, science had transformed in the eyes of the public from
noble savior to troublesome spoiler or worse—a tool of oppression.
Funding began to fall off, and science, which had been relatively silent
for nearly two generations, began to lose its standing.

By 1994, the year the "science wars" broke into the open and the culture wars were the topic du jour, the value of science had fallen so
steeply in the eyes of policy makers that newly elected House Speaker
Newt Gingrich, a self-professed science supporter, felt compelled to
propose eliminating funding for Congress's own science and technology advisory body, the Office of Technology Assessment (OTA), in
selling his budget-cutting package. The proposal was accepted. In fairness to Gingrich, he was able to protect and even increase science
funding in other areas, ultimately doubling the research budget of the
National Institutes of Health with the help of then-congressman John
Porter (R-IL). But the elimination of the OTA was a strategic blow that
would, many argued, affect science policy for the next twenty years. In
saving a relative pittance, Congress gave itself "a lobotomy," as congressman and physicist Rush Holt put it, with congressional staffers
turning instead to lobbyists, interns, and the Internet for science information of dubious quality when crafting major policy provisions. The
truth, forever of a nebulous quality in the eyes of baby boomers,
became even more fungible.

Michael Halpern of the Union of Concerned Scientists says that as a
result of losing the OTA, Congress has wasted billions of dollars on policies like the fence along the Mexican border that OTA scientists could
have told them would not work. "It was penny-wise and pound-foolish,"
Halpern says. "Without the OTA, it all became rhetoric."[10]

THE NEW NEWS CRISIS

Meanwhile, the move to free online news was being driven by the founding principle of the Internet: the hyperlink, a trail of context-dependent endnotes, based on a concept first developed by Vannevar Bush in a 1945 *Atlantic* magazine article in which he presciently describes the modern personal computer.[11] The model of free and linkable news stories completely upended the revenue structure of newspapers, making it very difficult to remain solvent and propelling them to jettison costs. But putting news stories behind a pay wall destroyed their linkability and so eliminated them from the burgeoning national dialogue occurring on the Internet. To save money, newspapers began by cutting expensive endeavors like investigative reporting and specialty divisions like science. Thus, an important faculty of the nation—its ability for broad critical self-assessment and data-based reflection—was suddenly eliminated.

As newspapers were grappling with obsolescence, yellow journalism spread from AM talk radio into TV with the advent of cable news. This trifecta combined to devalue the factual reporting and reason that once kept the country balanced, supplanting it with the battling opinion war-lords of the new media. Having trained at postmodernist universities, many emerging leaders in journalism didn't recognize this as a problem. It *wasn't their role* to discern the reality of things, they believed. Truth was subjective, its story a matter of one's perspective. Thus a reporter's role was to provide both sides of the story fairly and with balance, but also without judgment, which, in a world cut loose from knowledge, could be deemed politically motivated, and no one would be educated enough to refute the charge.

Marcia McNutt, director of the United States Geological Survey (USGS), describes the mind-numbing, tail-chasing public policy effects this "marketplace of ideas" produces.

> There doesn't seem to be much accountability. Remember when the "sand berm" issue for Louisiana was such a big deal? [It was proposed as a way to prevent the flood of oil from the 2010 Deepwater Horizon oil spill from reaching the shore.] Governor [Bobby] Jindal and the parish presidents got pretty much 24/7 coverage on CNN,

Fox, and everywhere else, saying what a bunch of losers anyone and everyone in the government was who wasn't rushing to put in their project, even though the USGS scientists and the Louisiana scientists said it was a stupid idea. But did the scientists get any airtime? No, and in fact with the way the media played it up, they were fearing for their lives. So the government reluctantly granted the permits and tried to make the best of it, because the public outcry in favor of the berms was so loud after all of the media attention. Then after the fact, when every prediction of the scientists proved to be right, where was CNN? Fox? Oh yeah—they did show up—to say what on earth was the government thinking to build the sand berms! "Another example of hasty government action!"[12]

The result is a crisis throughout American journalism. Serious journalism is being forced into small outlets on the Web, many of them nonprofit, or is being wrapped in the guise of one of the four horsemen of entertainment. Thus we see serious news being covered on Comedy Central's *The Daily Show* and *The Colbert Report* and Canada's *Naked News* and in theatrical documentaries such as *A Sea Change* and *Inside Job,* while cable news peddles scandal and outrage. Public broadcasting, one of the few news sources left that attempts to avoid the new conception of "fair and balanced" and instead stick to reporting the facts, is attacked as being politically motivated. As Hollywood screenwriting expert Robert McKee says, "Story is not about facts. It's not about intellect. It's about values and emotion."[13] Forcing journalists to deliver the high degree of values and emotion necessary to win in a crowded popular entertainment marketplace forces them to become yellow journalists, substituting emotions for facts and depriving Americans of the government accountability for which the fourth estate is relied upon in a democracy.

POISON II: THE HORROR OF THE VACCINE

By the late 1990s, Americans had become increasingly unable to discern between knowledge and opinion. With a niche for everything and emotion delivering the big money, the American public dialogue was getting hopelessly confused, and the invisible hand seemed increasingly balled into a fist, ready to strike out—at what, we didn't know, but *something*. Something had gone terribly wrong. The world was a mess. And science was poisoning us. Ruining our environment, causing cancer, and now—*this:*

Children—whom baby boomers protect with a level of fear and ferocity never before seen, perhaps recalling the cold-war danger of loss in their own early years—were in grave danger. Now science was taking *them* from us too, by giving them autism. But there was a lone scientist on our side, telling the truth, fighting against the man for us: a British surgeon named Andrew Wakefield.

In 1998 Wakefield published in the prestigious British medical journal *The Lancet* a scientific paper that linked childhood MMR vaccines to autism.[14] The paper gave Wakefield nearly instant celebrity because it crystallized the amorphous public fear of poison by science in another sort of Sputnik or Kuhn moment. He became a media star and began touring and speaking on the subject, and he coauthored a 2010 book about it. The only problem was that it wasn't so. Wakefield, as it later turned out, had doctored his evidence to fit his a priori conclusion, and the paper has since been discredited as "fraudulent" and was withdrawn by the journal.[15] The General Medical Council, which oversees British doctors, said Wakefield's "conduct in this regard was dishonest and irresponsible."[16]

But the damage was done and scared parents bought into the "science is poisoning us" meme that had been building all their lives since *Silent Spring.* They began refusing the MMR and other vaccines for their children. The MMR vaccine contained thimerosal, a preservative that contains mercury. It had been used since the 1930s to prevent contamination of vaccines by bacteria and fungi, after a horrible 1928 tragedy in which a multiuse vial of the diphtheria vaccine became contaminated with bacteria and killed twelve of the twenty-one children who were injected with doses from the vial just days after others had been safely vaccinated.[17] Thimerosal had helped ensure public safety for the nearly seven decades since. But now, suddenly, public opinion turned against the preservative.

The situation was greatly exacerbated when talk show hosts Oprah Winfrey and Larry King gave antivaccine advocate and charismatic former *Playboy* model Jenny McCarthy a platform on their shows. McCarthy is like many of us—well-meaning, passionate, and concerned for her child—but she has no background in science, and, with an a priori conclusion and a skepticism of science itself that impaired her ability to gain knowledge, she unintentionally did harm by promoting the nonexistent link. She was not alone. Robert F. Kennedy Jr., another nonscientist celebrity on the political left, authored a widely distributed, well-intentioned 2005 article in *Rolling Stone* and on Salon.com (since removed from the site), arguing that he was "convinced that the link between thimerosal and the epidemic of childhood neurological disorders is real."[18] In an age when truth is subjective, this sort of public pronouncement could seem reasonable.

Thimerosal had been removed from vaccines in 2001 as a precautionary move after the release of new findings about mercury and health, despite there being no scientific evidence of negative health effects. Still, the antivaccine movement continued to gain momentum, fueled by the same level of skepticism of science that had propelled the electromagnetic field scare, the cell phone scare, postmodernism, and the antiscience backlash as a whole. This backlash is not related to a lesser level of education, race, or economic status. On the contrary, according to a 2004 study from the Centers for Disease Control and Prevention (CDC), in the United States, unvaccinated children tend to be white and to have upper-middle-class, college-educated parents[19] who express concerns about the safety of the vaccines. These are organic-food shoppers. They know enough about science to be skeptical, but not enough to allay their fears, and they have too little trust in science to take the risk.

According to the same study, some seventeen thousand American children were going without childhood vaccines annually. The largest numbers were in certain counties in California, Illinois, New York, Washington, Pennsylvania, Texas, Oklahoma, Colorado, Utah, and Michigan. States that allow "philosophical exemptions" to laws that mandate certain vaccinations for children entering school had more communities with unvaccinated children, creating pockets of risk. The danger of this is that unvaccinated pockets of population form living petri dishes where nearly eradicated viruses can gain new footholds and endanger the wider public's health, similar to the way that cholera metastasized in the broader Haitian population in 2010. This is, in fact, what has happened. In 2000, the CDC declared that measles had been eliminated in the United States. But by 2008, after ten years of antivaccine misinformation, the United States had more measles cases than it had since 1996, two years before the Wakefield paper was published.

But the antivaccine scare also creates pockets of disease in certain religious and immigrant communities. In December of 2010, a now unlicensed and discredited Wakefield visited with members of the Minnesota Somali community, one of three such visits he held with them, reportedly telling the immigrants to "get vaccinated, but do your homework and know the risks."[20] The Somalis were concerned because autism appears to run at twice the national average in their community. As a result of the fear, vaccinations have fallen very low, and community members invited Wakefield to speak.

Then in February of 2011, an unvaccinated Somali infant returned from a trip to Kenya infected with measles. By April, fourteen cases had been reported, all but one traceable back to the unvaccinated infant. Half were Somali children, six of whom were unvaccinated and one who was not old enough for shots. The state had reported zero or one case of measles a year for most of the past decade.

Abdirahman Mohamed, a Somali physician in Minneapolis, told the Associated Press that Wakefield has caused global hysteria that has cost lives. He said he has warned the Somali community to stay away from him. "He's using a vulnerable population here, mothers looking for answers. He's providing a fake hope."

Antivaccine scares are not new; their history is almost as long as that of vaccination. Former congressman John Porter lobbies Congress for medical research investments as chair of Research!America, a health research advocacy group. Porter was raised as a Christian Scientist. His father walked with leg braces because of polio, and when the Salk vaccine was licensed in 1955, several of Porter's Christian Science classmates refused to get it. Porter thought of what the vaccine would have meant to his father's life, and he cut his ties with the church. "That is not what I would call moral thinking," he says.[21] The vaccine's success was profound. In 1952, the year Jonas Salk first began to test his polio vaccine in humans, the United States had 57,879 cases. By 1957, two years after the vaccine was approved for broad use, the number had fallen to 5,485, and by 1964, it was 122.[22]

DOES AMERICA COMPETE?

Porter says the silence by scientists over the last two generations has become a big problem in American policy making. "What members of Congress need to hear is passion," he says. "When scientists and engineers talk to a member they need to say, 'Look, this is really important.' The scientific community is very, very timid about asserting themselves. When President Bush was undermining scientific integrity and rewriting scientific studies to make religion preeminent, the science community did nothing publicly except for the Union of Concerned Scientists, and that silence was a shameful travesty. It's not just about 'Give me funding.' They've got to step up and say, 'This is wrong!' They need to impact the public in a way that really highlights how critical science policy is to our lives."[23]

By 2005, former Lockheed Martin CEO Norman Augustine was concerned enough to sign on to lead an all-star committee tasked with assessing America's economic competitiveness for the early twenty-first century for the National Academy of Sciences. Augustine is a former undersecretary of the army, chairman of the NASA Space Systems and Technical Advisory Board, and member of Bill Clinton's and George W. Bush's President's Council of Advisors on Science and Technology. His panel included a virtual who's who of American science and technology, and it issued a report called *Rising Above the Gathering Storm* that showed American economic competitiveness was slipping.[24] The landmark document formed the foundation of the America COMPETES Act, which was signed into law in 2007.

The study found that through a mixture of underfunding, misplaced priorities, and science illiteracy, the United States was falling behind competitors in the science-driven world economy in several critical measures. It served as a wake-up call for Congress, cutting through the fundamentalist rhetoric that had fallen over them like a poppy haze.

At about the same time, other reports were coming out showing similar declines, among them the National Science Board's *Science and Engineering Indicators 2006* annual report and reports on the international Organization for Economic Cooperation and Development's Program for International Student Assessment (PISA) tests, which measure and compare the performance of high school students around the world in science and math. In science, US students fell from seventh in a 1972 ranking known as the First International Science Study[25] to twenty-ninth out of forty-five countries measured in the 2006 PISA report,[26] below countries like Hungary and Poland and just above the Slovak Republic.

There are incomplete and inconsistent data sets in these international rankings, especially those predating 2000, but the overall trends were alarming. Members of Congress woke up long enough to pass the America COMPETES Act, but drifted back into the fog before funding it, leaving many of its initiatives to languish.

THE APPLE TREE

What could the science enterprise have done differently, given the same set of circumstances? It could have realized that science is and always will be political and communicated openly and transparently with that in mind. It could have done a better job of placing itself within the greater context of society—its ecosystem. It could have continued to sell itself to the public and make the case why its advances are good, moral, and important. It could have recognized that a democracy requires a plurality of voices, and that the voice of reason is a key part of that plurality, one that is independent of partisanship. It could have policed itself better, and more aggressively and more quickly exposed "junk" or fraudulent science before it could mislead many people. It could have stood up for what we know and how we know it, publicly refuted nonsense, and asserted the moral value that if you care about people, you must care about science, facts, and reason as the best tools and expressions of that love and caring. It could have thought about not just what it could harvest from the garden of government funding, but also what nutrients it must put back in.

Science in a democracy is like an apple tree. It needs to provide not only the fruits of knowledge and innovation, but also the strength of trust; the fertilizer of civic involvement, investment, and inspiration; and the shade of hope and charity. In this way science can create a virtuous circle that fosters a society that is both maximally responsive to the needs of science and well positioned to take maximum advantage of what science has to offer. It must make certain *the people are well informed.*

A REENGAGEMENT

This requires public engagement on a massive scale—but not just the whizbang of classroom or science museum demonstrations, and not just at the K–12 level. It requires scientists everywhere to come out of the laboratories and talk about what they know in the public square. It requires them to provide a constant reminder for all citizens of *the process* of how it is that we build knowledge, why this process is important, how facts are different from belief and opinion, why citizens should care, and how science intersects with their own lives and families, and thus *our policies.* It requires a reengagement.

That doesn't mean scientists should impose their personal values on the country—other than the values of knowledge and science. "Where science begins to appear political is when scientists, who unfortunately are also people, come to impose or even voice their own values, conflated with the data and their interpretation," cautions Alan Leshner, CEO of the American Association for the Advancement of Science.[27] It simply means presenting their work fairly and publicly. Learning how to do that is another matter.

There is a difference between being *political* and being *partisan* that Leshner's perspective does not tease out. Science is *always* political. It is a top-wing antiauthoritarian activity. But it is not partisan, left or right. America currently has no public forum whatsoever for candidate-public engagement on science, even though science has huge political and public policy implications. While scientists still rank among the most highly regarded of professionals in public opinion polls, with only teachers and members of the military ranking higher,[28] the majority of Americans can't name a living scientist[29]—unlike their local pastors and doctors. Science is distant, something done by nameless, faceless others in anonymous white lab coats.

KEEPING SCIENCE VITAL

Science itself holds the key to progress that is pro-health, pro-environment, pro-prosperity, pro-tolerance, pro-education, pro-skepticism, and pro-freedom. This top-wing political space is where science is naturally situated, and it cuts across the left-right political spectrum. Casting scientists as impartial and dispassionate authority figures works against what science naturally is, which is antiauthoritarian, creative, curious, progressive, conservative, open, and skeptical. Scientists need to be seen as they really are: passionate, curious, adventuresome, and creative people who care about other people and about making life better through creating new knowledge and the power that that effort yields.

Because new knowledge often challenges us to refine our ethics and to redefine how power can be used, science will always be inevitably and intrinsically political. The communication and participation enabled by citizen science initiatives and science debates that bring politicians together with scientists and the public are essential parts of keeping science vital in a society. It was bush league of the science community not to recognize this, but it's not too late to correct the oversight: Scientists simply need to speak up and play their part in the national discussion and to listen to—and sell the value of their work to—the public that so often pays their salaries.

It's the Process, Baby

How to do that? By making the scientific process, not authoritative pronouncements, the focus of communication with the public. What authority scientific claims have lies not with a practitioner, institution, or enterprise, but with each individual human's capacity to understand reason, observation, experimentation, and measurement. When scientists try to abbreviate this by saying, "Just trust us" or "It's too complicated to explain in plain English," they are furthering antiscience thinking. Scientists may indeed become authorities on their subjects, but such authority carries a burden of humility and the obligation to reach out and engage by translating what they know into simple, clear, and accessible language and disseminating it. Otherwise, they are separate from the people and just one of many competing authorities, along with kings, popes, policy czars, presidents, and pundits. This makes science into just another opinion that is thus politically suspect.

If instead of placing the emphasis on authority scientists focus on the *process*, they can safely engage in the public discussion in a way that is consistent with the framers' vision of the role of science in a democracy. They will have protected themselves *and* science, and also freed themselves to be the human beings they are.

PART III

TODAY'S SCIENCE
POLITICS

CHAPTER 9

TEACHING EVOLUTION: THE VALUES BATTLE

I don't believe in evolution. The dinosaurs live [sic] with Adam and Eve in the Garden of Eden 3,000 years ago. And then the Aliens came and gave us fire and the wheel.

—VISITOR'S WRITTEN COMMENT
at the Explore Evolution exhibition, University of Nebraska State Museum, Lincoln, 2005[1]

DON'T LET SATAN FOOL YOU

To understand how this is all coming to bear in America's culture wars, and why it is threatening America's national interests, let's look at the two largest fronts in the assault against science, the one being fought over values, and the other over money.

The values battle is being waged by religious conservatives on three major fronts: the nature and age of Earth and the universe, the theory of evolution, and the origin and nature of life and reproduction. It's important because our answers to these three questions lie at the center of physical science, biology, and the health sciences, and of our capacity to make policy decisions that will set up our children to compete in the emerging global science economy.

One of the most contentious of these issues is the teaching of evolution.

Fierce court battles that have been fought in numerous states since the 1925 Scopes trial have made contestants on both sides famous, with some of the cases going all the way to the US Supreme Court. Each time evolution wins, authoritarian religious conservatives adjust their language and redouble their efforts.

Classroom teachers often choose to simply skip the subject altogether rather than fight with creationist parents. One science museum director described overhearing a group of homeschooled children about to enter the paleontology exhibit being pulled aside by their parents and told, "Now, remember, those bones were put there by Satan to fool you."*

The question of whether evolution is real is dynamic because both views are uncompromising. One is based on knowledge, which is antiauthoritarian, the other on a literal reading of the Bible, which is authoritarian. They are thus opposites on the vertical political axis.

Modern medicine and biology are based on evolution—biology is essentially applied chemistry and physics in the context of evolution. It is the most fundamental principle in biology, the one that unified biology into an organized science. It connects and provides a framework for understanding all the various disciplines within the life sciences, from genetics to virology to oncology to organic chemistry. It is, at its most basic, simply the understanding of how life changes over time in relationship to its environment.

This helps us be better farmers, better environmental stewards, better computer programmers, and better doctors, improving health and productivity and saving trillions of dollars and millions of lives. For example, we have ten times the number of bacteria in our bodies as we do human cells, most of them working away in a symbiotic relationship that helps us digest food, fight off nastier bacteria, and stay healthy. But every year, pathogens try to attack us and take us down. Diseases like strep throat, meningitis, mononucleosis, gonorrhea, cholera, tuberculosis, Lyme disease, syphilis, tetanus, diphtheria, typhus, and pertussis are all

---

* An administrator at the Science Museum of Minnesota relayed this story, and similar stories have been relayed to the author by other science museum administrators in other states.

bacterial diseases, and they're just the beginning! It used to be that a course of antibiotics would cure most any of them, but over the last two decades these bacteria, which reproduce very quickly, have started to evolve into antibiotic-resistant forms that we can't kill in the same old ways. The theory of evolution is helping us to understand new strategies to slow this evolution before these killers get out of control again. For example, we know that antibiotics don't affect viruses, so we're working to get doctors to stop prescribing them across the board. When they do prescribe them, we want them to avoid giving low doses over long periods of time, because they're not high enough to kill all the bacteria, and the ones that remain evolve within our bodies into resistant strains. We now tell doctors to counsel patients to take all their pills, even after they start to feel better, or the knocked-down remnants of the bacterial army will regroup and evolve into resistant strains and come back with a vengeance. We need to kill them all, not leave the most resistant around to reproduce. And we've learned to stop the "preventive" use of antibiotics in farming and ranching, because widespread use just helps resistant strains get a foothold while keeping nonresistant, weaker strains out of the gene pool, helping to create superbugs. It's healthier for everyone if we treat the disease when it occurs, and we treat it overwhelmingly, rather than dribbling treatment around everywhere and encouraging the bacterial insurgency to adapt and spread. The theory of evolution is the only reason we understand this.

Politicians like Representative Michele Bachmann (R-MN) and former Alaska governor Sarah Palin suggest we should "teach the controversy," as if there were a legitimate scientific controversy over whether evolution is true. For the record: There simply isn't. But what most reason-minded critics of creationism don't understand is that the conflict is only tangentially related to the facts.

Since the time of Aimee Semple McPherson and the Scopes Monkey Trial, evolution has been painted by fundamentalist Christians as the cause of most social maladies. Consider this cautionary conservative Christian cartoon: It shows two castles, each on a tiny island, firing at one another with cannons. The castle on the left is on the island of "Evolution (Satan)" and flies the flag of Humanism, while the one on the right rests atop the island of "Creation (Christ)" and flies the flag of Christianity.[2]

In the balloons above the Evolution castle, one can see all the social ills caused by the theory of evolution: euthanasia, homosexuality, abortion, racism, divorce, and pornography. The misguided priests on the island of Christ are stupidly firing at these mere symptoms or hammering away at their own foundation, while the lone grim scientist-pirate-Satanist is blasting away at the bedrock of Christ himself, creationism.

With this type of emotional portrayal being incorporated into the context of the education of children, one can begin to understand why some on the religious right oppose evolution and think of creationism as godly, just as William Jennings Bryan and Aimee Semple McPherson did in the 1920s.

TEACHING THE CONTROVERSY

Michele Bachmann is such a woman. Bachmann was elected to Congress in 2006 and became a national standard-bearer for this type of religion-based politics. Before that she spent six years in the Minnesota Senate.

> I have no problem with teaching the various theories . . . of origins of life. . . . But, I think there's one . . . philosophy . . . that says only one could be taught and that one would be evolution. And because the scientific community has found that there are flaws in abiding by that dogma, I think it's important to teach that controversy. [Lawmakers and educators should not] censor information out of discussion because it doesn't meet within someone's dogmatic beliefs. Something that I think sometimes people don't like to hear is that secular people can be sometimes even more dogmatic in beliefs than people who are not secular. . . . In some ways, to believe in evolution is almost like a following; a cult following if you don't believe in evolution, you're considered completely backward. That seems to me very indicative of bias as well.[3]

This thinking is classic creationism and shows a postmodernist anti-science background, where science is simply another "way of knowing" that is on an equal footing with "but faith, or opinion." Bachmann said that because there are within the scientific community "eminent, reasonable minds" that disagree with the theory of evolution, "I would expect that teachers would disagree, and students would disagree, and the public would, certainly."

A Scientific Theory Is the Next Closest Thing to a Fact

Scientists will tell you that most of the conceptual problem lies in the general public's misunderstanding of the word "theory." A scientific theory is not a hypothesis or guess, as the word commonly means when used in casual conversation. A scientific theory is the one explanation that is confirmed by all the known and validated experiments performed to date. Experiments involving evolution have numbered in the hundreds of thousands over the past 150 years. A theory is thus among the *most certain* forms of scientific knowledge, and evolution is among the most certain of theories. But because science is inductive, scientists recognize that there is still a chance that it could be wrong.

In the case of evolution, the chance that it is wrong is somewhat smaller than that of, say, Earth being destroyed by a meteor within the next five minutes. We can see it working with our own eyes by watching viruses and bacteria evolve under the microscope. Any biologist can show it to you so you can see it with your own eyes too, and when you do, it becomes difficult to see how anyone could construe it as a matter of "belief." It's like saying "I don't believe in gravity" or "I don't believe Dick Cheney actually shot that guy in the face." It's ignorance of the evidence of observation.

Dogs of Intelligent Design

Like other creationists,[4] Bachmann says that many people confuse evolution with natural selection,

> and natural selection is not the same thing as evolution. No one that
> I know disagrees with natural selection, that you can take various

breeds of dogs . . . breed them, you get different kinds of dogs. . . . It's just a fact of life. . . . Where there's controversy is, Where do we say that a cell became a blade of grass, which became a starfish, which became a cat, which became a donkey, which became a human being? There's a real lack of evidence from change from actual species to a different type of species. That's where it's difficult to prove.[5]

This is another classic misconception of creationists. Darwin coined the term "natural selection" in order to distinguish it from the *artificial* selection done by breeders. The theory of evolution is not about selective breeding at all, which is the *opposite* of natural selection, and no evidence has ever suggested that human beings are descended from donkeys or blades of grass.

Creationists often refer to the writings of Michael Behe, a biochemistry professor and creationist and the author of *Darwin's Black Box,* a book arguing that some structures, such as the human eye, are just too complex to be the result of evolution and thus must be evidence of "intelligent design," a more recent version of creationism.[6] Like Harlow Shapley, Behe has made the mistake of clinging to an a priori first principle rather than building his understanding with observational evidence, and so his conclusions are not science; they're what Bacon called "science as one would," full of examples of "the vulgar Induction." In other words, rhetoric.

This was pointed out by the other professors in Behe's department of biological sciences at Lehigh University, who felt compelled to publish a statement clarifying their position. "While we respect Prof. Behe's right to express his views, they are his alone and are in no way endorsed by the department. It is our collective position that intelligent design has no basis in science, has not been tested experimentally, and should not be regarded as scientific."[7]

But some professors at religious colleges, Bachmann said, teach that "the Earth was created by an intelligent being—God, if you will—and that there are scripture passages that say that a day is as a thousand years and a thousand years is a day, and that therefore, over time, God could have created all this." Sure. Why not? But it's got nothing to do with knowledge, which flows, as Locke demonstrated, from *observation*. The rest "*with what assurance soever embraced, is but faith, or opinion,* but not knowledge, at least in all general truths."[8]

Let's not be mistaken. This sort of speculation is altogether human and entirely appropriate. But it's not science. Science, by its very foundation on Bacon and Locke and empiricism, limits itself to probabilistic conclusions that explain observations made of the physical world. Science makes no statements about the ultimate reality outside of these limitations, and teaching that it can is a corruption of the clarity of thought that is required of anyone who wants to excel in the science-dominated world of the twenty-first century. Confusing the matter for our children not only foolishly trades our ideological comfort for their ability to compete, but also, by unmooring them from knowledge, dooms them to the political paralysis and inability to problem solve and create a new future that has so plagued our republic in the first decade of the new century of science.

Bachmann is a classic example of how this kind of thinking can muddle one's logic. The more she examines the "universe and the natural world," she says, "the more convinced I am, personally, that this world was created by an intelligent being. I see evidence of intelligence everywhere. And if the scientific world points its finger in that direction, well, put it on the table."

Except the scientific world is not pointing its finger in that direction, Behe is, and he's not using science to do it. Bachmann continues:

> I read things about how carefully the world is made, that if the Earth was tilted just a fraction of a degree a certain way, or if the sun was just a little bit more beyond where it is . . . life could not exist on the planet. . . . And you see that and you say, How could it just be a big bang . . . that [made] everything come out so perfectly, to be perfectly conducive to life on this planet. It is just impossible, to me, that it could have been created just by random time and chance.[9]

Here we find another argument popular with creationist policy makers, which cosmologists call "the anthropic principle." It is related to the problems we have estimating probabilities and statistics in such things as gambling, life, and economic situations that involve insurance, investment, and risk. Imagine, for instance, the feelings Bachmann's blade of grass would have when the great holy golf ball landed upon it. "What are the chances," the blade might wonder, "out of all the millions and billions of blades of grass, that the great holy golf ball should choose *me*? I am special." And yet, with every stroke, the golf ball must land *somewhere*.

MEET THE FLINTSTONES

The prevalence of Americans who have creationist views has been well documented by numerous surveys, most prominently the *Science and Engineering Indicators* reports, a biennial issued by the National Science Board (NSB), the body charged with overseeing the National Science Foundation, and in Gallup polls. Both of these sources try to capture respondents' views with short conceptual questions. At the beginning of the millennium, 45 percent of Americans agreed with the statement that "God created human beings pretty much in their present form at one time within the last 10,000 years or so," while 52 percent of Americans were unaware that the last dinosaur died before the first humans arose.[10] That's to be expected. After all, we've seen them together on *The Flintstones*.

Fifty-four percent knew that Earth orbits the sun, and that it takes a year to do it. Fifty-one percent knew that antibiotics don't kill viruses. And 53 percent knew that "human beings, as we know them today, developed from earlier species of animals."

THE SWEET SPOT IN AMERICAN POLITICS

But by 2006, just *five years later,* something remarkable had happened. The last number had fallen significantly—from 53 percent knowing that "human beings, as we know them today, developed from earlier species of animals" to just *43 percent.*[11] Why? Did 10 percent of Americans get dumber about evolution over this time? Were evangelical creationists like Michele Bachmann and Michael Behe converting people in droves?

The answer may be suggested by considering the times. On September 11, 2001, al-Qaeda operatives flew planes into the World Trade Center and the Pentagon, and another hijacked aircraft, pointed toward Washington, DC, crashed in western Pennsylvania. America was shocked and angered, and its lawmakers responded, as traditionally happens in times of uncertainty, by increasing the power of the federal government and curtailing civil liberties—a swing toward authoritarianism. This is when the Bush presidency moved to a full bottom-wing position and the word "liberal" became an epithet to silence liberal (top-wing, antiauthoritarian) Republicans.

Between 2001 and 2006, the White House and both houses of Congress were largely controlled by Republicans, and the Republican Party was controlled to a large degree by authoritarian religious fundamentalists. George W. Bush had become the most profoundly evangelical president in American history. The Bush White House and reelection campaign worked as one to target and mobilize evangelicals as a voting block in the 2004 presidential election. Bush political adviser Karl Rove, himself a nonbeliever, famously walked the halls of the west wing whistling "Onward, Christian Soldiers" while preparing the 2004 Bush-Cheney reelection campaign.[12] Christian Coalition founding executive director Ralph Reed was hired as a campaign official and developed a strategy that included campaign activities in evangelical churches, endorsements and voter registration drives by pastors,[13] and mobilization of church congregations.[14]

The "civil war decade" was bearing cultural fruit, and the same security of strict biblical moral authority that had stiffened the wills of those venturing into the Wild West, that had emboldened the temperance movement of the post–Civil War 1800s, and that had given the conviction of courage to psyches unsettled by evolution, immigration, materialism, and other influences in the postwar Jazz Age now gave security and shelter in yet another time of American upheaval.

Viewed in this light, there emerges a possible correlation between this 10 percent of the electorate that had changed their minds on the question of evolution and the American swing voters who went overwhelmingly for Bush in the 2004 presidential election. Many 2004 surveys suggested that the number of swing voters in the 2004 election was about 13 percent—close to the number of new creationists. Other studies have shown this number to have remained relatively stable at between 9 percent and 13 percent in elections dating back to 1972.[15]

In 2004 Bush followed Reed's strategy of church outreach to unify political support within suburban congregations and to use those peer groups to capture politically unaffiliated voters who might swing from one cycle to the next, such as college-educated married white women, who were split almost evenly between the parties in 1992, 1996, and 2000 but broke sharply for Bush in 2004.[16]

"You are going to do what we used to call a friends and family program," Reed said, "where you had people take their Rotary Club or church or synagogue, or temple membership, or neighborhood directory, or even a tennis or garden club list, match it up against a voter file, and you will be able to do customized messages just to those people."[17] Members of congregations did indeed send in their church directories and formed "moral action teams" to fight what they increasingly saw as "a spiritual battle."[18]

Reed knew he was pushing churches into a moral and ethical gray area that could cause them to violate the law,[19] but Rove had told an audience at the American Enterprise Institute in December 2001 that only about fifteen million of nineteen million voters on the Christian right had turned out in 2000.[20] It was critical to Bush's reelection that the remaining four million be motivated to go to the polls, and Reed appeared to relish any negative publicity his strategy of church mobilization created as an opportunity to further reinforce the campaign's connection to faith and motivate true believers, saying that "Christians should not be treated as second class citizens."[21]

At the height of this overheated and often righteously angry national dialogue, the school board in Dover, Pennsylvania, voted on October 18, 2004, to require its ninth-grade science teachers to read a statement to their classes questioning the validity of evolution.[22] "Because Darwin's Theory is a theory," the teachers were required to tell students, "it continues to be tested as new evidence is discovered. The Theory is not a fact. Gaps in the Theory exist for which there is no evidence."[23] The district's science teachers defied the order, countering that it breached their ethical obligations to students "to provide them with scientific knowledge that is supported by recognized scientific proof or theory."[24] Several parents, including lead plaintiff Tammy Kitzmiller, agreed and sued the school district. John Jones, a Republican federal judge appointed by George W. Bush,[25] ruled that teaching intelligent design in public school science classes was unconstitutional.[26] After the decision, fundamentalist activist Phyllis Schlafly said that Jones had "stuck the knife in the backs of those who brought him to the dance." He was accused of being an "activist judge" and threatened to the point that he required the protection of US Marshals.[27] Later, in describing the job of a federal judge, he said that, unlike Congress and the presidency, which were created to be responsive to public sentiment, the founders "in their almost infinite wisdom" created the judiciary "to be a bulwark against public will at any given time—but to be responsible to the Constitution and the laws of the United States."[28]

As Bush, Bachmann, and other conservative evangelical congressional and state legislative candidates across the nation were campaigning in churches, pastors were warned by the IRS against making explicit endorsements of candidates from the pulpit,[29] so instead "moral values" were cited as the defining issue of the election.[30] Candidates were invited to speak about these moral values, and millions of voter guides were distributed[31]—in churches and by congregation members into the broader community[32,33]—on the moral issues said to be at stake, including abortion, same-sex marriage, faith-based initiatives, stem cell research, human cloning, and the teaching of creationism in public school science classes, all of which certain candidates were on the right side of, and others were not. A November 19, 2004, Gallup poll found that "the lowest levels of belief that Darwin's theory is supported by the scientific evidence is found among those with the least education, older Americans, . . . frequent church attendees, conservatives, Protestants, those living in the middle of the country, and Republicans,"[34] all voting blocks that went for Bush,[35] who heavily won the creationist vote,[36] which Gallup then measured at 25 percent of the US population.

Science, by contrast, had no voices speaking on its behalf from the pulpit or almost anywhere else, so it's not hard to imagine how such a vast swing in public opinion on the question of evolution could have happened.

Political consultants call this ideologically malleable electorate the "sweet spot" of American politics. Clearly, if something like political speech can influence substantial numbers of Americans' answers to factual science questions, something beyond education must be entering into the equation.

WHEN FACTS DON'T MATTER

Scientists will tell you this is just plain wrong. Facts are facts. They're not fungible, and these people are just poorly educated. The problem is that the public doesn't view science as a collection of facts that can't be argued with. In fact, some studies suggest that the brain processes facts and beliefs *in essentially the same way.*[37] Thus, when a fundamentalist Christian says she believes as a fact that God created people literally as described in the Bible, she means it, in the same way that a scientist believes a factual statement about evolution or geology. Whether this is a matter of science education not going far enough to eradicate superstition or something else is open to debate, but it underscores how critical the distinction is that Locke made between knowledge and "but faith, or opinion," and how hopelessly entangled the conflict between science and values politics sometimes seems.

That conflict came to a new head in April 2010, when the NSB deleted the questions and responses about evolution and the big bang (the origins of man and the universe) from its biennial *Science and Engineering Indicators* report.

Human beings, as we know them today, developed from earlier species of animals.

The universe began with a big explosion.

These two questions had appeared in the report since 1983. The NSB had already released a draft version with the responses, which showed that just 45 percent of Americans had answered "true" to the first question, a percentage much lower than that in Japan (78 percent), Europe (70 percent), China (69 percent), and South Korea (64 percent); and that only 33 percent of Americans had answered "true" to the second question about the big bang, compared with, for example, 67 percent in South Korea and 63 percent in Japan.*

The man responsible for the deletion was the chapter's lead reviewer, John Bruer, a science philosopher focusing on neuroscience, cognitive psychology, and education. The decision took the White House, which had been shown the earlier, unedited draft, by surprise. Scientists were outraged. A staff writer at *Science* magazine, Yudhijit Bhattacharjee, wrote about how stunning a development this was.

> Board members say the answers don't properly reflect what Americans know about science and, thus, are misleading. But the authors of the survey disagree, and those struggling to keep evolution in the classroom say the omission could hurt their efforts. "Discussing American science literacy without mentioning evolution is intellectual malpractice," says Joshua Rosenau of the National Center for Science Education, an Oakland, California–based nonprofit that has fought to keep creationism out of the science classroom. "It downplays the controversy."[38]

But did it? Consider the way the question was asked in a 2004 survey, the year of Bush's reelection when creation fervor was running high:

> True or false, *according to the theory of evolution*, human beings as we know them today, developed from earlier species of animals. [emphasis added][39]

---

* The deleted section of the draft can be viewed here: www.shawnotto.com/downloads/seind2010deletion.pdf.

When asked this way, not 45 percent, but *74 percent* of respondents knew the correct answer, a figure much more in line with the responses in other countries. This suggests that it's not that Americans are ignorant. The "science under siege" narrative that scientists use to gin up concern is not necessarily true. Americans *do* seem to understand. They just sometimes disagree anyway. Because—and here's the shocker—science is *political*.

FRAMING SCIENCE

If it's true that our brains process facts and Locke's "but faith, or opinion" in essentially the same way, how can one ever hope to break through? The key lies in *emphasizing the process,* which granulates the frame from an authoritarian assertion to an antiauthoritarian exploration of the senses and intellect: "Look, see it yourself?" This has the same effect as Locke's careful definition of knowledge: It removes science from a rhetorical frame conflict and refocuses the mind on observable reality, causing cognitive dissonance and questioning. When the evolution question is worded with the qualifier "according to the theory of evolution," that emphasizes process. We could also ask it with the qualifier "according to observations of the fossil record" and would likely get a similar result. This is because science is a physical, objective subset of the broader worldviews that it was carved out from, and that's okay. We are comfortable with its making limited assertions, which is all it claims to do because we cannot observe the whole universe at once. This is why it speaks in terms of probability, statistics, and math. This is the nature of inductive reasoning.

Prominent social scientist Matthew Nisbet and others have built up evidence that supports this. He suggests that people think about science as a "way of knowing" or "worldview" or "frame" and they think about religion as another "way of knowing" or "worldview" or "frame."[40] When people are forced to choose between these two frames, the conflict will skew the apparent level of science literacy lower. Scientists will say this doesn't make sense. The facts are the facts. But people do not consider only facts when making decisions, when investing in the stock market, when choosing a mate, when buying a house—or when voting.

This is an important difference between scientists—who are trained to set aside emotions and to adjust their worldview to a careful, detailed consideration of the evidence—and everybody else. Nobel Prize–winning physicist Richard Feynman put it very well when he said, "The only way to have real success in science . . . is to describe the evidence very carefully without regard to the way you feel it should be. If you have a theory, you must try to explain what's good about it and what's bad about it equally."[41] Bacon said the same thing, essentially, in *Novum Organum*.

This instills the values of honesty and integrity, which are impossible to fully adhere to when making a rhetorical argument whose purpose is to only present the information necessary to win. While many scientists make observations, think about the implications, and then respond emotionally to those insights about the world, a majority of Americans approach the world differently. They use persuasion to influence others to get what they want and need in life. They apply similar strategies to themselves, forming summary beliefs and principles that help them navigate difficult situations and relieve stress, both of which are, at their cores, rhetorical strategies. While a scientist seeks power through knowledge, a nonscientist seeks it through persuasion, which Francis Bacon called "the vulgar Induction."[42]

DRAWING A LINE IN THE SAND

Head-on challenges to these strategies will inevitably increase stress and raise ire—and Nisbet argues that it can lead to even further polarization. This in turn alters perceptions and starts a feedback loop. There is "lots of research," Nisbet says, that shows that people who have strong feelings about an issue tend to have difficulty estimating the proportion of people on their side versus the opposition and see any news coverage as hostile to their own interests. Journalists play on this, he says, using the "culture war" narrative, which plays to journalists' sense of story. As screenwriting guru Robert McKee says, "Conflict lies at the heart of all stories."[43] Journalists are trained to tell stories. They look for the conflict narrative to find the angle on a story. The "culture war" narrative is conflict on a grand scale. It's journalistic porn—cheap and easy to write, and it stimulates visceral responses that run pretty much the same way through any "culture war" story but carry little intrinsic meaning.

Nisbet offers as evidence a 2010 *New York Times* article[44] about how anti-evolution advocates are gaining new momentum by linking creationism with opposition to global warming science and human cloning, wrapping them all up in a "tarball" of antiscience thinking. The *New York Times,* he says, tends to savor these culture-war stories, which are usually heavily e-mailed among scientists and liberals. A story like that gives us no indication of its subject's real strength or threat as a movement, but it sets the reader up for an instant leap in logic: Science is under siege and we need to respond!

> This inevitably follows a pattern: Exaggerated perception of the threat. Chatter. Online discussion. People will write opinion pieces. And that gets the other side going. One of the interesting things is that this constant conflict narrative has a quality of a self-fulfilling prophecy. People look to news coverage to make sense of their own opinions. If evolution is constantly portrayed as a conflict of religion versus science, people take that as a cue on what to believe. The more you perceive this conflict in news coverage, and use it as a key to make your own opinion, to that extent your own opinion tends to become more extreme.

This is, of course, exactly what occurs on the right with AM talk radio and political-adrenaline cable "news" shows. Another example, this one on the center-left, comes from PBS's *Frontline* series. In 2010 the program produced a show on the antivaccine movement and called it "The Vaccine War."[45] There's no war. The press release about the show cast it as doctors and scientists against parents. This says there are only two positions: health advocates versus parents. What would many parents tuning in naturally choose? They would tend to side with the antivaxers because they don't identify with the scientists. They would be driven to a more extreme position.

Could it be that we enjoy this conflict? Clearly it sells newspapers, radio shows, television dramas, and political careers. What is so enticing about it is that it breaks the world down into us versus them and reaffirms us. We are right, they are wrong, and knowing that I am not alone in this difficult time makes me feel better. This is of course the baby boomer's life mantra, but it is by no means limited to one generation. The problem is that it is a sideshow that doesn't actually involve any leadership or an attempt to truly deal with the overarching situation by envisioning and going to a new future.

DEFUSING THE DEBATE: A GENERAL'S PERSPECTIVE

Science education advocate Eugenie Scott deals with that kind of polarization on a daily basis. Scott is executive director of the National Center for Science Education and has been at the forefront of the legal and policy battles against teaching creationism in public school science classes for twenty-five years. She pulls out a huge map of the roughly fourteen thousand independent school districts in the United States.

"Now realize," she says, "that each of those fourteen thousand school districts is controlled by a locally elected school board, whose members belong to the local churches and the local chambers of commerce, and who usually set curricula for science classes, usually with little knowledge except what they hear in church or on the radio or in newspapers." They are often committed and caring community leaders who lead busy lives, but typically few are scientists—partially because scientists have, for two generations, been civically disengaged.

Because of this, there are rarely science-educated voices guiding these discussions, and they can get wildly off the mark. Scott sees no gain in the "culture wars" narrative, even though she is on what could be termed the front line, because of its tendency to polarize and cause people to dig in. Instead, she seeks to build bridges by speaking about human-origins thinking not as either-or, but rather as falling along a spectrum from evolutionary atheists at one end to young-Earth creationists at the other and saying that most people fall in the middle.

"I think that people should understand science well enough that they don't confuse it with antireligiousness," Scott disarmingly tells her audiences. "Science is a really good way of knowing about the natural world. It doesn't make any pronouncements about how you should view ultimate reality."[46]

Nisbet says this kind of bridging language helps to break down polarization. He suggests that the responsibility for the heightened sense of conflict is not limited to evangelical creationists. Within the community of scientists there's a small social movement of militant atheists, including authors Richard Dawkins and Sam Harris, science blogger P. Z. Myers, and others whom Nisbet says have contributed to the polarization of the issue.

THE BATTLE'S FRONT LINES

Myers disagrees. In 2006, the science journal *Nature* listed Paul Zachary Myers's blog *Pharyngula* as the top-ranked blog written by a scientist. His following has grown since then. As of this writing, Myers says *Pharyngula* has some 2.5 million views per month[47] from about one million unique visitors, rivaling traffic on the Web sites of many major metropolitan newspapers.[48] The blog is named after Myers's favorite stage in embryonic development, the pharyngula stage, in which all vertebrates, including humans, have close similarities, such as a tail and gill arches. A brilliant and popular biology professor at the University of Minnesota, Morris, Myers is known for his sarcastic, incisive, and entertaining style. And it is precisely his outspokenness in giving readers tools to dispense with the arguments of antiscience evangelicals that has made him so hugely popular.

"[Science writer and Science Debate 2008 cofounder] Chris Mooney calls what I do 'the conflict frame,' and he thinks it's bad for science," says Myers. "But conflict is something to talk about. A novel that did not have a narrative, a central conflict, would be boring. Just because I draw a line in the sand, it doesn't mean I expect 100 percent of the people should abandon their faith. That line just gives us a focus."[49]

There is some truth to this. As noted earlier, drama is one of the four horsemen of entertainment, and the narrative or rhetorical argument is the strategy most people use to navigate the world. History shows that social progress is often made through a culture war of conflicting narratives, whether the movement is emancipation, women's suffrage, civil rights, women's rights, reproductive rights, or gay, lesbian, bisexual, and transgendered rights. This is confirmed by biologist Simon Levin's findings on opinion dynamics, which in some ways confirm the natural law basis for democracy. As mentioned in Chapter 4, Levin studies grazing patterns in herd animals, who vote with their feet—or their fins. "It's not the fish that are moving in the right direction that are able to win; in some cases it's those who are the most stubborn, and not responsive to others, that end up controlling the opinion dynamic," Levin says. "Look at George Bush."[50] The opinion tends to move in the direction of those who are the *least* flexible.

This suggests that Myers's principled stand may be the most effective approach if it is broadly communicated and articulated with conviction and leadership. The status quo never goes down without a fight. Authoritarian and intolerant tendencies like racism, for instance, are held in check in mainstream society largely through sanction and stigma, often with legal reinforcement—in other words, the peer review of the crowd. It is our relationship to other people that helps to keep us rational—an insight that groups like Alcoholics Anonymous, and the process of scientific peer review, use to great effect. But to move the crowd, one needs to be vocal, stubborn, and convincing.

Nisbet argues, however, that this only further polarizes debate. As the most politically vocal opponents of creationism, militant atheists have become stand-ins for the broader science community and contribute to the conflict narrative in news coverage. "The same type of identity formation and opinion extremity has happened on climate change as on evolution. It's always stated in binary terms: You're either a supporter or a climate denier. No in-between. It's become wrapped up in Democrat versus Republican in terms of political identity, as a result of activists using these issues to promote their own goals and media using the issues to tell the conflict narrative and sell newspapers and radio ads."[51]

While Levin's work supports Myers's approach when it comes to moving opinion, Myers is working against an overwhelming tradition and organization—religion—and is extending science to do what Scott says it doesn't do—make pronouncements about ultimate reality—by saying there is no God. This is beyond the realm of what it is possible for inductive reasoning to determine, and so it is outside the scope of science. This reduces the argument to competing claims—once again, all the way back to Locke—and thus ultimately the argument leads to brutality.

To Fight or Not to Fight

Myers argues that the conflict is already there in society and he is simply providing a much-needed countermessage to that of the radical authoritarian fundamentalists who have taken over much of the national political conversation. "Saying that there's not a conflict in the face of that is putting your head in the sand," he says, and much of the recent history of science bears him out.[52] "Right now the authority of science lies largely in it being sort of arcane and difficult to understand and I don't think that's healthy for us. Science should be something that everybody has access to and everybody can understand. Most scientists are perfectly willing and interested in engaging the public. Lots of them are actually pretty good at it, but there is no incentive in the university system to do that—tenure relies on grants, and then on teaching. Outreach is way down at the bottom and does not influence tenure or promotion decisions. People who do outreach are regarded as second-class citizens in the science world." Writing in a 2011 editorial, AAAS CEO Alan Leshner echoed this concern.

Unfortunately, traditional reward systems only emphasize publication and grant-getting, at the expense of efforts to promote increased
participation in science and engineering. Nationally, university
presidents, provosts, deans, and department chairs must find ways
to reward faculty members for reaching out to a wider community
of potential students and their families . . . Support like that is essential for innovation because increasing the diversity of the scientific
human-resource pool will inevitably enhance the diversity of scientific ideas. By definition, innovation requires the ability to think in
new and transformative ways.[53]

Diversity of ideas is precisely the opposite of what occurs with a strict
adherence to an ideological doctrine, be it the Bible or some other guide
for living, which creates conformity and "dittoheads"—the opposite of
what America needs to succeed in the new science-driven global economy. With that Nisbet agrees, but he argues that it's important to cast
outreach in terms of trust in relationships. Remove the perceived threat
or conflict. "People want to feel like you are listening to them," he says.
"If you tell versus listen, you promote mistrust."

Nisbet says we should identify topics of discussion that create interest, draw attention, and are relevant to people's lives to try to bring
doubters around to evolution. One sensible example is the immense
contributions evolution has made to agriculture or medical science. So
that's the frame. Then, who's the best communicator? Probably a medical scientist who has local community ties, plus a local teacher or religious leader. Go to places people are familiar with and feel comfortable
in, like churches or schools or malls, and sponsor opportunities for
discussion. It's not about promoting information or thinking about
communication as transmission under this approach. It's about having
a conversation. Communication is context dependent and political—a
valuable insight from postmodernism that scientists have fallen woe-
fully short in understanding.

This is all true, Myers says. But there is also a reasoned argument to
be made for what he does. "If I am loud enough and clear enough and

convincing enough, some people may—and I repeat, may—begin to reevaluate their thinking. Look at what Rush Limbaugh or Pat Robertson does. You can't tell me they haven't had influence. If you agree that they have, don't criticize me."

A PASTOR'S PERSPECTIVE

Lutheran pastor Scott Westphal ministers to a congregation in Scandia, Minnesota, in Michele Bachmann's congressional district. "I can't tell you how frustrating it is when parishioners listen to this divisive nonsense on AM talk radio instead of the teachings of their own pastor," Westphal says. "It doesn't get us anywhere as a people or a country. All I see it producing is anger and fear. Jesus questioned. He wanted people to open their hearts and their minds and let go of fear." Westphal has had scientists and science advocates lead discussions about evolution, science, and the big bang at his church. "Protestants started out by questioning. These things don't have to be in conflict. By meeting scientists and talking openly about these things, people can let go of their misconceptions and see why scientists think the way they do—and that they are our friends and neighbors, not threats or opponents."[54]

WHY THE FIGHT MATTERS

Eugenie Scott says she herself is "nontheistic," but that others in her office, such as her colleague Peter Hess, are "theistic" but nevertheless hold strong convictions that creationism must not be taught in science class. It can be taught at home or in a church setting, but government-sponsored schools need to teach science so that we have an informed citizenry. "If we're teaching creationism, we're not teaching science," Scott says. "The assumption of creationism is that natural phenomena require supernatural explanations. I'm not saying science is atheistic about ultimate reality. It isn't. To say that you can explain something using natural causes is not the same thing as saying there are no supernatural causes. Science is atheistic in the sense that plumbing is atheistic. It limits itself to the study of natural causes."[55]

This is critically important. The United States has gotten as far as it has in terms of technology and dominance because of science. Because of our understanding that even if you haven't figured something out, you can just keep plugging away, looking for those natural causes and sooner or later you'll find them. Teaching creationism in school science classes is teaching a habit of mind that is toxic to that problem-solving method. It teaches you to just throw up your hands and declare that the problem is unsolvable, particularly if that problem is tough or might have consequences for a particular religious belief. It teaches you to value not diversity of ideas, but conformity. If you do that, you're basically giving up on science and on the probability of finding those answers. That is not going to take America where we need to go.

CHAPTER 10

CLIMATE CHANGE: THE MONEY BATTLE

This is nothing more than a giant, global redistribution-of-wealth scheme. Man-made global warming, the hoax that it is, has always been nothing but that—with the accompanying gigantic growth-of-government from nation to nation occurring at the same time—and the loss of individual liberty and freedom. Tell the people in Denver! Tell the people in Fargo! I woke up the other day looked at the weather, and it was 35 degrees in New York! And you know, more and more people are starting to ask, "What is this global warming?" More people are starting to consider the notion that we actually may be in a cooling phase, 'cause there hasn't been any significant warming in years.

—RUSH LIMBAUGH, MARCH 27, 2009[1]

If the IPCC [Intergovernmental Panel on Climate Change] had been done by Japanese scientists, there's not enough knives on planet Earth for hara-kiri that should have occurred. I mean, these guys have so dishonored themselves, so dishonored scientists.

—GLENN BECK, FEBRUARY 10, 2010[2]

THE GREAT FALLACY OF SCIENCE

For two generations, scientists have labored under the notion that science is not political. They removed themselves from America's national dialogue, and Americans became less interested in science as a result. And over those two generations, as the public and even academia grew more skeptical of science, the public policy implications of scientists' insights into nature grew at the same time that their ability to communicate with society atrophied.

Meanwhile, using the fruits of science, the world progressed. Population grew exponentially. Development and technology exploded. Oil use sky-rocketed, largely in response to the broadly dispersed, low-density land development model the United States had instituted as a defense against nuclear attack. A new economic model evolved based on sustained opti-mum production, low-density development, high consumer turnover, and perpetual market expansion. Science continued to leverage new efficiencies and engineering to produce new conveniences. People responded to the incentives of this bounty and consumption skyrocketed. Wealth exploded.

Science-driven market economics began to change the planet and pro-gressed to the point that the majority of the policy challenges facing the United States and other leading governments revolved around science. For convenience we can call them SEEP challenges, for they always exist at the intersection of science, economics, environment, and population. And nowhere did SEEP challenges come into more urgent focus than on the issue of climate change.

THE SCIENCE OF THE CLIMATE

Our understanding that increasing carbon dioxide levels in the atmo-sphere could change the climate is not new. The relationship between carbon dioxide, water vapor, and climate was first laid out in detail in 1896 by the Swedish physicist Svante Arrhenius, who estimated that a doubling of carbon dioxide levels would cause global warming of 4.9° to 6.1°C. In his landmark paper, Arrhenius reported that "a simple calcula-tion shows that the temperature in the arctic regions would rise about 8° to 9° C, if the carbonic acid increased to 2.5 or 3 times its present value," and he suggested that over time the burning of fossil fuels—then largely five hundred million tons annually of coal—could have an influence.[3]

This influence was first documented experimentally by Charles Keeling in the late 1950s as the US economy was coasting off its postwar boom. Keeling was a young postdoctoral researcher in geochemistry at the California Institute of Technology in 1955 when he figured out how to make a device that would reliably measure the level of carbon dioxide in the atmosphere. He spent three weeks that summer camping out at Big Sur, with his wife and newborn son, taking samples and recording measurements.

Roger Revelle, the director of the Scripps Institution of Oceanography in San Diego, thought this work might be important. Public transit was dying and auto use was exploding, and, along with it, suburban home energy use. Scientists had debated since Arrhenius's time whether the extra carbon dioxide from burning fossil fuels might have any effect. Keeling's precise measurements, if taken in pristine air away from major sources of fossil-fuel emissions, might be able to answer that question. Revelle persuaded Keeling to join Scripps, and Scripps and the US Weather Bureau (now the National Weather Service) funded a measuring station at Mauna Loa Observatory, high in the mountains of Hawaii.

Keeling's measurements showed that carbon dioxide levels fall as plants take it in during the Northern Hemisphere's growing season, then rise during the fall and winter months as vegetation dies and decomposes, releasing it back into the atmosphere. The planet is essentially breathing. But over time, he also found something troubling: The annual levels, year after year, were rising.

Keeling's first measurements in 1958 showed an average concentration of 315 parts per million (ppm). Every year thereafter, the levels rose by a little more than they had the previous fall and winter, but they fell by only the usual amount caused by growing vegetation in the spring and summer, creating an overall upward trend. Today, the number is more than 390 parts per million. Keeling's chart of the annual rise and fall in CO_2 levels and their inexorable climb has become one of the most famous images of science, one that is etched into a wall at the National Academy of Sciences in Washington, DC. It is called the Keeling curve.

When scientists saw the CO_2 increase, they realized it would have major public policy implications for how we generate and use energy, but they didn't have enough data. They knew CO_2 helps trap heat in the atmosphere, keeping the planet warm enough for life, but they didn't know if the extra CO_2 Keeling's measurements showed was enough to affect global temperatures. They began expanding the field of geophysics to get a handle on just how big the problem might be.

Over the next five decades, thousands of scientists went into the field and collected massive amounts of data. What they found, using many different methods, was that the increasing CO_2 levels were causing global warming, just as Arrhenius had predicted, and we were approaching levels that could change climate patterns in ways that destabilize our economy, our national security, and our environment. We had to stop burning fossil fuels.

As had happened to Galileo, making those simple observations once again bumped science up against an established political and economic power elite that viewed it as a threat to their authority. This time it wasn't the church, it was the energy industry. And like Galileo had, scientists badly misjudged the situation.

THE POLITICS OF THE CLIMATE

The costs of scientists' disengagement had come home to roost. Early in the Bush administration, the writing was already on the wall. By 2001, scientific conclusions were being held back or altered by ideologically motivated appointees. Still, the climate scientists held out hope. Former president George H. W. Bush had publicly acknowledged climate change and the need to do something about it back in 1990. So, during the administration of President George W. Bush, they appealed to Congress and urged the president to sign the Kyoto Protocol, an international treaty first adopted by members in 1997 that is designed to limit carbon emissions.

THE NATIONAL ACADEMIES' VIEW

Instead of signing on right away, President Bush asked the conservative National Academies to look into the matter. They did, and in 2001 concluded that:

- Climate change is real.

- Satellites have detected no increase in solar energy reaching Earth since they were placed in orbit in the 1970s, yet the global climate is warming, so the warming cannot be attributed to the sun.

- By analyzing the isotopes of the hydrogen and oxygen atoms that make up the mountain and polar ice caps, we can infer past temperatures in the region.

- By analyzing tree-ring density and calibrating it with past temperature records and with ice-cap isotopes, we can estimate annual temperatures going back more than a thousand years.

- We can observe that carbon dioxide is a greenhouse gas that traps heat.

- We can measure the amount of carbon dioxide created by burning fossil fuels because it is a different isotope from environmental carbon dioxide.

- Atmospheric and oceanic carbon dioxide levels are increasing from burning fossil fuels, and the rate of this increase correlates with the warming temperatures. By the year 2000, when the report was being researched, the levels had increased from the 315 ppm measured in 1958 to 370 ppm.

- By analyzing air bubbles trapped in ancient ice, we can observe that carbon dioxide levels are at their highest level in at least 400,000 years and are continuing to rise.

- A warmer atmosphere holds more moisture and is stormier. Water vapor is also a very powerful greenhouse gas, leading to a positive feedback loop that can accelerate warming.

- We would logically expect this to result in a drier climate as more moisture remains trapped in the atmosphere and in more violent storms because of the extra moisture and heat. Heavy rainfalls tend to run off before they can be absorbed, so we would logically expect less moisture in the soil and more flooding.

- The most recent decade [then the 1990s], when global temperatures are averaged, is the warmest in at least a thousand years.

- As logically expected, there is more rapid warming over land and the north polar region than over the oceans, which take longer to warm. We also expect some areas will have local cooling as weather patterns change. Weather should not be confused with climate.

- Vast increases in methane gas production by large-scale ranching and farming, coal mining, landfills, and natural gas handling, are also adding to the warming, as methane is a very potent greenhouse gas. Increases in emissions of nitrous oxide, a by-product of the use of nitrogen fertilizers, are also contributing.[4]

Bush got the report and responded June 11 with a major speech in which he said he would not sign on to Kyoto because China and India, as developing countries, didn't have to reduce their emissions and because we still didn't know enough.[5] Like much of science during the bottom-wing Bush administration, the exhaustive report was virtually ignored. Subsequent National Academies reports and testimony delivered to President Bush and to Congress in every subsequent year of the Bush presidency further developed these data with new research, showing, for example, that the climate was warmer than at any time in the last 650,000 years,[6] and that increased warming may lead to the release of the large amounts of methane that are currently frozen under the ocean and in northern Russia,[7] possibly inducing a positive feedback loop of global warming that we may not be able to stop. These reports met with a similar response.

OF POLAR BEARS AND PROFITS: A CASE STUDY IN ANTISCIENCE PROPAGANDA

What wasn't ignored was polar bears. One of the many data sets scientists had been tracking during this time was wildlife counts, to see if and how wildlife were reacting to climate change, particularly in the polar regions, where its effects are most pronounced. By 2005, scientists had ten years of data showing declining numbers of polar bears. Because polar ice was retreating earlier in the season, polar bears had a narrower time frame in which to hunt during the spring when seal pups are born. The seals are a primary spring food source for both adult bears and their cubs. The retreating ice also made it more difficult for the bears to hunt for other prey in their diet. The polar bears observed were thinner and producing less milk, causing bear cubs to die and reproduction rates to decline.[8] As a result, overall polar bear populations in the Hudson Bay area were down by about 20 percent. Based on this data, a young environmental lawyer named Kassie Siegel at the Tucson, Arizona–based Center for Biological Diversity petitioned[9] the US Fish and Wildlife Service (FWS) to list the polar bear as threatened under the Endangered Species Act due to global warming.[10]

"We wanted to force the Bush administration to acknowledge the reality of climate change, specifically via its effects on the polar bear," says Siegel. "The Endangered Species Act requires the government to judge cases only on the basis of the best available science, not economic impact, not political pressures, not industry compromises. We could use that to say, 'Okay, Bush administration, either go ahead and protect the polar bear, or deny the petition and we can go to court and litigate the best available standard on climate change, and you don't want that either.'"[11]

The petition generated considerable publicity, most of it sympathetic to the bears, which the public thinks of as cute and fuzzy and somehow majestic. They were even Christmas season mascots for Coca-Cola and Marshall Field's. FWS director H. Dale Hall said the agency received some six hundred thousand public comments.[12] The FWS asked the US Geological Survey to prepare nine detailed studies of the polar bear population to better understand what was happening.[13]

But then a strange thing happened. Scientists began to speak up and challenge the idea that polar bears were threatened. It began with David Legates, then the state climatologist for Delaware. Legates published a May 2006 study that critiqued a report by scientists from eight arctic nations and six indigenous peoples documenting how climate change was causing melting of the polar ice cap and concluded that the study's claims were "not supported by the evidence."[14] The only problem is that Legates's "study," which was widely quoted in conservative publications, was an unscientific policy paper prepared and published by the National Center for Policy Analysis, a nonprofit think tank in Dallas whose "mission is to seek innovative private sector solutions to public policy problems."[15] Legates was listed as a senior fellow[16] and was also working for numerous other climate change denial outfits, many funded by Exxon-Mobil and the Koch family foundations.[17] The paper offered a rhetorical argument couched as science.

The same day, a polar bear biologist named Mitch Taylor from the Department of the Environment of the Government of Nunavut, Canada, published an opinion piece in the *Toronto Star* arguing that the polar bear population was increasing, not decreasing, and this was all hype.[18] The population, Taylor argued, had grown by 25 percent during the previous decade. Energy-industry groups and conservative media outlets reported that the overall polar bear populations had exploded—from five thousand in 1972 to twenty-five thousand in 2007.[19] The actual estimate was a low of five thousand to ten thousand and a 2007 population of twenty thousand to twenty-five thousand,[20] and the report neglected to mention that polar bears were recovering from a period of overhunting in which sport hunters took 85 to 90 percent of the polar bear kill, a decimation that stopped only when Canada established quotas in 1968 and Alaska banned sport hunting in 1972.[21] The government of Nunavut had just increased its polar bear sport hunting quota by 20 percent in January,[22] and Taylor, their polar bear expert, was "concerned that listing the bears as threatened could lead to a ban on sports hunting."[23] The sport hunting of polar bears brings about $2 million to Nunavut communities each year.[24]

Next came a paper in an obscure science journal named *Ecological Complexity,* coauthored by Legates, Harvard University professor Willie Soon, Markus Dyck of Nunavut Arctic College, and others.[25] It argued: Wait a minute! Polar bears aren't declining in population because of loss of Arctic ice due to climate change—Arctic temperatures aren't changing that much. And if polar bears are declining at all it's probably because of their interactions with humans or because they are competing with other polar bears. Any claim that the decline has to do with climate change is irresponsible science done by people with a political agenda.[26]

Conservative bloggers pounced on the paper and mainstream journalists became interested. Here was another case of science saying something and then someone else doing a more thorough study, and just the opposite is revealed to be the *real* case, like other times science has reversed itself, about, for example, butter being good or bad for you or the good then bad then good again health effects of red wine, and so on.

But there was a problem that few of the journalists covering the story noticed: This "scientific" paper actually *wasn't*. Like Legates's policy paper, it did no original research at all. Instead, the paper's authors selectively cherry-picked bits and pieces of research from other papers to support an a priori conclusion. Thus the paper was posing a *rhetorical* argument as a scientific one, a classic "vulgar Induction," as Bacon called it, producing "a case, or several cases, wherein their proposition holds."[27] But *it was not science.*

Not being science, there was nothing to peer-review, so it was accepted for publication the day after it was submitted and run in the journal as a "Viewpoint" article—scientists' term for something that's interesting and perhaps informative, but is "but faith, or opinion."

The paper cited 101 sources and so *appeared* to be scientific. Willie Soon had authored other scientifically discredited, energy-industry-funded climate change denial papers in the past, including with Legates.[28] He is not a climate scientist or a polar bear expert. He is an astrophysicist. He is educated enough to understand that what he is doing is not science, but rather the opinion-based rhetoric of "vulgar Induction," or pseudoscience, and that he is using the credibility of his position at Harvard to fool people. Soon is a frequent speaker at climate change denial conferences, such as those sponsored by the Heartland Institute. *Ecological Complexity* has ethical rules about disclosure of conflicts of interest.[29] Because of this, Soon acknowledged receiving funding from Exxon-Mobil, as well as the Charles G. Koch Charitable Foundation and the American Petroleum Institute, in the back matter of the polar bear paper. But perhaps journalists didn't read that far.

Combined, these and similar obfuscations were promoted by groups and bloggers aligned with Exxon-Mobil, the Koch family foundations, and the conservative think tanks they donated to,[30] which pushed them to mainstream media outlets. Many news outlets reprinted versions of the stories without critical examination, their reporters using their control-C keys in a way BBC business journalist Waseem Zakir refers to as "churnalism."[31]

AM talk radio hosts picked up the new "findings" from "science" in a second wave of denialism.[32] On the basis of the articles' and radio shows' pronouncements, conservative and industry groups that said there were simply no data to support the decision criticized the federal government for its unfounded and unreasonable activism in listing the polar bear under the Endangered Species Act. These groups included the American Enterprise Institute for Public Policy Research,[33] the Heritage Foundation,[34,35] the George C. Marshall Institute,[36] and the National Association of Manufacturers.

A FIVE-PRONG PROPAGANDA STRATEGY TO MANIPULATE DEMOCRACY

The method has now become an oft-repeated five-prong propaganda strategy of cloaking rhetorical arguments in the clothing of scientific legitimacy in order to effect a desired policy objective. It begins with phony science, in the form of either a pseudoscience paper or another event that casts doubt on the accepted views of mainstream science, followed by canned and slanted press material[37] prepared for and fed to the press by industry-affiliated nonprofits and bloggers promoting a narrative about the supposed controversy. Once laundered in the legitimate press, the story is then picked up by partisan talk radio and cable news purveyors, who reference these mainstream sources, react with outrage, and call for policy action. These three prongs of the propaganda onslaught lay down political cover in a sort of suppressive fire for the fourth prong of the attack: legislative or other action by partisan allies in government. In this case, the fourth prong began with a major speech by Senator James Inhofe[38] attacking the petition followed by Alaska governor Sarah Palin's effort to prevent the polar bear's "threatened" listing.

All of this conspires to support the fifth prong. Now seemingly supported by science, the press, the government, and sometimes religion—in other words by all the other major houses of power in society—the real actors can step safely out from behind the curtain for the main act of the culture-war drama. Five industry organizations* filed suit in August 2008 against then interior secretary Dirk Kempthorne and FWS director Dale Hall, joining Palin's effort to reverse the listing.[39]

Thus the five-prong strategy is:

1. Phony science is promulgated.

2. Canned blog and press stories that create a controversy narrative based on the phony science are fed to the mainstream media.

3. Partisan talk radio and cable news shows react to the laundered narrative with outrage and call for policy change.

4. Governmental allies propose reasoned corrective responses to the outrage.

5. The aggrieved, patriotic main actors take the stage and reluctantly plead their case, either to policy makers or in advertisements to their constituents, or both

---

* The American Petroleum Institute, the National Association of Manufacturers, the US Chamber of Commerce, the National Mining Association, and the American Iron and Steel Institute

The strategy is designed to erode the primacy of knowledge and slowly move public opinion toward accepting the defined objective as the truly reasonable position.

The coordination between energy-industry companies and their allies in this five-prong attack on the polar bear listing was well orchestrated: The documentation accompanying Palin's April 9, 2007, letter to Kempthorne referenced Willie Soon's polar bear paper seven days *before* it was actually published.[40]

THE PART THE PUBLIC DIDN'T HEAR

Siegel later received copies of e-mails between Hall and Kempthorne over the decision of whether or not to list the polar bears. "Hall says in one e-mail 'I got this call from Lyle Laverty,' who was Kempthorne's assistant, 'and he told me the secretary has decided not to list the polar bear and I told him of course that's his decision to make but I don't want to be at the press conference because I can't support it and I won't be helpful to you.' Then they have all these notes back and forth, reasons to not list: We want to be in the driver's seat on policy; why not let the courts decide. There was really a lot of deliberation, they didn't want to do it, but they realized they would lose because the legal standard was the science. Then on May 15, 2008, Kempthorne gets up there and says he's listing the polar bear."

Kempthorne made it clear he wasn't pleased. "While the legal standards under the ESA compel me to list the polar bear as threatened," he said, "I want to make clear that this listing will not stop global climate change or prevent any sea ice from melting. Any real solution requires action by all major economies for it to be effective. That is why I am taking administrative and regulatory action to make certain the ESA isn't abused to make global warming policies."[41]

Hall—who had been appointed by President George W. Bush—testified before Congress that numerous scientific studies showed there was no significant scientific uncertainty that polar bears were endangered by global warming.[42] But the public didn't hear *that* story.

THE HOCKEY STICK ATTACK

After the polar bear attack, energy companies stepped up their invest-ments in the infrastructure supporting this five-prong strategy, from funding anti-climate-change nonprofits and political activist groups to paying scientists to cast doubt on the real science, all the way up to mas-sive lobbying, political advertising, and political contribution campaigns. Public data from the Securities and Exchange Commission and charitable organizations' reports to the IRS show that between 2005 and 2008, Exxon-Mobil gave about $9 million to groups linked to climate change denial, while foundations associated with the private oil giant Koch Industries gave nearly $25 million.[43] The third major funder was the American Petroleum Institute (API).[44] Between them they created and funded dozens of groups engaged in producing pseudoscience, activism, organizing, media outreach, and lobbying and implemented an all-out propaganda campaign. But this was a drop in the bucket compared to what they would spend overall. Between 1999 and 2010, the energy indus-try spent more than $2 *billion* fighting climate change legislation, more than $500 million of it from January 2009 to June 2010, or almost $1,900 *per day* in lobby expenditures for every US senator and representative in Washington—and those numbers don't include nonreportable expenses like publicity, earned media, rallies, and polling.[45] They spent an esti-mated $73 million more on anti-clean-energy ads from January through October 2010,[46] and Koch family foundations gave an overall $48 million to groups engaged in climate change denial between 1997 and 2008.[47]

It was an assault that climate scientists were wholly unprepared for, even though it had been building for some time. In 2003, Willie Soon had coau-thored another controversial phony-science paper, this one published by conservative think tank the George C. Marshall Institute.[48] That paper attacked the work of prominent climate scientist Michael Mann, developer of the famous "hockey stick" graph. Mann's graph charted average tem-peratures for the last thousand years, and the sharp increase it shows in the last century makes it resemble a hockey stick. It was used prominently in the 2001 report issued by the United Nations' Intergovernmental Panel on Climate Change (IPCC).[49] It was also used, more famously, by a scissor-lift-riding Al Gore in the documentary film *An Inconvenient Truth*.

Tree scientists have correlated tree-ring density measurements with annual average temperature data. Mann used this correlation to analyze tree rings going back a thousand years and infer unknown temperatures. He then plotted them on a chart using blue and, being a scientist, he added gray bars to represent the statistical probability of error in his estimates, which increased with time. Finally, he added current known temperature measurements in red.

The findings by Mann and two other climate scientists were originally reported in a paper published in *Nature* in 1998,[50] and further refined the following year. They represented classical Baconian and Lockean observational empiricism. By proceeding logically from observations, they were able to create knowledge that could be tested and verified by others, which it was. The results of all of the various methods and observations overlapped in relative agreement, creating confidence that the temperature estimates were reliable. This is *science*, versus the top-down approach of rhetoric. It had the simple, iconic power of Hubble's four-page paper placing Andromeda outside the Milky Way. That kind of power would soon make it the target of a much, much larger attack.

SCIENCE'S RIGHTFUL PLACE

After the election of President Barack Obama in November 2008, many scientists hoped that a corner had been turned and America was entering a new era in which science would be restored, as the president promised in his inaugural address, "to its rightful place," which was the mission statement of Science Debate 2008.

Obama's science advisor, John Holdren, was outspoken on population and climate change. Energy secretary Steven Chu is a Nobel prize–winning nuclear physicist and a prominent climate scientist who supports transitioning from fossil fuels to a low-carbon economy.

NOAA administrator and undersecretary of commerce for oceans and atmosphere Jane Lubchenco is a renowned marine biologist and past American Association for the Advancement of Science (AAAS) president who is also vocal about dealing with climate change. US Geological Survey director Marcia McNutt had run Monterey Bay Aquarium Research Institute and is similarly outspoken. The oceans are currently absorbing about half of the extra carbon dioxide being produced by humans. This is making them far more acidic, inhibiting the formation of corals and shells and threatening the foundation of the entire aquatic food chain. Holdren, Chu, Lubchenco, and McNutt all stated that the president needed to prioritize the major science issues, and all had backed the call for a science debate during the campaign.[51]

But once in office, Obama took a more cautious approach. The economy had collapsed and, like President Kennedy early in the space race, Obama found himself caught between his broader vision and financial problems, this time caused by the Great Recession. He pulled back, prompting some to call him timid.[52] He had to choose which initiatives he would put his full force behind. Democracy being what it is, he chose the one he thought would have the most immediate positive impact for voters and the economy: health care reform.

It proved a fateful decision. Anti-cap-and-trade opponents were gearing up for a "civil war" and the "greatest part of that battlefield is the global warming battle," according to energy-industry-shill Lord Christopher Monckton, a British journalist with no particular expertise in climate science who travels America eruditely calling global warming "bullshit."[53] Obama's delay gave opponents the time and ammunition they needed to regroup.

CLIMATEGATE

On November 17, 2009, the battle was rejoined. Days before the start of the Copenhagen climate summit, an unidentified hacker posted on a Russian FTP server a sixty-one-megabyte file containing e-mails stolen from servers at England's University of East Anglia Climatic Research Unit (CRU). The hacker then posted a link to the file on the climate skeptic blogs *The Air Vent*[54] and *Watts Up with That?*[55] as well as the blog *RealClimate*, which is run by several leading climate scientists, including Michael Mann.[56] The CRU is one of the world's leading centers for climate research and a hub of global climate science communication. The file contained thousands of private e-mails exchanged by top climate scientists over more than thirteen years. The Air Vent blogger, Jeff Id, quickly began highlighting ones that made it appear that scientists were cooking their data to reach an a priori conclusion—which, ironically, was exactly what energy-industry-funded scientists like Willie Soon were actually doing. Blogger Anthony Watts co-broke the story on *Watts Up with That,* and even though he was traveling he too managed to identify key e-mails and quickly posted them on his blog.

The leak was then broadly publicized, with reporters' attention being drawn to the most damning and easily reframed e-mails. One of these, reproduced by Watts in his "breaking news" post on the hack, was between CRU director Phil Jones and the three original authors of the hockey stick graph—Michael Mann, Raymond Bradley, and Malcolm Hughes.

From: Phil Jones <p.jones@xxxxxxxxx.xxx>

To: ray bradley <rbradley@xxxxxxxxx.xxx>, mann@xxxxxxxxx.xxx, mhughes@xxxxxxxxx.xxx

Subject: Diagram for WMO Statement

Date: Tue, 16 Nov 1999 13:31:15 +0000

Cc: k.briffa@xxxxxxxxx.xxx,t.osborn@xxxxxxxxx.xxx

Dear Ray, Mike and Malcolm,

Once Tim's got a diagram here we'll send that either later today or first thing tomorrow.

I've just completed Mike's Nature trick of adding in the real temps to each series for the last 20 years (i.e. from 1981 onwards) amd [sic] from 1961 for Keith's to hide the decline. Mike's series got the

annual land and marine values while the other two got April-Sept
for NH land N of 20N. The latter two are real for 1999, while the
estimate for 1999 for NH combined is +0.44C wrt 61-90. The Global
estimate for 1999 with data through Oct is +0.35C cf. 0.57 for 1998.

Thanks for the comments, Ray.

Cheers

Phil

Prof. Phi l Jones

Climatic Research Unit Telephone +44 (0) 1603 592090

School of Environmental Sciences Fax +44 (0) 1603 507784

University of East Anglia

Norwich Email p.jones@xxxxxxxxx.xxx

NR4 7TJ

UK

As in the polar bear case, a five-prong propaganda attack was
employed: phony science; canned stories via bloggers for the press; AM
talk radio and partisan cable news reaction and outrage; government
intervention; and hand-wringing by the real actors, our much-maligned
and patriotic heroes. Suddenly, scientists found themselves not just fend-
ing off junk science, but directly in the crosshairs.

"Phil Jones has gone on record saying that he was using the term 'trick'
in the sense of 'a clever way of solving a problem,' like when you divide both
sides of a math problem to isolate a variable. It's a way to make things
apparent, like a magic trick," Mann says. "In referring to our 1998 *Nature*
article, he was simply pointing out that the proxy record of tree-ring data
we used to estimate historical annual temperatures ended in 1980, so it
didn't include the warming of the past few decades. So in our *Nature* article
we also showed the thermometer data that was available after 1980 and
through 1995, which we clearly labeled in a different color so that the reconstruction could be viewed in the context of recent instrumental

temperatures."[57] The validity of Mann's temperature reconstruction was subsequently independently verified by the National Academy of Sciences.[58]

Mann says Jones's reference to adding data to "hide the decline" referred to work by their colleague Keith Briffa, who was being copied on the e-mail. "It had nothing to do with hiding a supposed decline in global temperatures," he says. That was a reframing of it by antiscience opponents. The "decline" referred to a well-known decline in the response of high-latitude tree-ring density to temperatures after about 1960.

In an article published in *Nature* later in 1998 that correlated tree-ring density to temperature, Briffa and his colleagues explained that the close correlation began to decline after about 1950 and was unreliable after 1960.[59] "Decline" was being used in that specific scientific sense.

"Hide" probably hadn't been the best word choice, since the existence of the decline was a major point in the original Briffa *Nature* paper. Jones was simply saying he was substituting instrumental data to [pick your word here] the decline. "The irony," says Mann, "is that Phil was trying to be precise and careful, and was criticized as if he were trying to be just the opposite."

Exploiting the different meanings of a word—scientific and colloquial— to foment public mistrust was becoming a familiar strategy, one taken from the playbook of evolution deniers and their talk about "theory."

CHURNALISM AND THE CONFLICT FRAME

Journalists who don't understand science can easily be manipulated as tools of propaganda campaigns in a science-dominated world. To understand both the difference between journalistic and scientific communication and the care journalists must take, consider the year 2008. News headlines about global warming could accurately say:

A. "2008 Warmer Than Any Year in the 20th Century Except 1998"

B. "Global Temperatures Cooled over the Last Decade" (because 1998 was warmer)

C. "2008 the Coldest Year of the Century" (which started in 2001, with 2001 through 2007 all being warmer than 2008)

D. "8 of 9 Warmest Years Ever Recorded Since 2000" (1998 is the ninth)

Or, as Rush Limbaugh put it on his March 27, 2009, show: "More people are starting to consider the notion that we actually may be in a cooling phase, 'cause there hasn't been any significant warming in years."[60]

Each of these headlines is true, but each gives a very different impression about global warming. As the saying goes, "figures don't lie, but liars can figure." And therein lies the problem. Language, the tool of the press, is imprecise.

Media outlets ran the peddled angle that the e-mails seemed to show that a small cadre of politically motivated climate scientists close to Al Gore was not using "the real temps" and was trying to "trick" the public and "hide the decline" in global temperatures.

Citing this reframing, Sarah Palin stepped up her criticism of climate science, calling it "snake-oil science" and "junk science."[61] Instead of investigating the truth or falsehood of the claims, which could have been easily established, mainstream media outlets lapped up the prepackaged "controversy" and called it "Climategate."

"Too many lazy journalists simply uncritically parroted what they had read from dubious sources, such as climate change denial outfits and blogs," said Mann.[62]

The story highlights the problem that arises when most of the journalists in a democratic society whose major policy challenges revolve around science have no training in it. It's simply not good enough to write he-said-she-said conflict narratives or to churn out prepared press releases after adding a few quotes. Life in a democracy depends on the press informing citizens about what's *really* going on based on knowledge, not on warring opinions that can have little or nothing to do with the underlying science driving the issues that policy makers ultimately have to grapple with. If one side presents knowledge and the other "but faith, or opinion, but not knowledge," simply reporting both sides is not balanced journalism and constitutes malfeasance by the press.

THE TALK RADIO PRONG

After the "phony science" prong was in place and the packaged stories were in the media, the AM talk radio prong stepped into the attack. Rush Limbaugh's November 24, 2009, show laid down the outrage and wrapped all the talking points together.

> The people who have been preaching to us about global warming have been doing so, as the left usually does, from the "crisis mode" standpoint. "We've got twenty years." "We got ten years." Remember Ted Danson in 1988? "Ten years to save the oceans." Ten years to this; 20 years for that. "We're killing ourselves. We're killing the polar bears!" Except it hasn't warmed in ten years. And now we've got the hoax fully exposed. Wouldn't you think that people genuinely believing in manmade global warming and its destructive results would be happy that it isn't happening? They're not. They are distressed, and they're trying to cover up the hoax, and they're going to try to weather the storm—'cause it isn't about global warming like health care is not about health care, like cap and tax is not about cap and tax, like Obama is not who he is. They're all frauds. They are all liars. They are skunks, and they ought to be held up for public ridicule. Obama said he wants to "restore science" to its rightful whatever? Then he ought to be leading the way to find out who these people are, what they've done, who they've infected, who went along with them—calling them out by name—making sure that every scientist at every university in this country that's been involved in this is named and fired, drawn and quartered or whatever it is. Because this is a worldwide hoax, and it's primary target was you. The people of the United States of America.[63]

The five-prong propaganda deployment had all the characteristics of a successful psyops campaign and became "one of the best-funded, most highly orchestrated attacks against science we have ever witnessed," says Mann, and the more than $2 billion in anti-climate-science spending proves he's right.[64] "The evidence for the reality of human-caused climate change gets stronger with each additional year. Greenhouse gases don't care whether you're a Democrat or a Republican. Nor do the ice sheets. Unfortunately, some have found it convenient to politicize the science, as others have with the science of tobacco, acid rain, ozone depletion, stem cell research, human health, environmental contaminants, etc."

A FAILURE OF AUTHORITY

On December 4, 2009, twenty-nine of the country's most prominent climate scientists, led by the fairly conservative and highly respected then–AAAS board chairman James McCarthy of Harvard, sent a joint letter to Congress saying:

> In the last few weeks, opponents of taking action on climate change have misrepresented both the content and the significance of stolen emails to obscure public understanding of climate science and the scientific process. . . . Observations throughout the world make it clear that climate change is occurring, and rigorous scientific research demonstrates that the greenhouse gases emitted by human activities are the primary driver. These conclusions are based on multiple independent lines of evidence, and contrary assertions are inconsistent with an objective assessment of the vast body of peer-reviewed science. . . . If we are to avoid the most severe impacts of climate change, emissions of greenhouse gases must be dramatically reduced.[65]

This was a noble statement, and prominent scientists frequently use this "overwhelming authority" approach of issuing open letters from groups of top scientists. It is very impressive, and to people who trust science it is influential in the same way that instruction from a group of bishops can be influential for devout Catholics. But such an approach has two problems that doomed it to probable failure in this instance.

First, it relies on people having respect for the unquestioned authority of science leaders who are putting their reputations on the line, but as we've seen, that authority had been called into question, so simply appealing to it again carried little weight with "nonbelievers." Thus, Palin was able to flippantly but effectively preempt the appeal with a Facebook posting calling concerns over global warming "doomsday scare tactics pushed by an environmental priesthood."[66]

Second, unquestioned authority isn't the message of science in the first place. Science has never rested on "trust me" or "take it on faith"—that is, until *Sputnik*. Authority was the message of kings and popes, the thing Locke was trying to get away from. Science's strong suit has always been its focus on observation and process: "See, look for yourself. Don't take my word for it." The authority response suggested that science was just another "expert opinion."

In some ways, the open letter played right into the message box of the energy industry, which was casting science as a political opinion instead of knowledge, and not surprisingly, it fell on deaf ears among those who didn't "believe" in global warming. Scientists were losing the debate because they didn't understand the terms of the battle. They rightly saw global warming as a fact and missed the political ramifications of a high-profile politician delivering the message of science in a postmodern world: *It made the facts partisan.*

There was the "Al Gore, environmentalist, proscience, pro-Hollywood, Prius-driving progressive" frame versus the "Watch out for your wallet, liberals and their government-paid scientists are fleecing you with a hoax and junk science" frame, which was, ironically, being aggressively promoted by oil-industry front groups who were in fact trying to do just what they accused the scientists of doing. It's one of the oldest ambushes in both politics and propaganda: Hide by ascribing your own worst motives to your opponent *before* they expose *you* to the public. Bingo! You're inoculated with the defense of sour grapes: Anything they accuse you of is just sour grapes over your accusation of them. If your accusation seems true, theirs seems like false payback. If your accusation seems false, theirs still seems like false payback. You can't lose as long as they fall into the tempting trap of playing in your frame. "It's not about science literacy," says social scientist Matthew Nisbet. "It's about competing mental models."[67] Once those models become polarized, they get wrapped up in special interest framing and science gets tossed around like any other opinion. How to avoid that? Emphasize the process.

THE FOXGATE E-MAIL SCANDAL: A CRACK IN PRONG THREE

That this was a full-blown battle between climate scientists and propagandists became abundantly clear later, when the curtain was lifted for a brief instant to expose the workings of the third prong—slanted cable news. On December 8, 2009, Fox News's White House correspondent Wendell Goler delivered a live report from Copenhagen on the daytime news show *Happening Now*. Goler was asked by host Jon Scott about "UN scientists issuing a new report today saying this decade is on track to be the warmest on record."

Goler said yes, 2000 to 2009 was "expected to turn out to be the warmest decade on record." He said it was a "trend that has scientists concerned because 2000 to 2009 [was] warmer than the 1990s, which were warmer than the 1980s." He then said, "Ironically, 2009 was a cooler than average year in the US and Canada," which, he said, was "politically troubling because Americans are among the most skeptical about global warming."

When Scott brought up the Climategate e-mails, Goler said that although people had raised questions about the CRU data, "the data also comes from the National Oceanic and Atmospheric Administration and from NASA. And scientists say the data across, across all three sources is pretty consistent." Goler then quoted a statement from UN Secretary-General Ban Ki-moon, saying, "Nothing in the data quote casts doubt on the basic scientific message that climate change is happening much faster than we realized and that human beings are the primary cause."

This unbiased reporting must have caused an irate phone call from *someone* to chew out Fox News's Washington managing editor Bill

Sammon, because within fifteen minutes, Sammon had fired off a strident e-mail to the staffs of *Special Report,* the Fox White House Unit, *Fox News Sunday,* FoxNews.com, and several other reporters, producers, and executives:

> From: Sammon, Bill
>
> To: 169 -SPECIAL REPORT; 036 -FOX.WHU; 054 -FNSunday; 030 -Root (FoxNews.Com); 050 -Senior Producers; 051 -Producers; 069 -Politics; 005 -Washington
>
> Cc: Clemente, Michael; Stack, John; Wallace, Jay; Smith, Sean
>
> Sent: Tue Dec 08 12:49:51 2009
>
> Subject: Given the controversy over the veracity of climate change data . . .
>
> . . . we should refrain from asserting that the planet has warmed (or cooled) in any given period without IMMEDIATELY pointing out that such theories are based upon data that critics have called into question. It is not our place as journalists to assert such notions as facts, especially as this debate intensifies.[68]

The directive ignored the fact that the *e-mails* had no bearing on the *data* Goler mentioned, which had been accumulated over *fifty years* of research, a fact that had already been established by a wide variety of the nation's leading scientists, academies, and journal editorial boards. Over the previous and coming few days, the *New York Times,*[69] Politifact.com,[70] FactCheck.org,[71] the Associated Press,[72] and McClatchy Newspapers[73] released statements saying that the e-mail exchanges did not change the credibility of the fundamental science, which was based on a wide variety of sources, whereas Sammon's statement was an ideological directive that placed knowledge on an equal footing with opinion and thus was not in keeping with the standards of good journalism. It would set the tone of junk science skepticism for all Fox News reportage and GOP messaging on climate change that were to follow.

FOR NOW WE SEE THROUGH
A GLASS, DARKLY

Enter prong four of the propaganda strategy: the government allies. On December 7, 2009, Congressman Jim Sensenbrenner Jr. of Wisconsin, the ranking Republican on the House Select Committee on Energy Independence and Global Warming and a longtime global warming critic, sent a letter to Rajendra Pachauri, the chair of the IPCC, demanding that Mann and the other top climate scientists included in the e-mails be barred from participating in the next IPCC assessment of climate change for having "caused grave damage to the public trust in climate science in general, and to the IPCC, in particular."[74] Concern appropriation, the act of appearing to adopt the concerns of your opponents while in fact appropriating their language to use against them, is another classic propaganda and political campaign tactic that frames the appropriator as the reasoned centrist and his or her opponents as the fringe.

On December 9, the day after Sammon issued his directive, Palin also used concern appropriation in a major op-ed published in the *Washington Post* in which she said that the Climategate scandal

> exposes a highly politicized scientific circle—the same circle whose work underlies efforts at the Copenhagen climate change conference. . . . The e-mails reveal that leading climate "experts" deliberately destroyed records, manipulated data to "hide the decline" in global temperatures, and tried to silence their critics by preventing them from publishing in peer-reviewed journals. What's more, the documents show that there was no real consensus even within the CRU crowd. Some scientists had strong doubts about the accuracy of estimates of temperatures from centuries ago, estimates used to back claims that more recent temperatures are rising at an alarming rate. This scandal obviously calls into question the proposals being pushed in Copenhagen. I've always believed that policy should be based on sound science, not politics.[75]

Except, of course, when it comes to the teaching of creationism in science class, or to the funding of "fruit fly research in Paris, France—I kid you not!"

WHEN YOU'RE 'SPLAININ', YOU'RE NOT GAININ'

The propaganda campaign wasn't the only problem facing climate scientists. On December 1, another story had broken that further undermined Copenhagen, climate science, and the IPCC. Graham Cogley, a geologist at Trent University in Peterborough, Ontario, who apparently was unrelated to the propaganda campaign, noticed that the IPCC's fourth assessment contained an error. The assessment, which was released in 2007, the year the IPCC and Al Gore had won the Nobel Prize, said that "Glaciers in the Himalayas are receding faster than in any other part of the world," and that all the glaciers in the central and eastern Himalayas could disappear by 2035.[76] The IPCC classed the statement as "very likely," meaning that it had a probability of being true of 90 to 99 percent, implying that it was the result of measurement and observation and was supported by inductive reasoning and statistics,[77] and said the glaciers' "total area will likely shrink from the present 500,000 to 100,000 km^2 by the year 2035." But it was wrong.

The statement was sourced to a World Wildlife Fund report, which in turn had drawn it from a 1999 article published by science journalist Fred Pearce in *New Scientist*.[78] But the scientist who was quoted in the *New Scientist* article, Syed Hasnain of Jawaharlal Nehru University in New Delhi, was speaking offhandedly, offering an opinion, and had never repeated the prediction in a peer-reviewed journal. He has since called the comment "speculative." When Cogley went back to check previous statements, he found a separate 1996 study by a leading hydrologist, V. M. Kotlyakov, that said, "the extrapolar glaciation of the Earth will be decaying at rapid, catastrophic rates—its total area will shrink from 500,000 to 100,000 square kilometres by the year 2350."[79] It now appeared the IPCC error was the result of a simple transposition—2350 to 2035—that no one had caught, even when the 2035 prediction had garnered worldwide media attention going into Copenhagen.

This error occurred concurrently with Climategate. The press smelled blood in the water, and coverage of the error went worldwide. The retreating glaciers' iconic beauty had inspired press stories, but so had the fact that an estimated one billion people depend on Himalayan glacial melt for water. That the dramatic and dire prediction was not supported seemed to confirm the worst of the right's characterizations of climate scientists as doomsayers and shattered the public's confidence in their credibility. This was classic Chicken Little.

This situation was worsened by a study by India's environment minister, Jairam Ramesh, that similarly suggested that Himalayan glaciers might not be melting as much as was widely feared and accused the IPCC of being "alarmist." Instead of exercising caution and scientific open-mindedness, the embattled Pachauri dismissed the Indian study as "voodoo science," recalling Shapley's response to Humason.[80]

From a political tactician's perspective there is only one way out of this trap: Set ego aside, admit the mistake as quickly as possible, accept full responsibility, and take immediate steps to make things right. The row with India further damaged the IPCC's and Pachauri's worn credibility since the IPCC's statement was unsupported. If anything, *it* was voodoo science, and Pachauri came off as arrogant and out of touch.

By now the mainstream press was actually gathering news, combing through the report's three thousand pages themselves, looking for other possible biases and inconsistencies. They had an angle. The meme that an elite and arrogant group of scientists close to Al Gore was working a political agenda had begun to stick.

A few days later, this conscientious investigation brought more errors to light. The report also wrongly stated that 55 percent of the Netherlands was below sea level because the IPCC had failed to independently confirm information supplied by a Dutch government agency.

It was an opposing campaign manager's dream, the gift of a scandal that keeps on giving. It couldn't have come at a worse time or been more poorly handled by the IPCC, which should have pulled the report and taken charge of combing it themselves, found the errors themselves, made the corrections themselves, and then thanked their opponents for being such good citizens to assist them in the process of strengthening the report's credibility. Instead, over the rest of the month, several more minor errors were identified one after another in a "death by a thousand cuts" nightmare that is a classic campaign strategy: Dribble out the bad news to make your opponent explain him- or herself over and over until all credibility with the public has been lost. The adage in politics is "when you're 'splainin', you're not gainin'," and so it was.

After all the publicity, here's the most shocking thing: The errors were not in the *scientific* section of the IPCC report, published separately as *Climate Change 2007: The Physical Science Basis,* which is subject to peer review. They were in the section published as *Climate Change 2007: Impacts, Adaptation and Vulnerability,* which speculated about possible impacts of climate change and was not subject to peer review. So they were not purported to be scientific. But journalists and the public didn't seem to notice that distinction.

"Neither of these statements were important enough to even make it to the technical summaries or the summary for policy makers, so they really had no effect whatsoever. And let's not forget," Mann says, "we're talking about a couple mistaken sentences in thousands of pages. What's amazing to me is that these are the only errors that were found in a thousands-of-pages-long document. Were that most sources had such a low error rate."

All that is true—but it didn't matter. The IPCC errors had nothing to do with Mann or the science, but the whole thing was being conflated with the CRU e-mails and Mann's hockey stick graph, and the narrative of climate science began shifting under his feet.

THE NEW MCCARTHYISM

Recognizing what was happening, Mann took the bull by the horns and embarked on an aggressive publicity campaign, appearing on CNN and other news outlets to try to counter the massive propaganda attack by doing brilliantly what scientists should have been doing for decades: breaking it down and talking process.

In response, climate change deniers renewed their criticism of the hockey stick graph, which they argued was an inaccurate depiction based on faulty science (prong one). The graph had a self-evident power of the sort scientists often dream of but rarely achieve. Invalidating it was the holy grail of climate change denial, the mortal blow they badly wanted to inflict upon the public image of climate science.

With a climate change bill finally taking shape in Congress, the attacks escalated. At the same time, the energy industry was spending half a billion dollars lobbying Congress, trying to discredit Mann and other climate scientists, and running television ads and pushing news stories to influence public opinion and cow policy makers. In keeping with the idea that politics is narrative, they needed a bad guy to tar and feather. Having appeared on the news programs, Mann fit the bill.

In February 2010, Senator James Inhofe of Oklahoma, the ranking Republican on the Senate Committee on Environment and Public Works and a longtime climate change denier, attacked under prong four. Following Rush Limbaugh's on-air suggestion of November 24, 2009, Inhofe went after Mann personally in a Senate minority report.

> The CRU controversy is about far more than just scientists who lack interpersonal skills, or a little email squabble. It's about unethical and potentially illegal behavior by some of the world's leading climate scientists. The report also shows the world's leading climate scientists acting like political scientists, with an agenda disconnected from the principles of good science. And it shows that there is no consensus—except that there are significant gaps in what scientists know about the climate system. It's time for the Obama

Administration to recognize this. Its endangerment finding for greenhouse gases rests on bad science. It should throw out that finding and abandon greenhouse gas regulation under the Clean Air Act—a policy that will mean fewer jobs, higher taxes and economic decline.[81]

Inhofe's next act should be morally repugnant to every American. In an act of McCarthyism, he publicly named the scientists he wanted investigated for possible referral to the US Justice Department for prosecution: Raymond Bradley, Keith Briffa, Timothy Carter, Edward Cook, Malcolm Hughes, Phil Jones, Thomas Karl, Michael Mann, Michael Oppenheimer, Jonathan Overpeck, Benjamin Santer, Gavin Schmidt, Stephen Schneider, Susan Solomon, Peter Stott, Kevin Trenberth, and Thomas Wigley.

The bullying of scientists by a sitting US senator was picked up by and retweeted in the right-wing-media echo chamber. The scientists began getting hundreds of attack e-mails accusing them of falsifying science for political purposes—and then death threats. Unless you've been through that kind of public shaming and national shunning, it's hard to imagine the fear; the negative power of harsh, malevolent "peer" review; and the gross sense of injustice. It strikes against every value that Americans hold dear. "People said I should go and kill myself," Phil Jones said. "They said they knew where I lived. I did think about it, yes. About suicide. I thought about it several times, but I think I've got past that stage now."[82]

Jones, who has since been exonerated by the British House of Commons, said the worldwide scandal—over *nothing*, it would turn out—was something he was totally unprepared for. "I am just a scientist. I have no training in PR or dealing with crises," he said. And that lack of training, the result of two generations of scientists having devalued communication and political involvement, is a major problem now that they find themselves increasingly involved in political battles about scientific findings that vested interests see as challenging their power and fight against. Climate change is just the beginning.

Stephen Schneider, another of Inhofe's targets, was already familiar with the oil companies' tactics. He wrote the book *Science as a Contact Sport*.[83] He shared one of the e-mails he received during this period.

> You communistic dupe of the U.N. who wants to impose world government on us and take away American freedom of religion and economy—you are a traitor to the U.S., belong in jail and should be executed.[84]

A HUNDRED-YARD DASH
INTO THE WEEDS

Ralph Cicerone, president of the National Academy of Sciences, talked to several top scientists about what to do. He worried that the collapse of the public's confidence in climate science could spill over into other sciences. It was a legitimate concern, with legislatures in three US states—Louisiana, South Dakota, and Texas—telling their schools to teach "alternatives" to the scientific consensus on climate change.[85]

South Dakota's resolution "urges" schools to take a "balanced approach" to teaching about climate change, because the science is "unresolved" and has been "complicated and prejudiced" by "political and philosophical viewpoints." Whose?

> The South Dakota Legislature urges that instruction in the public schools relating to global warming include the following:
>
> (1) That global warming is a scientific theory rather than a proven fact;
>
> (2) That there are a variety of climatological, meteorological, astrological, thermological, cosmological, and ecological dynamics that can effect [sic] world weather phenomena and that the significance and interrelativity of these factors is largely speculative.
>
> (3) That the debate on global warming has subsumed political and philosophical viewpoints which have complicated and prejudiced the scientific investigation of global warming phenomena.[86]

The embarrassment of a formal South Dakota state resolution that pronounces that *astrology,* as in horoscopes, can "effect" weather is over-shadowed by the tragedy of a collapse of public trust in science that threatens the future of democracy itself, as the state legislature of South Dakota, nervously clutching its Bibles and consulting its astrological charts, slips back into antiscience, authoritarianism, ignorance, and superstition, angrily shaking its pitchforks at whatever scapegoat its attention is next directed to.

CLIMATE SCIENCE AS FRAUD AGAINST THE TAXPAYERS

This prong-four attack took a more sinister turn with the act of Virginia attorney general Ken Cuccinelli. On April 23, 2010, Cuccinelli subpoenaed the papers and e-mails of Michael Mann and the University of Virginia, where Mann once worked, demanding all "data, materials and communications" of Mann's related to any grants he may have sought or received in which any funds came from the Commonwealth of Virginia.[87] He cited a 2002 law, the Fraud Against Taxpayers Act, that gives the attorney general the right to demand documents and testimony in cases in which tax dollars have allegedly been obtained falsely by state employees.[88]

An engineer and an attorney, Cuccinelli has said he doesn't believe in climate change. He argued that he is simply exploring whether there were any "knowing inconsistencies" made by Mann when applying for grant money.[89] "The revelations of Climate-gate indicate that some climate data may have been deliberately manipulated. . . . The legal standards for the misuse of taxpayer dollars apply the same at universities as they do at any other agency of state government. This is about rooting out possible fraud and not about infringing upon academic freedom."[90]

"It's totally unacceptable for Cuccinelli to put forward allegations of fraud against Dr. Mann simply because he doesn't agree with the results of the research," says Francesca Grifo, director of the Scientific Integrity Program at the Union of Concerned Scientists.[91] In fact, Grifo argues, it can have a chilling effect on research and academic freedom across America.

That is, of course, the intent, which is what makes Cuccinelli's abuse of power so pernicious. Who will be the next target when an elected official-cum-regent wants to quash the results of research he or she doesn't agree with? Stem cell researchers? Geneticists? Nutritionists when an official has ties to the food industry? How does that get America where we need to go in a highly competitive, science-driven global economy?

INTO IGNORANCE

Republicans' mad dash into the intellectual wilderness picked up speed in March of 2011 when a Republican-sponsored bill intended to prevent the EPA from regulating greenhouse-gas emissions passed out of committee. The agency in 2009 had issued an endangerment finding declaring the gases a threat to public welfare, which served as the EPA's legal basis for regulation. Repealing the finding would eliminate its authority over greenhouse gases.

The finding was scientifically sound, but at the hearing the Republicans on the House Energy and Commerce Committee's Subcommittee on Energy and Power directed anger and distrust at scientists and respected scientific societies. The British journal *Nature*, one of the two leading periodicals of science, grieved the day as a new low point for science in America. "Misinformation was presented as fact, truth was twisted and nobody showed any inclination to listen to scientists, let alone learn from them. It has been an embarrassing display, not just for the Republican Party but also for Congress and the US citizens it represents."[92]

Scientists were described at the hearing as "elitist" and "arrogant" and said to be hiding behind "discredited" institutions, words that recalled Eisenhower's departing warning against the scientific-technological elite. But in fact, it is elitist and arrogant to doom future generations for one's own personal comfort. One lawmaker cited melting ice caps on Mars as evidence contradicting anthropogenic warming on Earth. Antarctica was falsely reported to be gaining ice. Also mentioned was the myth that in the 1970s the scientific community had warned of an imminent ice age. Not surprisingly, subcommittee chairman Ed Whitfield (R-KY), a former oil distributor,[93] touts his home state's $3.5 billion coal industry on his official Web site and says he "will continue to be a leader in the development and responsible use of new coal technologies and innovations."[94]

At the request of Democrats on the subcommittee, several scientists were on hand to answer questions, but the Republican lawmakers weren't interested in facts, only in their a priori conclusions and the rhetorical arguments that supported them. This was a fact-free decision-making process that rammed through a bill despite all the evidence to the contrary. In effect, it severed the fundamental relationship of democracy to knowledge, and thus ran counter to Locke's foundational ideas contained in the Declaration of Independence. If knowledge does not have primacy in public decision making, then no truth can be said to be self-evident, and we are left with the tyranny of ideology enforced by might. These Republican lawmakers were behaving in as un-American a manner as King George.

The story is by no means unique. In 2011 it was happening across the country, as a new wave of freshman Republican and Tea Party state legislators found themselves in power. In Minnesota, for example, legislators concerned about accommodating industries whose discharge could kill that state's famed wild rice harvest passed a bill that arbitrarily increased the environmental standard for sulfates in water from 10 p.p.m. to 50,[95] against the scientific basis for the prior standard, which dated to 1944 research by state fisheries scientist John B. Moyle. "No large stands of rice occur in waters having a SO_4 [sulfate] content greater than 10 p.p.m.," Moyle wrote in the *Journal of Wildlife Management*, "and rice generally is absent from water with more than 5 p.p.m."[96]

Science is inherently political. If those who disagree with its findings are allowed to harass and intimidate scientists or quash those results, all Americans lose, and democracy loses. Cuccinelli's, Inhofe's, and the Republican House Caucuses' acts were cowardly, tyrannical, and deplorable. They should be condemned in the same terms used to castigate anyone who seeks to suppress freedom. Thomas Jefferson, who founded the University of Virginia in 1819, would have been appalled.

Knowledge is power, the saying goes, so it follows that suppression of knowledge for ideological reasons weakens the powerful. It weakened the Roman Catholic Church's credibility when it indicted Galileo, and Italy lost its place in the intellectual and economic vanguard.

Suppression of knowledge weakened Russia in the Lysenko affair, in which a political ideologue and former peasant named Trofim Lysenko ingratiated himself to communist leaders and was placed in charge of national agriculture because of his ideological conformity. He denounced and suppressed scientists who questioned his odd schemes as "fly lovers and people haters"[97] (because geneticists were doing fruit fly research—I kid you not!) and his uneducated methods decimated Soviet agriculture. Soviet scientists who opposed him were persecuted, jailed, and shot. Similar to modern Republican characterizations of climate science as "junk science" by an "environmental priesthood," Soviet geneticists, physicists, and chemists were characterized as "caste priests of ivory-tower bourgeois pseudoscience."[98] Soviet agriculture, biology, and genetics were held back for forty years, weakening the Soviet Union and helping lead to its eventual downfall.

Suppression of knowledge similarly weakened China during Mao Zedong's Great Leap Forward. Mao prided himself on his peasant roots and considered intellectuals arrogant, dangerous antirevolutionaries, similar to the modern characterization of them by Rush Limbaugh. Similar to Eisenhower, Mao was concerned that scientists could take over as a "technical elite," so he demanded that ideology take precedence over science, effectively silencing scientists.[99] In 1957 he set forth a plan to transform China into a modern industrialized society. It would overtake Britain in fifteen years while simultaneously feeding its own people and exporting grain to other nations. Mao had no knowledge of metallurgy, but based on a single demonstration he mobilized millions of peasants to smelt steel in "backyard furnaces." They burned trees, doors, and furniture as fuel and melted scrap metal like their pots and pans. At the same time, peasants were given outrageously optimistic grain production quotas based on Lysenko's assumptions. Because ideology and appearance of success mattered more than the facts of their meager harvest, the peasants gave more grain to the state than they could spare. Meanwhile, millions of other peasants were diverted off the farms into large-scale public works projects needed to industrialize the country, and grain crops were left to rot in the fields. Scientists and others who suggested that Mao's plans were unrealistic were "struggled against," sentenced to hard labor, and often executed.[100] The furnaces failed, the steel was unusable, and forty million Chinese people died in the greatest famine in human history.

Knowledge is power. Now its suppression is weakening America.

THE NEW DENIALISM

In November 2010, the prong-four energy-industry financing of the Tea Party movement through FreedomWorks and other front groups paid off, and Inhofe and other congressional outliers gained new allies who placed ideology ahead of knowledge. The political tide swept in a new class of eighty-seven freshmen GOP representatives and thirteen senators. Ninety-four of these one hundred new lawmakers either explicitly denied climate change, signed the FreedomWorks "Contract from America" to reject cap and trade,[101] or signed the "No Climate Tax" pledge promoted by Americans for Prosperity (AFP),[102] another front group that was started by David Koch and Richard Fink, a member of Koch Industries' board.

AFP was another of the lead organizations behind the Tax Day Tea Party protests of 2009.[103] Koch Industries is America's largest private oil refiner, and in 2010, *Forbes* named it the second-largest privately held company in the United States after Cargill, with annual revenues of $100 billion.[104]

Not a single freshman Republican publicly accepted the scientific consensus that anthropogenic greenhouse gases are an immediate threat.

"I am vindicated," said Inhofe, who was ridiculed by environmentalists in 2003 when he declared that man-made global warming was the "greatest hoax ever perpetuated on the American people."[105] Forty-six of the freshmen had picked up on the Sammon-style messaging led by Fox News and explicitly denied anthropogenic climate change is even happening in public statements they made on the campaign trail.[106]

"Today's global warming doomsayers simply lack the scientific evidence to support their claims," said newly elected congressman Bill Huizenga (R-MI) in a typical newbie comment. "A host of leaders in the scientific community have recognized that the argument for drastic anthropogenic global warming is no longer based on science, but is being driven by irrational fanaticism."[107]

But the newbies were not alone. Forty-four incumbents professed what were sometimes even more strident antiscience views, always with the common themes of elevating the critics, casting doubt, and branding global warming a "hoax."[108] Incoming House Speaker John Boehner (R-OH) led the charge on national TV, making embarrassingly ignorant statements when speaking to George Stephanopolous on ABC's *This Week* in April 2009. "George, the idea that carbon dioxide is a carcinogen that is harmful to our environment is almost comical. Every time we exhale, we exhale carbon dioxide. Every cow in the world, you know, when they do what they do, you've got more carbon dioxide."[109]

This willful rejection of science as a Republican Party plank is lamented by Republican former congressmen like Vernon Ehlers, John Porter, and Sherwood Boehlert. New York's Boehlert, who was chair of the House Committee on Science, wrote of it in a *Washington Post* piece shortly after the 2010 election.

> I call on my fellow Republicans to open their minds to rethinking what has largely become our party's line: denying that climate change and global warming are occurring and that they are largely due to human activities.
>
> There is a natural aversion to more government regulation. But that should be included in the debate about how to respond to climate change, not as an excuse to deny the problem's existence. The current practice of disparaging the science and the scientists only clouds our understanding and delays a solution. . . .
>
> We shouldn't stand by while the reputations of scientists are dragged through the mud in order to win a political argument. And no member of any party should look the other way when the basic operating parameters of scientific inquiry—the need to question,

express doubt, replicate research and encourage curiosity—are exploited for the sake of political expediency. My fellow Republicans should understand that wholesale, ideologically based or special-interest-driven rejection of science is bad policy. And that in the long run, it's also bad politics.[110]

EXPERIMENTING WITH THE ECONOMY

By 2010 the climate bill was long and full of compromises, directives, and exceptions that irked conservatives[111]—at more than 1,200 pages it was roughly double the length of the Clean Air Act Amendments of 1990 on which it was modeled. As it moved forward some pundits didn't deny the science, but rather argued that "cap and tax," as cap and trade came to be called, was risky "experimenting with the economy." They argued that the government should not make major changes in people's lives and the economy on the basis of a few years of research. That kind of precipitous action doesn't really give time for the debate to get organized, they said, especially when long-term trends need to be studied. And, they cautioned, new research methods can give unexpected results. For example, remember "vegetables will prevent cancer, wait, sorry, they won't."

This argument forgets that we have been accumulating data for not just a few years but for fifty years, and that we experiment with the economy constantly. Any time we do a tax cut for the wealthy on the grounds that it will trickle down, that's an experiment. Deregulating banks was an experiment that didn't have the best results. Bailing out those same banks was another experiment. We experimented with the economy plus American lives by unilaterally declaring war on Iraq. America is an ongoing experiment; that is the nature of our legislative process. In fact, we know a lot about bringing externalities into the market with fees and taxes. We can quantify those things better than the schemes some economists have proposed for Social Security reform.

But these arguments fail to move the new Tea Party activists, who for all their embrace of anarchy, libertarianism, and freedom appear to be interested only in lockstep authoritarian adherence to the team message. By contrast, conservative economists generally support cap and trade. One of these is Douglas Holtz-Eakin, the former head of the Council of Economic Advisers to President George W. Bush. Holtz-Eakin went on to become the director of the Congressional Budget Office, putting him at the top in terms of Republican economic credentials. In 2008 he was director of domestic and economic policy for John McCain's presidential bid. After that, Senate minority leader Mitch McConnell (R-KY) appointed him to the Financial Crisis Inquiry Commission, and as president of the conservative think tank American Action Network, he actively opposed the 2010 health care reform bill.

Holtz-Eakin says the entire cap and trade argument is misplaced. "There needs to be some reeducation," he says. "Conservatives *invented* cap and trade, to battle acid rain. They were leaders in overthrowing liberals on it under Reagan. Before that it was all command-and-control approaches, and *we* brought market forces in to bear through cap and trade and it saved a *ton* of money." In fact, what was expected to cost about $4 billion to $6 billion annually wound up costing a quarter of that and has saved an estimated $70 billion annually in quantifiable health care expenses—a return on investment of more than forty to one.[112] Holtz-Eakin says conservatives have forgotten that, and lost their roots. "They've taken positions that are divorced from any reality on the policy and from their own history."[113]

TRIUMPH OF THE EPITHET

By calling the approach "cap and tax" and turning that into an epithet, the energy-industry special interest groups poisoned the water. This is a common propaganda tactic because it destroys critical thinking. In a democracy, authority is brought to bear at the ballot box, and the fastest way to shape that outcome is to use epithets to demonize the opposing side. Thus we see the rise of negative campaigning and the use of epithets and talking points to chant the groupthink message until it is believed. "Liberal." "Cap and tax." "Socialism." When 2012 GOP presidential candidate Mitt Romney said in June 2011 that he believed the world is getting warmer,[114] Rush Limbaugh's response was "bye-bye nomination."[115] After that level of emotional distaste has been reached, effecting a reeducation by a rationalist conservative like Holtz-Eakin is not going to be easy. "You have to explain how it really works. These are ideas that matter to conservatives and previous conservatives embraced them, so you use some strength by affiliation."

Many of the Tea Partyers were propelled into office by a genuine fear of big government overreach. "That's not unjustified," says Holtz-Eakin. "There was a lot of overreach. The health care bill was enormous overreach. But it doesn't mean there shouldn't be some reach, and there's a conservative principle that says government can do good by intervening with appropriate regulations, like cap and trade. It's gotten to be such a loaded word in Washington you can't even say it, but *it's a conservative free market idea.* You either pick the price or set the quantity, and let the price fall out. Let the top-line policy drive the outcome and you let private business innovate and get out of the line completely. What that means is there should be a serious effort to clear out the regulatory underbrush at the same time. We don't need the EPA using the Clean Air Act to regulate carbon on top of it. You have to have something for people who want to be against government. This is what went wrong with the energy bill in 2009. We put it in and people said we don't believe it because there's another ream of command and control language right alongside it."

Siegel says the command and control approach works, and if Republicans have rejected cap and trade, that's their loss. "We already have all these laws set up and ready to go. The Clean Water Act, the Clean Air Act, the Endangered Species Act. Maybe they're not perfect but they're the world's best, most successful laws. A theoretical cap and trade system is not incompatible with them." She says the only way to counter special interest money is for the administration to use the bully pulpit. "We need the government involved, and the Obama administration has really dropped the ball on climate. They should have been out there with a full-court educational press, a World War II type mobilization to counter the misinformation and educate the public, and they haven't done it."

Demonizing cap and trade, Holtz-Eakin says, threatens conservative principles because if it becomes completely discredited, conservatives won't have it in their toolbox for managing environmental regulations in the future. "We don't want to go back to the old, big-government approach of one size fits all of the 1970s," he argues. What's really happening, he offers, is that a decade of marketing and $500 million spent by the petroleum industry in the prior *eighteen months alone* to lobby Congress was being brought to bear. "Conservatives need to figure out what they stand for. Is it just the highest bidder, or are there principles in there that mean something? I've always felt it's in their political interest to not deny the science, that's where the votes of the future are. But it awaits a teachable moment."

ARE WE THINKING ABOUT THIS ALL WRONG?

That teachable moment may never come. Emerging from beyond the ashes of the cap and trade battle is a new idea that is threatening to derail the entire climate change debate. It could rightfully be considered the second wave in the anti-climate-change propaganda campaign, in which a counternarrative is presented to confute and divide the thinking of one's opponents. The counternarrative in the climate wars is called "geoengineering," and it may just be the most dangerous idea yet.

Two years into his presidency, the president was briefed on ways to tackle climate change, and the report didn't even mention carbon taxes, or cap and trade, or even reducing greenhouse-gas emissions; it recommended geoengineering—using technology to directly intervene in Earth's climate system by spreading chemicals across millions of square miles of the Atlantic Ocean to reflect sunlight into space and force down Earth's temperature.

But the president wasn't Barack Obama. It wasn't George Bush. It wasn't even Ronald Reagan. It was Lyndon Johnson, and the year was 1965.[116] "This generation has altered the composition of the atmosphere on a global scale through radioactive materials and a steady increase in carbon dioxide from the burning of fossil fuels," Johnson wrote in a letter to Congress that year.[117]

The problem of anthropogenic climate change has dogged every president since, but the political will to do anything about it hasn't improved in nearly fifty years. Now the argument for this scheme is being presented again, and its divisive effect is working. Increasingly desperate for a solution as they watch policy makers losing touch with reality, leading scientists are beginning to take the once-fringe idea more seriously. Some American conservatives like it as a way to tackle the problem while saving face politically and requiring little change in energy use. And most attractively, it's cheap. Crazy cheap.

"My estimates are a couple billion dollars a year," says leading climate scientist Alan Robock.[118] "I never thought I'd hear myself saying it about that amount of money, but that's really cheap compared to these other things we're considering." Robock is editor of *Reviews of Geophysics*, a publication of the American Geophysical Union, the leading international professional association in Earth and space sciences, and is a lead author of an upcoming IPCC report on climate change. He shared the 2007 Nobel Peace Prize for his work on prior IPCC reports. A cheerful guy with a bearded, cherubic face, he's not what one would imagine as a "climate doomsayer."

What worries Robock is that the low cost puts the possibility in the hands of a small nation or even a wealthy individual. "There are several billionaires in the world who could afford to become Greenfinger," he says, referring to the megalomaniac Goldfinger of the James Bond film. "Or what if one day Bangladesh decided they were tired of being overheated? They could afford to do it all on their own."

AN EXPLOSIVE IDEA

There are several ways scientists and engineers could take command of the planetary thermostat with technology, but the leading plan is to spray thousands of tons of sulfur dioxide into the upper atmosphere in order to reflect sunlight back into space, essentially installing a global sunshade to cut down on the amount of light reaching the surface. This is referred to as modifying Earth's albedo, or reflectivity.

Called sulfur aerosol injection, the plan comes from observations of the Philippines' Mount Pinatubo eruption in June of 1991. The volcano blasted nearly twenty million tons of sulfur dioxide into the upper atmosphere.[119] It spread into a haze that reflected enough light back into space to decrease the amount hitting the planet's surface by about 10 percent. Over the next eighteen months, global temperatures fell by about 1°F. The sulfur stayed in the stratosphere for about three years, creating spectacular sunsets, but also slowly destroying parts of the ozone layer, before gradually falling back to Earth as acid rain.

The same end result—a high-atmosphere sulfur dioxide haze—can be achieved technologically, and various methods have been proposed, from adding sulfur dioxide to commercial jet fuel to pumping it up through 65,000-foot-long fire hoses running from the ground to enormous zeppelins that would spray it into the stratosphere.

DIVIDE AND CONQUER

That the counternarrative was dividing scientists became apparent shortly after Obama's inauguration. John Holdren, in his first media interview as Obama's science adviser, confessed as much to AP science reporter Seth Borenstein. Borenstein's article stunned DC policy circles, reporting that "tinkering with Earth's climate to chill runaway global warming—a radical idea once dismissed out of hand—is being discussed by the White House as a potential emergency option, the president's new science adviser said Wednesday."[120]

A frustrated Holdren shot off a clarifying e-mail to scientists:

> I said that the approaches that have been surfaced so far seem prob-
> lematic in terms of both efficacy and side effects, but we have to look
> at the possibilities and understand them because if we get desperate
> enough it will be considered. I also made clear that this was my per-
> sonal view, not Administration policy.
>
> Asked whether I had mentioned geo-engineering in any White
> House discussions, though, I said that I had. This is NOT the same
> thing as saying the White House is giving serious consideration to
> geo-engineering—which it isn't—and I am disappointed that the
> headline and the text of the article suggest otherwise.[121]

National Academy of Sciences president Ralph Cicerone, himself a climate scientist, also supports further investigation. Three years earlier he
had said of similarly exotic ideas for combating climate change, "We
should treat these ideas like any other research and get into the mind-set
of taking them seriously."[122] His wording suggests a certain resignation.
Cicerone even took steps to help Nobel laureate and atmospheric chemist
Paul Crutzen publish a paper on sulfur injection.[123]

Crutzen won his Nobel for showing how chlorofluorocarbons and
other industrial gases were damaging the ozone layer. Robock asked him
what he had intended with the sulfur injection paper, which seemed such
a divergence from his past work. Crutzen said he was despairing that
there would ever be a move toward mitigation. He thought this might be
our last, most desperate hope. "I asked him if when he wrote the paper
he thought maybe it would be so scary, it would make people more serious about mitigation, and he said yes," Robock said.[124]

You Got to Be
Freakin' Kidding Me

The political momentum the radical approach enjoys has been helped along
by its destigmatization by University of Chicago economist Steven Levitt
and journalist Stephen Dubner in their bestseller *SuperFreakonomics*. The

book presents an engaging account of some of the world's top minds, wizards led by former Microsoft chief technology officer and billionaire Nathan Myhrvold, their "Harry Potter,"[125] hunkering down in a converted Harley-Davidson factory, smirking at Al Gore's histrionic tones and setting out to solve the problem once and for all, like men.

> When *An Inconvenient Truth* is mentioned, the table erupts in a sea of groans. The film's purpose, Myhrvold believes, was "to scare the crap out of people." Although Al Gore "isn't technically lying," he says, some of the nightmare scenarios Gore describes—the state of Florida disappearing under rising seas, for instance—"don't have any basis in physical reality in any reasonable time frame. No climate model shows them happening." . . . It isn't that current climate models should be ignored, [astrophysicist Lowell] Wood says—but, when considering the fate of the planet, one should properly appreciate their limited nature. . . . Then there's this little discussed fact about global warming: while the drumbeat of doom has grown louder over the past several years, the average global temperature during that time has in fact decreased.[126]

But the book is chock-full of misinformation and propaganda. For example, *SuperFreakonomics* cheekily complains that "When Al Gore urges the citizenry to sacrifice their plastic shopping bags, their air-conditioning, their extraneous travel, the agnostics grumble that human activities account for just 2% of global carbon dioxide emissions, with the remainder generated by natural processes like plant decay."

Well, *that* certainly takes the air out of the global warming argument. All this over *just 2 percent?!* Maybe that Al Gore really is overreacting. But what the book doesn't say is that, as described earlier, the global carbon cycle is relatively stable, with roughly equivalent amounts of carbon dioxide being absorbed and emitted by growing and decaying vegetation each year, as the roughly equal rise and fall in the Keeling curve each year illustrates. The plants grow in the spring and summer, soaking up carbon dioxide, and they decompose in the fall and winter, putting about the same amount back into the atmosphere. The 2 percent is the extra amount that humans add *each year,* and *it stays in the atmosphere.* Then, the next year, another 2 percent gets added, and the next 2 percent more. This is what makes the Keeling curve climb every year. After fifty years, 2 percent each year can become a big, big problem.

THE 2 PERCENT SOLUTION

Levitt and Dubner adopt the conspiratorial tones of Atlantic City wise guys, writing dismissively about the "religious fervor" of climate scientists. Al Gore "isn't technically lying," and "It isn't that current climate models should be ignored," they tell us, but they're just saying, don't take these lefty pantywaists too seriously. If it's a problem, if it bothers you, fine, we'll take care of it. Don't get your undies in a bunch.

The second half of the chapter tells how. The wizards, conducted by Myhrvold, "fueled by an astonishing amount of diet soda," admit that it is a problem after all and then go about solving it for $250 million and change. Wham bam, thank you Mother Nature. Here's what we do, see? We erect an extremely long hose. We shoot stuff up it at an extremely high volume. Then we ejaculate it into the stratosphere. Problem solved.

Other economists and scientists will tell you that this approach ignores the risks. A true comparison, economic or otherwise, would include a risk-risk analysis, as opposed to the rosier cost-benefit analysis. There are geophysical risks, climatological risks, political risks, economic risks, defense risks, and others that the wizards can't even begin to quantify, so instead they set them to zero. But that's not science. It's not even good economics. It's propaganda.

Climatologist Raymond Pierrehumbert puts it more simply. "They're idiots," he says, "and you can quote me." This is surprising because Pierrehumbert works just two blocks from Levitt. He is a lead author of the IPCC's Third Assessment Report, published in 2001. He is also working with several top conservative economists at the University of Chicago's famed Milton Friedman Institute for Research in Economics, including some Nobel laureates, who agree that climate change is real,

human caused, and a problem that needs to be dealt with and are looking for ways to use free market economics to break the logjam.

"Unlike the other economists in the Friedman school, Levitt is unwilling to use his intelligence to actually see whether what he's saying makes
any sense," Pierrehumbert says. "And the sad thing is, if he didn't want to
do the math himself he could have walked nine minutes over to my office,
or even sat in on one of the talks I give in the business school. I just analyzed one tiny piece of what he says and did some basic middle school
math and showed that Nathan Myhrvold was completely wrong, for
example, about solar panels. It's just sloppy, politically motivated thinking. Levitt says the problem is because environmentalists don't think like
economists and don't like cheap solutions, and sulfate aerosol injection is
cheap. But he glosses over all the known risks, and he gets his math wrong
in the most basic middle school level ways. This is a well-known economist at a top university. You'd think he would hold himself to the same
standards he'd expect of his own students. The chapter is wrong on so
many points that it should be a major intellectual embarrassment."[127]

RISKS? WHAT RISKS?

The first known risk of sulfur aerosol injection is the weakening of the
monsoons. Monsoons happen because the Asian landmass heats and
cools much faster than the Indian Ocean. The sharp temperature contrast causes wind to blow in off the ocean, carrying moisture. It hits the
mountains and spawns the monsoon season—a period of heavy rains
that provide water for crops that feed about three billion people across
Pakistan, India, China, and the rest of South Asia.

If we reduce the amount of sunlight reaching Earth's surface, there's
less evaporation off the oceans. The land won't heat as much, so there is
also less wind. Together the effects could vastly reduce or even end the
monsoons. "It's been observed after every major volcanic eruption we
have data for," says Robock.

A second known risk is ozone depletion. Ozone blocks ultraviolet rays, which have enough energy to ionize DNA, thus potentially causing cancer. The hole in the ozone layer worsened after the Pinatubo eruption, but it's located over Antarctica. The problem is that a lot of the geoengineering work would have to be done in the Northern Hemisphere, since that's where the effects of climate change are most felt. So we'd see a substantial weakening of the ozone layer there. With most of Earth's population in this hemisphere, the rates of death from skin cancer and blindness from glaucoma would skyrocket.

The third known risk—the mother of all known risks, one might say—is that once you start geoengineering, you can't stop. If we stabilize the temperature but keep putting carbon into the atmosphere, the chemistry of the atmosphere will get more and more out of balance, and like any addict, we'll have to keep upping our dose of sulfur to maintain the same effect. "We are committing the next ten thousand years to keeping that up," says Pierrehumbert. "That's what these guys aren't telling you."

KOCHED UP

This idea is easily the riskiest suggestion in the history of human civilization. Carbon dioxide would continue to accumulate, but we wouldn't be feeling the effects because we'd be sedating the climate system with sulfur. If somehow we lost the will or the ability to maintain and continually increase the sedation, the accumulated effects of deferred climate change would come crashing down on us in the course of a few years, ending the world as we know it. *SuperFreakonomics* is a very aptly titled book.

Let's look at a very crude example of why this is a bad idea. Let's say we're injecting sulfur aerosol and everything's great, temperatures are under control. Feeling good. But then let's say there's a severe drought in India. Millions of people begin to die. Work at the world's customer service and information processing companies freezes. Capital stops flowing. The economy tanks. The Indian government loses control of the situation because the global power structure has flattened. Nation-states aren't what they used to be. Terrorists blow up Western dams to score political points. The most massive of the Indian outsourcing companies tries to get a handle on things, but Maoist rebels destroy its server farm.

In the northern latitudes the skies are no longer blue, they're milky white from the sulfate injection. It's perpetually cloudy. The lake fish are mostly gone from acid rain. Because of the lower light conditions, people are eating antidepressants like candy, becoming chemically dependent in yet another way. Drug and alcohol use skyrockets, as does the prevalence of obesity.

The drought spreads into China as the monsoons dry up there too, threatening the food supply. China joins India in telling the United Nations that America and the United Kingdom must stop geoengineering. The president goes on television and refuses, explaining what could happen to the entire planet, that we have no choice. However, China and India are under immense internal pressures and threaten nuclear action. An international crisis ensues. The world goes to DEFCON 1, the highest defense readiness condition. Britain caves and stops geoengineering. Facing a thermonuclear showdown, America backs down too, on the condition that China revalue its currency and put carbon dioxide scrubbers on all its factory chimneys. China says okay to the scrubbers but not the currency, and America agrees. Nuclear war is averted.

After a year or so, the chemicals come out of the atmosphere and the monsoons resume. It is a great rebirth. Dancing in the streets. We've kicked the habit. The US president is reelected in a landslide. But all this time, we've been pumping more carbon dioxide into the atmosphere. Scientists have been warning about global warming, but that's just their *opinion*. Liberal hand-wringing. Global socialist scheming. Redistribution of wealth. We're at 490 ppm of carbon dioxide, so what?

But global temperatures, which have begun to creep up, now shoot up 2°F in the first few weeks after the sulfur is all gone. Just a warm spell. Then 4°. Then ten. Then twenty. Alarmed, we again start injecting sulfate aerosol, but the heating cycle has too much momentum. Methane starts bubbling up from the oceans and outgassing from the melting Siberian permafrost, and we can't catch up. People die by the millions in unprecedented heat waves across central Asia and the United States. The beneficial, nourishing rains dry up, but now they dry up everywhere and instead there are violent thunderstorms, severe floods, hurricanes and wildfires, and mass extinctions.

It's the *rate* of change that's hard to adapt to, not the absolute level, and this rate is much too fast. The world would become a *superfreak*—a sulfur huffer—and experience deadly withdrawal symptoms if it were ever to even try rehab.

WHO GETS TO BE GREENFINGER?

Fiddlesticks, say geoengineering supporters, that's just more alarmist Al Gore style fear mongering. We'll just do it for a little while, until we get past this crisis and can transition our economy in a more orderly way.

Okay, let's say we buy that argument and go ahead with it. There are other political risks, the first being identifying which government bureaucrat we want to be in charge of Earth's climate. Holtz-Eakin, the former Bush economic adviser, argues that "there is going to be the fear of government getting it wrong, of overreach. Right now there's no faith in any institutions—government, Wall Street, insurance companies, they're all bad. You have to re-create faith in institutions before you can get the institutions to work. But even then, do you really want to give government *that much power*?"[128]

There are also major foreign policy concerns, beginning with the international ENMOD treaty—the Convention on the Prohibition of Military or Any Other Hostile Use of Environmental Modification Techniques, ratified by the United Nations in 1978 in response to the United States' use of silver iodide cloud seeding to alter the weather over the Ho Chi Minh Trail during the Vietnam War. Operation Popeye was an effort to "make mud, not war" to slow North Vietnamese supply lines by lengthening the monsoon season.[129] The treaty outlaws "deliberate manipulation" of "the dynamics, composition or structure of the Earth, including its biota, lithosphere, hydrosphere and atmosphere, or of outer space."[130] Each of its seventy-four signatory countries agreed not to "engage in military or any other hostile use of environmental modification techniques having widespread, long-lasting, or severe effects." So if a country interpreted either geoengineering or its cessation as a hostile act, we could find ourselves at war over the climate.

Setting that aside, suppose we call a UN convention to discuss having the IPCC run it, assuming that conservatives can get over their distaste of the United Nations and the IPCC and their fears about world government, and liberals over their fears about globalization. Russia sees the great breadbasket potential in a warmer Siberia, but India wants it cooler. How do we get agreement *then*? The political negotiations and potentially enormous economic concessions that hotter, more populous countries involved in climate control talks might demand of cooler, more developed countries would likely make nuclear and Arab-Israeli negotiations look like tiddledywinks.

If we can't get to agreement through diplomacy, what about our favorite market economics idea—privatize it? What if we give Exxon-Mobil's geoengineering division a stake in this? Or let's say it's BP Geoengineering—Beyond Petroleum. It comes in and says, Look, we know you can't agree, we'll do a multiparty contract, monetize the whole thing, and give you a guarantee. Just sign here, problem solved. The world would, literally, be its oyster, and ultimately the corporation that supplied the world's sulfur jones would become the dominant force on Earth.

Geoengineering may seem on the surface like an idea that we should keep in our back pocket in case the real problem of carbon control can't be solved, but it is a bit like taking up a crack habit to counter the effects of your drinking problem, and it will lead to equal ruin.

SEPARATING THE SCIENCE FROM THE POLITICS

Winning the global warming battle will require separating the science from the politics. But is this possible, if all science is political? Yes. The key lies in confuting assumptions, framing science in ways that emphasize the process of science over its products, and focusing on solving shared problems.

Unlike questions about evolution, the politics of climate change have been cast in terms of money, freedom, and socialism. These are less about ideology than self-protection, and that is the value that the battle needs to be based on.

This is in no small part because the topic wasn't properly presented to the public in the first place. *An Inconvenient Truth* told us "nothing is scarier than the truth" and sought to scare people into action. Political strategists could have told the filmmakers that if you want to get people to vote against something, particularly Republicans, you need to get them angry about it, not scared of it, and if you want them to vote for something, you sell them hope and freedom. The irony of all this is that science informs a lot of successful psyops and propaganda strategies, but it isn't being used by most scientists even as a tactical defense. They instead sort of stumble blindly along and seem confused when they get bowled over.

Science has always been about the hope and freedom offered by creating knowledge and therefore power. Selling science with fear worked when we had a specific opponent in the days of the cold war, but that old standby only works if science is *the answer,* not *the messenger.* We mock what we are scared of. Scary movies make people laugh. In fact, recent studies show that climate "doom and gloom" messaging causes people to dig in their heels and become *more* committed to disbelief, not less.[131]

Politically speaking, the IPCC and *An Inconvenient Truth* may have inadvertently exacerbated the climate battle by casting science as the messenger of fear and associating it with a *partisan political authority figure,* former vice president Al Gore.

Raymond Pierrehumbert disagrees with this conclusion. "I think with Al Gore, the only way he could have been less polarizing on this would be if he had been less effective. Anyone who was as effective as Al Gore would have attracted the same level of attacks." But Pierrehumbert is a geophysicist, so he looks at climate change from a factual perspective. Politics, on the other hand, is about narrative and emotion—and in this case especially, it is a war of propaganda. Al Gore is a greathearted and brilliant man on a noble mission. But *An Inconvenient Truth* cast Al Gore as the principal *protagonist* in the climate change battle, a role that powerfully made it *his story,* not *our story,* with all of *his* qualities—left-leaning, Democratic, anti-Bush. "I used to be the next president of the United States," he tells audiences to laughter, unnecessarily alienating half of the United States. The decision created a dramatic frame that begged for opposition. The power of the film's very important message was thus undermined by the filmmakers' narrative choices, which made for good drama, but bad politics.

If scientists wish to communicate what they see from their vantage points, they would be wise to avoid this kind of identity politics and focus instead on inclusion and process. They would profit by organizing local conversations at churches and chambers of commerce and community centers across the country to bring community leaders together—not to tell them what to do, but to make it *our story*: to present what they know and then *listen* and be part of a conversation—something scientists in general haven't done for two generations.

They would be wise to form a rapid response team to immediately, loudly, and publicly debunk antiscience and pseudoscience propaganda before it can "set" in the public consciousness, which usually takes about twenty-four to forty-eight hours. The American Geophysical Union is organizing one such effort, as has University of St. Thomas scientist John Abraham's ClimateRapidResponse.org, and hopefully they will be able to adequately resource and maintain them.

Scientists would be wise to support the Science Debate effort, which provides another model for ideological-frame-confuting conversations among policy makers, scientists, and citizens—as well as an opportunity for accountability within the electoral process—in a time, place, and mental space in which people are used to thinking about policy and the future. "Somebody just has to stand up and do the right thing," says Kassie Siegel. "You got elected to make a difference in the world."

Michael Mann says that whether the science issue is climate change or something else, this last part will be critical. "We need to hold our policy makers and media accountable for distinguishing legitimate science from denialist propaganda and antiscience."[132] He's right.

FREEDOM, BABY

The biggest part of the problem is the money the energy industry is pouring into propaganda and lobbying. "We're just a bunch of scientists. How do we combat *that*?" asks Pierrehumbert.[133]

It's tough. But ultimately, as Mann pointed out, the facts will come out *if* scientists engage enough. One way that may produce better results is to stop pitching the doom-and-gloom scenarios, as research suggests. Optimistic statements more readily attract support. It should be about what you *get,* not what you might *lose.* This is a basic sales tactic.

"Actually, I have been doing that a lot more lately," says Pierrehumbert, "and it works. Imagine how wonderful it would be if you could get on a shiny new high-speed train in Chicago and visit your relatives in Madison without having to go to the airport, and all the hassle. What if we had a robust energy-supply system with a smart grid, so that someone with a bright idea about putting energy into the grid could hook up and make money on that. Think of the possibilities! All the ways we can make life *better, easier, and cheaper* by doing things that also happen to reduce our carbon footprint."

What climate scientists, politicians, economists, and engineers should be talking about is the problem and the concrete solutions: How do we plan communities and energy supply in ways that work better for people, that give them *more choices*—that *increase their liberty.* That is what science and America have always been about.

There's emerging evidence that this is a successful strategy. Residents of deeply conservative Salina, Kansas, by and large think that global warming is a big liberal hoax put on by Al Gore to get their money. "Don't mention global warming," warns Nancy Jackson, director of the Climate and Energy Project, a small nonprofit working to reduce fossil-fuel emissions.

"And don't mention Al Gore. People out here just hate him." But saving energy is another matter.[134]

The Climate and Energy Project decided to see if it could remove climate politics from the energy debate by focusing on thrift, patriotism, spiritual conviction, and economic prosperity—in other words, focusing on the process and the solutions to our problems—and it worked. Energy use in targeted towns declined by as much as 5 percent compared to neighboring towns. The Lawrence Berkeley National Laboratory featured the effort in its report on changing energy use behavior in America.[135]

SELLING EFFICIENCY TO CONSERVATIVES

Further north, along the heartland's Interstate 35 in Minnesota, state auditor Rebecca Otto in 2009 won the national Excellence in Accountability Award from the National State Auditors Association for a report[136] detailing how the 4,300 local governments in Minnesota could make wise investments to reduce their energy use.* As a bloc, US local governments are among the largest energy consumers in the world, far larger than most nations. But in the report, Otto never mentioned the word carbon, except when reproducing vendors' self-descriptions in the back matter.

"I get conservative Republicans coming up to me and saying, 'Isn't that some sort of liberal tree-hugger thing?' or 'Isn't that green stuff communism?' and I say, 'No, I'm about efficiency and saving money. It's about our wallets,' and they sort of nod, because I'm more conservative than they are on a lot of fiscal issues," says Otto. "I say to them 'Anything to save tax dollars and improve efficiency, and you know, it saves on health care costs too. I'm sure you have somebody in your family that has asthma, or maybe you have it. Some days you go out for a run and it hurts your lungs to breathe, it limits your freedom, and that's not right,' and

* Rebecca Otto is the wife of the author.

they get that. You have to put it in terms of individuals, what's in it for me, and maybe a little wider to their family. That resonates with people."[137]

The Minnesota state auditor and the economists are right. People respond first to self-interest and things that affect their families, which are the core units of the economy and of altruism in America. This is the most basic and universal level of moral and ethical development, and thus the broadest message point.

This can be an antiauthoritarian, top-wing position. Let's apply it to urban design as an example. If you have to commute a long distance in your car, is that really good for your wallet or your family? The market shows that people want to live in smaller, higher-quality communities where they have sidewalks, they can talk to their neighbors, and they have safe streets, clear air, and conveniences that free up more time. The only reason they don't is because of the way we used to design suburbs, which was about dispersing everybody to avoid the bomb during the cold war. But dwellings in newer developments that employ a more old-fashioned urban residential design have been big sellers. People like living like that. They save money and have a better lifestyle. That's not exactly rocket science, but it turns out that getting those sorts of communities takes planning because it requires changing the old cold war design that was imposed on suburbs in the 1950s. These days, if you're talking about planning anything, it invites critics to call it "socialism." But when you approach it with an emphasis on freedom, people see that it's about having and increasing liberty. Right now, because of the outdated cold war mentality, people who want to live in high-quality communities can't do it because we're not building those communities. But they should have that choice if they want to. They should have that freedom.

See how that works?

PART IV

TOMORROW'S SCIENCE POLITICS

CHAPTER 11

FREEDOM AND THE COMMONS

One death is a tragedy. A million deaths is a statistic.

—ATTRIBUTED TO JOSEPH STALIN*

If I look at the mass I will never act. If I look at the one, I will.

—MOTHER TERESA

SCIENCE'S CHALLENGE TO DEMOCRACY

Clearly we are reaching a crisis point. Science already affects nearly every aspect of life, yet Americans are increasingly unable to make decisions grounded in reality. Our quality of life, our economy, and the ongoing viability of the planet hang in the balance. Our paralysis is presenting perhaps the greatest moral and political dilemma in human history.

The problem is only going to accelerate as the century wears on. Like business, science has gone global. It owes allegiance to no nation, and knowledge is always a double-edged sword. Advanced science education of a world-class quality is increasingly accessible in many places outside

* Russian scholars have not verified this popular Western attribution. The comment may have come from the German writer and satirist Kurt Tucholsky, aka Kaspar Hauser. Tucholsky wrote a 1932 article titled "French Joke" that listed several painful jokes, including a comment by a French officer saying that World War I wasn't so bad because "The death of one man, that is a tragedy; a hundred thousand dead, that is a statistic" (Tucholsky, K. "French Joke" in *Learn to Laugh Without Crying*. Berlin: Ernst Rowohlt Verlag, 1932, p. 148).

the United States and Western Europe, and research labs rarely have particular geopolitical requirements beyond liberty. Scientists are connected with one another via the Internet and are sharing information at the speed of light in unprecedented volumes, across fields, and with unprecedented interconnectivity.

Technology is flattening power structures and weakening nation-states as knowledge spreads through broadly distributed communications and social media networks. New power structures based on antiauthoritarian leaderless groups are emerging even in nondemocratic societies, fueled in part by science's shifting of knowledge and power into the hands of individuals. This restructuring expands opportunity for freedom, but also increases terrorists' and anarchists' access to tools, necessitating increasing responsibility and regulation. The US Defense Department's 2010 *Quadrennial Defense Review Report* examines how US armed forces are adjusting to this changed environment. "Threats to our security in the decades to come are more likely to emanate from state weakness than from state strength," the review predicts.[1] One of the factors that undoubtedly will play into that is the very real danger of the striation of society in terms of the level of scientific and technological literacy.

In reaction to these unprecedented changes, we are seeing a staunch embrace of anti-intellectualism among those who favor a more authoritarian social order and power structure, a position naturally opposed by science's antiauthoritarianism. This is not just a US phenomenon; it is happening worldwide, in each of the major world religions and in many of its leading nations. At the same time, moral direction has also been lost by a class of American leaders steeped in postmodernism who believe that what defines reality is not knowledge, but rather the most forcefully articulated argument.

As a consequence, the United States increasingly finds itself lacking an effective framework for solving our most pressing problems. Lacking both the framework and the social mechanisms that can help the public and elected representatives understand and navigate this vast new world of challenges, we are increasingly paralyzed. We need a new set of political and ethical approaches for the twenty-first century. Among the most essential tools are science debates, which focus policy attention on knowledge-based discussion of the world's most pressing problems.

At the center of many of the arguments over proposed solutions to environmental problems associated with increasing population and industrial development are differing views of how individuals and freedom relate to regulation and the commons—the common property of humankind. These questions are central to the relationship between science and democracy itself.

THE TRAGEDY OF THE COMMONS

In December of 1968, a little-known University of California Santa Barbara biologist named Garrett Hardin published a paper in *Science* that would change the way we look at economics. The core dilemma it identified, which came to be called "the tragedy of the commons" after the paper's title, lies at the heart of the unresolved environmental challenges of the twentieth century, among them climate change, ocean acidification, overfishing, biodiversity loss, habitat fragmentation, overdevelopment, pollution, exploding population, and unsustainable energy use, to name a few. The dilemma suggests that politicians are paralyzed by a fundamental conflict between the environment and the economy that arises from the deeply held, mistaken belief that freedom and regulation are incompatible.

Hardin's paper was remarkable because it offered such a sound rebuttal to the ideas of the Scottish economist Adam Smith, whose collaborator and mentor was David Hume. In 1776 Smith argued in *The Wealth of Nations* that in a shared economy, an individual, who "intends only his own gain," was in effect "led by an invisible hand" to promote the greater public interest, since willing buyers and willing sellers will always arrive at a natural price for things, and the highest value and efficiency will be obtained. "Nor is it always the worse for the society that [the individual's intention to do social good] was no part of it. By pursuing his own interest he frequently promotes that of the society more effectually than when he really intends to promote it."[2] The argument of the invisible hand was so well made that it has become an axiom of economics: Just get out of the way and let the market work.

But, Hardin asked, did the same reasoning still hold true in the economics not of 1768, when the world seemed unlimited, but of 1968? Imagine a situation where village herdsmen share a common pasture, he offered. Over time, various factors—disease, war, poaching, periodic famine—keep both the herds and their herdsmen at well below the carrying capacity of the pasture. In this situation, we could consider the pasture limitless, as the world appeared to be in Adam Smith's time, and his economic argument would hold true.

But then one day, increased use of good farming practices permits social stability and these losses are minimized. At this very point of highest social good, the logic of the commons creates a tragedy. Each herdsman thinks, "What is the utility *to me* of adding one more animal to my herd?" Since the herdsman receives all the proceeds from the sale of the animal, the positive benefit is +1. However, because the loss of grass, more weeds, increased erosion, etc., caused by the additional animal's grazing is shared equally by all of the herdsmen, the detriment to the herdsman is only a fraction of –1.[3]

In this way, each herdsman is motivated by the only "rational" economic conclusion: Add an animal . . . and another . . . and another.[4] But what seems rational when the problem is looked at within the *individual's* frame becomes grotesquely irrational when the frame used is the *collective* of all the herdsmen, and indeed of their economy and society at large, who now must cope with an overgrazed pasture that can sustain only a fraction of the number of cattle it did prior to the bubble. The environment and the economy both collapse.

Hardin concluded that in this circumstance, "each man is locked into a system that compels him to increase his herd without limit—in a world that is limited. Ruin is the destination toward which all men rush, each pursuing his own best interest in a society that believes in the freedom of the commons. Freedom in a commons brings ruin to all."

FREEDOM VERSUS TYRANNY

The simple dilemma that drives the tragedy of the commons is writ large in the greatest political argument of our time: the clash between individualism and collectivism. In the political realm this first became a clash between capitalism and communism and more recently one between supercapitalist anarchy and democratic socialism. Politics is narrative, and every narrative argument has an underlying value that is at stake. In this case the value is *freedom* versus *tyranny*, which is essentially the same as the antiauthoritarianism versus authoritarianism that drives the science-politics debate. But which road leads to freedom, and which to tyranny?

Capitalism is a more moral system if it offers *more freedom* to individuals. What do we mean by freedom? Liberty means *freedom to choose*, as Benjamin Franklin's and Adam Smith's friend David Hume defined it.[5] But there are qualifications. In a capitalistic and democratic system, my freedom to do as I wish is moral and just to the degree that it does not reduce your freedom to do the same.

To mediate a fair compromise, each of us must accept limitations equally to receive the equal benefits of freedom from the tyranny of *might makes right*. Thus society has created titles to land and other private property, one of the most basic functions of Western governments. My freedom is bounded by the sanction of government-issued property titles, as is yours, based on how much treasure we can each muster for the purchase of land and the payment of the taxes that protect our freedom and provide for common services to the property. What we have each gained through this self-imposed limiting and taxation to pay for roads, a sheriff, and a judge is freedom from the conscienceless, who would use brutality to take our lands by force. Freedom from tyranny proceeds from laws and regulations. The alternative is the assertion of the mighty. Thus we hear talk of the rule of law. This seems just and reasonable to most people as the price of enjoying the benefits of living together in a democratic society.

But what happens when the argument is extended beyond private property to the use of the commons, such as the public parks and lands, lakes and rivers? To carrying concealed weapons in public? To smoking in public, or not using a seat belt in a car, or riding a motorcycle without a helmet? To the atmosphere, the oceans, the rain forests? To the planet? To paying or not paying taxes to maintain these common things? What is the proper equilibrium between my individual rights and the rights of the collective of everyone else—now *and into the future*? How do we balance freedom and our rights to the commons?

Today's Herdsmen

Because we have a limited planet, today's herdsmen—nation-states, supranational corporations, and individuals—all have powerful economic incentives to pursue their rational self-interests until the tragedy of the commons occurs on a global scale. This is the nature of a boom or a bubble, but the assumption of limitless growth is what our economic model is based on.

The purpose of democracy is to find a balance that protects the equal rights of all individuals by using the antiauthoritarian rule of law, regulation, and the vote to maximize the overall level of freedom and minimize the might makes right of tyranny. Thus, each herdsman's rational self-interest is moderated in a forum with all others. Democracy affords the opportunity through regulation to govern the engine before it races to ruin.

Understanding this relationship, teasing out the logical fallacy that underlies the tragedy of the commons, ties into the fundamental relationship between economics, democracy, and knowledge gained from science. Finding ways to create *more freedom* and *more wealth* that allow for progress and economic growth while also sustaining the environment is the great task of the twenty-first century.

Freedom Through Regulation

Marine geophysicist Marcia McNutt recently witnessed this dilemma firsthand. As director of the US Geological Survey, McNutt was in charge of measuring the flow rate of the Deepwater Horizon oil leak and attempting to assess the damage to the marine ecology of the Gulf of Mexico.

"In the Gulf I see this playing out very concretely every day . . . the short-term extraction issues of oil and gas versus the long-term restoration issues of the Gulf coastline," McNutt says. "And many of the policies that need to be put into place that would allow a long-term restoration plan that would restore wetlands, stop coastal erosion, allow for more prosperous fishing, more bountiful fishing, more abundant wetlands, better sediment supply to nourish beaches, and therefore better tourism, hunting, recreational values, quality of living, better human health, better water quality, better air quality . . . all of that is on the line because of concerns by the Gulf Coast citizens that the policy that will allow all that will come at the expense of the oil-extraction activities. For example, the channels that are cut through the marshes that allow wellhead access, and many of the other dredging policies that allow ships access for maintenance of oil wells and pipelines that allow this to continue, are done with the thought of 'one more pipe won't matter, one more ship channel won't matter, one more bulwark by the Army Corps of Engineers won't matter'—yet it all disrupts the ecological balance that protects the coastlines and coastal cities."[6]

Throughout history, the dilemma has repeated itself over and again and has the quality of a mirage: As we get close, it seems to evaporate. Is regulatory tyranny an illusion? The answer seems to be yes. The evidence shows that successful regulations that define a *fair* trade in the commons do not reduce freedom, *they increase it.*

Consider sewage disposal. In the Middle Ages we soiled the commons by throwing feces out into the streets and into the water sources from which we also drew drinking water—until science showed us that this is how cholera spreads. There was a time not many years ago when a factory owner whose land abutted a river thought it was well within his rights to dump factory by-products into what seemed an endless flow. The factory owner would have called antipollution regulations infringements of his freedom. A flowing river cleans itself every twenty miles, the old saying went. Today we'd view those acts as equally stupid and unconscionable. So have the stricter regulations and laws reduced or increased our freedom? To the extent that the common waters are cleaner, are we each deprived or *enriched?* It's pretty clear we are each healthier, wealthier, and *freer* because of fair and equal regulation based on the knowledge from science. As Hume pointed out, freedom is the liberty to *choose*. Its sources are knowledge, science, democracy, and fair and equitable regulation—all of which work to maximize the liberty to choose.

In a similar way, we accept limitations on our individual freedoms to gain greater freedom in the forms of regulations that reduce smog, acid rain, ozone destruction, the use of DDT, backyard burning of garbage, noise pollution, certain carcinogens, and, more recently, exposure to secondhand smoke and injuries caused by not wearing seat belts and texting while driving. Generally, most people appreciate these laws and regulations for the freedom they provide from the tyranny of others' decisions and actions imposing upon them bad health, environmental destruction, devastating injury, higher insurance rates, capricious death, and other *takings* that reduce quality and quantity of life.

We have not yet seen fit to extend our freedom to regulating other pollutants that limit our choices and those of our children, such as carbon dioxide; genetically modified organisms that are released into the environment with limited knowledge of how they will impact biocomplexity and natural processes in the long run; and nanotechnology that is being broadly adopted by industry and entering the environment and the food chain mostly unregulated, again with little knowledge of the long-term effects.

How can we give up freedom to roam land without fences; freedom to dump sewage where we please; freedom to dump by-products into the commons and the air, to use ever more advanced technology to promote agricultural productivity and industrial value and to reap the "limitless bounty of the sea"? These are the acts of our ignorant past, but once we build knowledge of environmental effects, once we "know better," they become the acts of tyrants and bullies that deprive everyone of freedom and coerce everyone by removing choices, including those of the actor; they impose a tyranny of trash—of ignorance. Ignorance is not bliss. It's tyranny. They take private property; they take health, life, and clean water; they take clean lungs and fresh air; they take fish by depleting the oceans, money by raising the cost of insurance. The shifting—or, as economists say, the externalizing—of private costs and risks onto the commons takes from everyone, and in fact *reduces* wealth.

TYRANNY ON THE COMMONS

All of this ties back in to the ideas of conservative economics, particularly those of its father, the American economist Milton Friedman. In every economic transaction there is a willing buyer and a willing seller, and they agree on a price that benefits both. But there are spillover effects in many economic transactions—costs and/or benefits that are transferred to third parties. Friedman called these spillovers "neighborhood effects." Today, most economists call them "externalities."

At their most basic, externalities don't have to involve buying and selling. If you smoke in a restaurant instead of stepping outside it's easier for you, but it's worse for everybody else. That's an externality. If you throw your McDonald's bag out the car window it makes life easier for you, but it's worse for everybody else. That's an externality, too. If a pretty girl walks by in a low-cut top she's probably a little colder, but the added happiness of the boys in the neighborhood is a positive externality. For some reason, economists seem to love that example.

Now let's add in money. If your power company sells you coal-fired electricity at three cents per kilowatt-hour, that's a good deal. But if it does it by burning cheap, dirty coal and doesn't have a smoke scrubber, the soot that's dumped on the neighboring town is a negative externality. The people in that town are subsidizing your cheap power by paying the cost in terms of property damage, extra cleaning, poor health, aggravation. And the *next generation* is also subsidizing your price break if the utility isn't removing the carbon dioxide from its emissions. In terms of freedom, you are forcing a *tyranny on the commons* because the third party (in this case, everyone else) is deprived of having a choice about whether or not to partake in the transaction that gives you cheap power. It's not fair or equal freedom.

Now let's say someone in your town starts a company to produce a new kind of biofuel. This biofuel is so efficient that everybody wants it, and the company's business skyrockets. It needs to hire more workers and expand its plant. As a result, the town starts to boom economically and your home's value soars. That's a positive externality: Some of the wealth the company is generating has spilled over onto you, even if you don't buy the biofuel or work at the plant. Friedman would say this is not fair either and that you should pay some of this wealth back to the plant to balance the freedom in the transaction.

A science-based example of a positive externality is vaccinations. Vaccinations work based on the number of people in the population who are vaccinated. Once a certain threshold is reached, the disease can't spread effectively and is essentially eliminated. As long as enough people are vaccinated, others who choose not to be still get to enjoy that positive externality at no cost. These are what economists call freeloaders. Economists like the golden rule: Do unto others as you would have them do unto you. They don't like externalities because they represent inefficiencies in the marketplace that skew the real value of transactions and so are implicitly unfair. By reducing choice, they reduce freedom. Those who suffer by exposure to external costs do so involuntarily, while those who enjoy external benefits do so at no cost. Friedman thought that overcoming neighborhood effects "widely regarded as sufficiently important to justify government intervention" was one of the key roles of government.[7]

Government can do this in two ways: by imposing taxes and laws (regulations). Taxes used punitively—for example, sin taxes levied on such things as alcohol and tobacco—force the generator of an externality to *internalize* the cost of making life worse for everyone else. The alcohol tax, for example, induces those who drink alcohol to internalize some of the increased costs and risks of bad behavior, such as the need for more police to protect the public. Laws against littering coerce fast-food patrons to properly dispose of the trash from their meals, and if they are caught breaking the law, they incur a fine that acts like a more painful tax and covers the cost of their irresponsibility to everyone else.

Emitting excess carbon dioxide may make life easier for you, but it makes it worse for everybody else. Using Friedman's logic, now that we are widely aware, through science, of the costs that carbon emission shifts onto society and future generations, "to overcome neighborhood effects widely regarded as sufficiently important to justify government intervention," it's incumbent upon us to find ways to regulate the market that will make emitters internalize that cost in order to maximize freedom and market efficiency. We can do this by imposing carbon taxes, implementing cap and trade schemes, or employing other methods of tax or law that lie at the economic-environmental nexus to get them to internalize the cost of what they are taking from our freedom, thereby reducing their tyranny and increasing our choice/freedom. Because ignorance equals tyranny, this task is also the fundamental thrust of science.

Okay, we might say, that is well and good. We understand that cap and trade is a market-driven, conservative economic idea, as George W. Bush's own chief economic adviser, Douglas Holtz-Eakin, pointed out. We know that market-driven environmental laws and regulations reduce soiling of the commons and increase individual freedom. But how do we get at Hardin's biggest concern? How do we manage the environment in a limited world when everyone cares about economic growth and prosperity and wants just one more cow? Are we not rushing toward ruin? Can we have economic growth and a clean environment in a limited world?

IT'S THE ECONOMY *AND* THE ENVIRONMENT, STUPID

National Oceanic and Atmospheric Administration (NOAA) administrator Jane Lubchenco thinks we can. Lubchenco is one of America's leading voices on tackling climate change, but she challenges the idea of a limited world, pointing out that science has consistently expanded the economy beyond the zero-sum days of Thomas Hobbes. "By creating new knowledge, science changed economics away from a zero-sum game," she says.[8]

Evolutionary biologist Simon Levin thinks that the way to tackle the problem is to develop an economic model that values the commons as capital. "I see a real split even among my students," he says, "many of whom think economic growth and environmental protection are antagonistic elements and therefore a primary goal must be to reduce consumption and economic growth. I don't believe that's the case, but the interplay is an example of the sorts of problems we have to wrestle with."[9]

Levin got together with 1972 Nobel Prize–winning economist Ken Arrow and ecologist Paul Ehrlich to look at the question of whether we're consuming too much, and he says they found that the answer was "not uniformly—[in] some areas [we] are and [in] some [we] are not." How do you evaluate that? It requires coming up with an acceptable definition of general welfare. If you take a comprehensive wealth measure that includes the ability to eat, to preserve the environment, to maintain your health, and other similar measures, in some areas we're underinvesting and underconsuming, Levin argues.

LIMITLESS GROWTH?

In broad strokes, there is truth to Levin's approach. The Droid cell phone I hold in my hand uses far fewer resources than the mainframe computer I used to program in BASIC over a teleterminal in junior high school. It has far more power and, more important, it adds far more *value* to my life. Better quality and utility combined with lower resource use and cost can be a formula for economic growth and wealth generation in other sectors as well, particularly as nanotechnology becomes more sophisticated.

But look closer. Yes, the per-capita resource use is lower now, but the market is vastly larger. The global population is becoming unsustainable, and the road from ENIAC—the first electronic digital computer, completed in 1945—to my Droid is littered with glass, plastic, steel, gold, silver, platinum, and toxic metals including barium, cadmium, chromium, copper, lead, nickel, and zinc. My Droid may weigh only six ounces, but I'm responsible for a burden of about 180 pounds of discarded computing technology, on average.* And these amounts are now being dwarfed by the e-waste being generated in the rest of the world.[10]

The solution is to be smarter and more conservative, to find ways to grow the economy without depleting the corpus of natural capital, and thus inducing neighborhood effects. In a business sense, this means without spending one's assets to finance current operations—i.e., don't burn cash to heat your home. This is largely a matter of perspective. Imagine, for example, that we live in a galactic economy and you are the CEO of the corporate planet Earth, Inc. Your job is to maximize the shareholder value of Earth. To keep a healthy balance sheet, you will want to account for, monitor, and restore assets both known and unknown in Earth's minerals and biodiversity, ranging from the sources of the new miracle drugs to those of the life-expanding high-tech resources of tomorrow. What makes sense for competition on this scale is different from what makes sense within a smaller frame of reference. If ignorance is tyranny because it removes choices, we need to develop a way to quantify the cost of the unknown wealth or choices we may be leaving on the table—in other words, our opportunity cost.

This opportunity cost, and arriving at a present value for the unknown economic bounty our current resource stock may contain, is the most fundamental value equation we need to solve. If we burn our capital, we won't be able to compete. Today's corporations currently mining these resources may be motivated to become their best protectors if we can find a way to define and value that future unknown economic opportunity. This shifts the business activity from *mining* resources to *farming* them. It seeks to maximize productivity and biodiversity to provide maximum biological creativity and the attendant economic potential for finding the next big thing. This seeks, in other words, to maximize freedom.

---

* The EPA estimates that in 2007, discarded electronics totaled about 2.25 million US tons. By 2010 it was about 3.3 million. Placing that on a very rough growth curve going back to 1980 gives a number of about 27 million US tons over that time. At two thousand pounds in a ton and a population of about 300 million people, each American is responsible for roughly 180 pounds of electronic waste. In China, e-waste in 2010 totaled about 2.5 million tons—fast approaching US levels—and that country has more than four times the population.

Ecosystem Services and Natural Public Capital

For an economic model to include environmental sustainability, value must be assigned to common commodities, be they biodiversity, a stable climate, clean water, rain forests, parks, mineral resources, topsoil, what have you. Because the economy is a human system that trades in human values, the first principle of valuing the commons is to state it in terms of human values. Some environmentalists argue that this approach is wrong, that the values transcend human purposes. That may be true, but for purposes of trading in the human economic system, it has no meaning. Any value we put on biodiversity has to relate to the way people value it, which comes down to a utility question. This is consistent if the objective of utility is to provide greater freedom. As a means for beginning to think about this, environmental scientists are now talking about ecosystem services—the services that ecosystems provide to humans.

Included in ecosystem services are fuels, foods, and other direct benefits like natural resources, pharmaceutical products, the filtering of surface water by wetlands, and the buffering of coastal cities from storm surges by coastal wetlands. Ecosystem services also encompass indirect benefits in terms of ethical values we choose to assign to the environment, such as the benefits of being able to visit the wilderness or to experience the transcendent beauty of nature. These services are the starting point. But to think of them as simply *services* is only part of the financial calculation. The services are provided by working capital—the *natural capital* owned by the business Earth, Inc. And since the economic system is a human one, we need to relate it to humans as "natural public capital." We are the shareholders in Earth, Inc. We have inherited it; we can invest it, even borrow from it, but we should never spend it down because it supports our ongoing operations and delimits our future growth.

So how do we arrive at values for natural public capital? How do we deal with uncertainty, inequity in distribution of wealth, intergenerational equity? And how do we put a value on things that don't have a market price, like our love for our children?

A FULL ACCOUNTING

In 1997, ecological economist Robert Costanza set out to answer those questions. He and a team of fellow researchers embarked on a quest for metrics that would put a value on the ecosystem services that nature provides to our economy. [11]

Published, appropriately, in the journal *Nature,* the paper had a powerful effect because it put a price tag on the commons, showing that the annual value of the world's ecosystem services to the economy was at least $33 trillion, or nearly twice as much as the $18 trillion that made up the world's combined gross national product—the sum total annual value of each nation's traded goods and services—and possibly as much as $54 trillion, depending on certain assumptions. For comparison, the US gross domestic product in 1996 was about $6.9 trillion. [12] The economics behind this valuation were hotly debated, and many economists protested because nobody had asked them, but Costanza and his group by and large got the numbers right, and the paper held up to the scrutiny it received.

Duke environmental scientist Stuart Pimm, who wrote the *News and Views* column for the issue, describes the moment as a watershed. "Within a decade you could go to meetings at the UN and we had a world that's saying, 'We need to raise $20 billion,' and that would be the price of stopping global deforestation, which is putting 1.5 billion tons of carbon into the atmosphere each year. The cheapest way to slow climate change is to stop tropical deforestation." [13] The argument now had *currency* in a way that made it discussable in public policy and economic terms.

Most policy makers outside the United States now understand that you can't treat the global commons as if they have no value. They are exceedingly valuable; in fact, they are the source of our biological existence and the foundation of our economy, and it's worth our while to protect them. "Globally, everybody except the United States now gets this," says Pimm. The political paralysis in the United States on both the science and the economics has become the largest single impediment that is reducing freedom and thus imposing a tyranny.

"When you're dealing with equity issues, it's got to involve conversations with social scientists, humanists, economists, and the public," says Levin. And it's complicated because it's not just about today's people, but also about tomorrow's: How do you value the right to spend down natural public capital that depletes the wealth and freedom of future lives—reducing their liberty—either by polluting it or by using it up? What is ethical, and what is "spending my children's inheritance," as the popular bumper sticker proclaims?

MIDNIGHT IN THE GARDEN OF GOOD AND EVIL

Clearly, transforming the frame from Ayn Rand's "rational self-interest" to the broader perspective of "rational self-interest in a shared and limited pasture" held by a farmer who surveys the field, does a little science, and realizes that if everyone adds one cow all will fall to ruin relies on political speech, if not altruism. Because of this, Stuart Pimm sees religious engagement as critical. He's focused much of his career on species extinction and how to prevent it and says that speaking with American Christian groups in particular is important. "You've got very large groups of Eastern Orthodox, Anglican Communion, Methodists, Unitarians, all of whom come down strongly on this notion [that] there needs to be creation care versus the group on the Cornwall Declaration saying, 'All of this is deeply evil and we need unfettered access to natural resources,'" says Pimm.

"The Cornwall Declaration on Environmental Stewardship" and its associated "An Evangelical Declaration on Global Warming" are public letters expressing the opinion of the Cornwall Alliance, a coalition of evangelical clergy, theologians, and policy experts, including Charles Colson, James Dobson, Jacob Neusner, and R. C. Sproul. With striking similarity to the indictment of Galileo, the group's position on climate change is:

> We deny . . . that Earth's climate system is vulnerable to dangerous alteration because of minuscule changes in atmospheric chemistry. . . . There is no convincing scientific evidence that human contribution to greenhouse gases is causing dangerous global warming. . . . We deny that carbon dioxide—essential to all plant growth—is a pollutant.

Reducing greenhouse gases cannot achieve significant reductions in future global temperatures, and the costs of the policies would far exceed the benefits. We deny that such policies, which amount to a regressive tax, comply with the Biblical requirement of protecting the poor from harm and oppression.[14]

The Cornwall Declaration's position on sustainability is:

Many are concerned that liberty, science, and technology are more a threat to the environment than a blessing to humanity and nature. . . . While some environmental concerns are well founded and serious, others are without foundation or greatly exaggerated. . . . Some unfounded or undue concerns include fears of destructive manmade global warming, overpopulation, and rampant species loss. . . . Public policies to combat exaggerated risks can dangerously delay or reverse the economic development necessary to improve not only human life but also human stewardship of the environment. . . . We aspire to a world in which the relationships between stewardship and private property are fully appreciated, allowing people's natural incentive to care for their own property to reduce the need for collective ownership and control of resources and enterprises, and in which collective action, when deemed necessary, takes place at the most local level possible.[15]

This split view of sustainability as either a church stewardship value or associated with socialism or communism (and thus totalitarianism) is uniquely American. In Europe, for example, few at either end of the political spectrum take any particular issue with the facts of climate change, and European countries have in many ways taken the lead on tackling the problem. After Tony Blair was elected British prime minister in 1997, the Labour Party's first conference was on climate change. Then chief scientific adviser Lord Robert May had prepared a report that Blair presented to all attendees, and it helped set a national course. "The wheels of government grind slowly," says May, but Britain's Climate Change Act, adopted a few years later, provided for specific carbon targets and established "a Climate Change Committee to recommend the targets and monitor progress towards them. The UK is the only country to have such formal apparatus in place." [16] In a more recent example,

Norway committed $1 billion to help combat deforestation, which some, like Pimm, say is the cheapest way to combat climate change. The amount was roughly seventy times more on a per capita basis than the United States promised in the Copenhagen Accord. The list goes on. Pimm sees the disparity between the United States and other nations as symptomatic.

'It speaks to a very deep division, and one only has to look at the political influence of people like Sarah Palin and James Inhofe," he says. "These are not politicians arguing on the edges, these represent very deep emotional issues where a good half of the US population isn't marching in step either with the science or with the religious, spiritual, and ethical motivations in other countries. So you can't just say we're doing a damn lousy job of explaining science in the media. It's way beyond that. Clearly the US has had a very long and powerful urge to be isolated from the rest of the world. If it was only the Atlantic seaboard that voted in presidential elections, we'd have a very different country. It's not surprising [that] you have Inhofe in the middle of the country. Fewer than 30 percent of Americans even have passports, and roughly half of those are only for travel to Canada and Mexico.[17] The size of the country has people in the middle very isolated from the rest of the world. They sort of admire the fact that Sarah Palin thinks that Africa is a country."[18]

The differences between American perspectives and those prevalent in other countries are of course not limited to climate change. University of Pennsylvania bioethicist Jonathan Moreno points out how concern about genetic pollution caused by genetically modified food crops draws large numbers of students to campus demonstrations in Europe, but in the United States, "you get maybe eight people. If it's about reproductive choice, the situation is reversed."[19]

Pimm says his greatest fear is that the "extraordinary isolation" of Americans is in the long run going to be "hugely damaging" not only to the world, but also to our own well-being. "This idea that we can go and tell the rest of the world to go f— themselves with impunity, that's a very stupid thing. We need resources, they cost an extravagant amount of money, and that's getting worse as China comes online competing with us—not just as a producer, but worse, as a consumer, particularly an energy consumer. I live in Florida, where the junior US senator, Marco Rubio, denies global warming, but many of his constituents are losing their homeowners' insurance[20] because insurance companies are worried about costly storms from global warming."[21]

The Insurance Cost of Climate Change

Richard Feely, senior scientist in charge of monitoring ocean chemistry for NOAA, estimates that the United States and other countries could stop climate change for an investment roughly equal to 2 percent of their gross domestic product,[22] which in 2011 was about $15 trillion.[23]

If that's correct, an annual US investment of roughly $300 billion, less than the country spends on imported oil in one year, could get the US economy off carbon. Instead of going to despotic foreign countries, much of that $300 billion would be spent on domestic businesses, stimulating the economy. The economics of this approach should attract the attention of anyone who has run a business. If you can improve your cash position and eliminate a number of long-term ongoing expenses while transitioning into the future with new technology and strengthening your balance sheet, all for 2 percent, you jump at the chance.

Insurance companies, naturally, are focused on these emerging economics. They are already transferring increased risk from climate-related disasters into policy premiums or dropping coverage altogether, particularly in coastal areas,[24] but they are seeing sharp upward trends in exposure across the board.

Ernst Rauch, head of the Corporate Climate Center at Munich Reinsurance America, a major reinsurer of US and global insurance companies, reported in the 2010 *Natural Catastrophe Year in Review* webinar[25] that, globally, disasters had risen from fewer than 400 events in 1980 to 950 in 2010, reflecting an increase, on average, of a factor of

more than two and a half. Geophysical events, such as earthquakes, tsunamis, and volcanic eruptions, were relatively unchanged. Most of the
increase was in atmospheric perils related to climate change, he reported,
including the flooding of up to one-quarter of Pakistan and the terrible
heat wave in Russia, which killed at least 56,000 people.

Robert Hartwig, president of the Insurance Information Institute,
reported that eight of the twelve costliest US natural disasters have
occurred since 2004, and that eight of the twelve top disasters affected
Florida. When you look at US losses from catastrophic events related to
the climate—storms, floods, droughts, and wildfires—in 2010 and compare them to 1980, their number has quadrupled, but remained
unchanged in other categories of loss. Carl Hedde, head of risk accumulation for Munich Re, said, "The majority of the events are storm related."
In fact, in the United States, 2010 had roughly double the events compared with the first half of the decade, a trend that appears to be in a
geometric growth curve relative to the Keeling curve.

At the 2010 midyear natural catastrophe webinar, Peter Höppe, head of
Munich Re's geo risks research and Corporate Climate Center, had pointed
out that the March, April, and May of that year were the warmest on
record, and that the North Atlantic was 2°C warmer than the long-term
average, suggesting increased hurricane activity in the coming years.[26]

The growing population is also a driver. Global population is currently
roughly seven billion, with roughly eighty million more people currently
being added every year. By 2050, it is expected to top nine billion. As
population increases and science makes society both more intercon-
nected and more high tech, the costs associated with each disaster
increase. A hurricane in Florida causes far more economic damage now
than it did in the less populated, less wired, and less wealthy Florida of
1980. Multiply the increasing costs by the increasing population by the
increasing number of catastrophic storms, and you can see why insurance premiums are skyrocketing. Over the decade of 2001 to 2010, US
losses from climate-related catastrophes were greater than the $300 bil-
lion Feely's numbers suggest would be the annual cost of conversion to a

carbon-free economy. Considering the growth curve in losses, one can predict that within the next decade economics will transform the climate change debate.

While insurance-industry modeling is limited in scope, it does provide some useful ways for thinking about the interplay between wealth and the environment. What makes it interesting is that it turns the tragedy of the commons on its head, functioning in reverse. Like science, wealth affords freedom by expanding the power of choice. Insurance is a commons, a shared pasture, if you will. It provides financial security by spreading individual risk to the group. In exchange, the individual gives up some small portion of freedom, represented by the monetary cost of the premium—a tax—and by policy terms that limit behavior—regulations. The small limitation on freedom is, generally, viewed as a worthwhile tradeoff for the greater freedom from financial catastrophe that insurance coverage provides. Insurance is essentially economic regulation that expands freedom by protecting against the brutality or tyranny of catastrophic events in the same way that property titling and sewage regulation expand freedom by protecting against might makes right and ignorance. John Hubble, the famed astronomer Edwin's father, a staunch Baptist, was an insurance underwriter. "The best definition we have found for civilization," he wrote in 1900, "is that a civilized man does what is best for all, while the savage does what is best for himself. Civilization is but a huge mutual insurance company against human selfishness."[27]

Insuring the Commons

That is also a useful way to think about protecting global ecosystem services and natural public capital. Once we establish their economic value, we can create an insurance vehicle that protects against the exposure to catastrophe. Costanza offers that the best way to handle it, at least for accidents such as the Deepwater Horizon oil spill, is by requiring an "assurance bond"—a sum of money large enough to repair the involved ecosystems if an accident occurs.[28] The face value of each bond would be based on the total value of the ecosystems at risk and be set by an independent agency or government-chartered body. Companies posting a $50 billion bond to drill in the Gulf of Mexico would be made aware that they were engaging in a very risky business and that investments like acoustic blowout preventers, at a cost of $500,000, are good deals and thus justifiable to shareholders.

This approach is not new. We regularly ask private parties to purchase insurance to cover the risks they pass on to the public. Purchasing automobile insurance is now mandatory in the United States. Building contractors must carry performance bonds in order to build sizable projects. In the film business, performance bonds protect investors from the risk of the director or producer failing to bring a high-quality picture in on budget.

These kinds of vehicles cause private interests to internalize the risks they are undertaking and recognize them ahead of the fact, instead of dealing with them solely through law and litigation afterward—an idea that has been developing since the early 1980s as a market-based, more efficient alternative to the "command and control" method of law and litigation.[29]

The latter system is the one we currently use for mitigating environmental damage. It's inefficient because it doesn't motivate prevention, it requires costly government monitoring of behavior, it shifts the burden of proof to the public, and it seeks reparations only after the damage is done. The *Exxon Valdez* oil spill is a case in point: Litigation took decades to make it through the courts and was ultimately resolved for a fraction of the actual loss. A year after the Deepwater Horizon spill, during which BP CEO Tony Hayward said, "We will make this right," the marshes were dead, dolphins were spontaneously aborting by the score, and the Gulf fishing industry was deeply threatened. Clearly, this route to accountability is ineffective, unjust, and inefficient. Whether the goal is to preserve the environment or maximize economic efficiency, or simply to solve problems and avoid costly litigation, shifting the economics of risk to the front end by being proactive instead of reactive is better business practice because managers are aware of them in advance, can measure them, and can control for them in their business models and account for the risk in their product pricing, thereby protecting the company against otherwise uncontrollable market punishment. Peer-reviewed, published scientific research by Costanza, Andrew Balmford, and others has now shown a return on investment of at least $100 for every $1 invested in preserving remaining wild nature[30]—a figure that will make any Wall Street trader's eyes light up if we can find a way to monetize it.

ON LEADERSHIP

To achieve that, the federal government, whose purpose is to administer the commons and maximize freedom, will need to be more involved in crafting policy, fair and equitable regulation based on science, and international treaties that level the playing field. This needs to be done in cooperation with the insurance industry, environmental groups, scientists, and the private sector, so that forward-looking, innovative companies, nations, and individuals—those focused on the future—aren't at a competitive disadvantage to those who tyrannize by taking natural public capital for private gain, as has been done in the more ignorant past. It's also going to take the involvement of Wall Street and financial agencies that work to develop accountability and well-informed decision making in public and private sector financial reporting, such as the Governmental Accounting Standards Board, the Federal Accounting Standards Advisory Board, and the Financial Accounting Standards Board. Currently, no laws require the government or private industry to use full cost accounting methods that include the value of ecosystem services, but many local governments have adopted a limited version of the methods to account for non-cash costs associated with solid waste management.[31] Such programs offer a conceptual base on which to build.

We are living in a time of supranational corporations and a global economy based on feudalism. These corporations can no longer be considered constituents of any one nation. Yet beyond the stock market, there is no global economic or regulatory system to govern their actions, and so it is a free-for-all that ultimately results in the freedoms of all individuals and all nations being reduced by these corporations' takings from the natural public capital.

This lack of a governing system is not the corporations' fault, but there is no reason to assume that this situation will change on its own; these corporations have been freed by the global market that has developed because of technology and the Internet. But without concordant global regulation they have become locked into a pattern of quarterly competition for returns that do not account for the burning of public capital and so, like the herdsmen of Hardin's commons, they are, by the Wild West structure of the supranational global marketplace, forced to consider only "what is the utility to *me* of adding one more animal to my herd?"

The way to free them from that tyranny of the unregulated marketplace is to adjust the market, to level the playing field with fair and equitable regulation, as Friedman suggested, and to refuse to let them burn public capital to artificially inflate their financial performance. That is a cheap and dirty way to see quarterly gain, but it is a great externality and thus a tyranny that is antithetical to free markets and to democracy, and so it cannot be tolerated. The Wild West was won by the rule of law. Corporations must be asked, on a fair and equitable basis, to internalize their takings. Those who cannot probably should not be in business anyway, because they are either too unresponsive to markets, are abetting bloody physical tyrannies for their gains, or have unrealistic cost structures and are thus bound to fail eventually.

And this is where leadership comes in. The United States is by far the world's largest economy. For all the talk of Exxon-Mobil and the Koch brothers spending $1 billion to scuttle cap and trade, ethical corporations could collectively swamp that to level the playing field and get it passed. For all the talk of China, the US economy is still three times its size.[32] It is the only single economy that wields enough power to do this in a way that the rest of the world will fall in line behind. If it is fair, corporations will thank the United States for exercising leadership and providing that fair and equitable regulation, that sheriff, and freeing them from the Wild West tyranny of the tragedy of the commons.

We don't lack ability; we lack leadership. The assumption that other economies will not follow the United States' lead is incorrect. Much of the world is well aware of the issues and places at least as much value on the environmental commons as the United States does, and in many cases more, but is even more powerless to do anything about it because of the momentum of the US economy. By exercising leadership, recognizing the shared need of the bounded pasture, and saying, "This is what we all have to do and this is where we are going to go as a world now," America would be in the unique position of being able to maintain and strengthen its moral leadership and prestige as a superpower and the leading economy. It would be in a position to rein in supranational corporate feudalism in a fair way; and to dictate the next era of foreign policy—one that rewards and punishes based on national economies' willingness to fall in line with the economic and moral policies that maximize sustained freedom on a global basis—the values that the United States was founded upon. This moral leadership will improve the US economy as well, by forcing the global players to adopt regulatory structures consistent with US environmental policy.

This is this generation's calling. But are we capable of leading? Do we have an understanding of science adequate to even see the issues clearly, or to base our political arguments on knowledge and not just warring opinions? Are we able to look up from the grist wheel of day-to-day performance long enough to even consider the vast economic and political opportunity that is within our grasp? That is the question this generation will ultimately be judged upon, and it remains unanswered.

PART V

THE SOLUTION

CHAPTER 12

TALKING ABOUT SCIENCE
IN AMERICA

The only way to have real success in science, the field I'm familiar with, is to describe the evidence very carefully without regard to the way you feel it should be. If you have a theory, you must try to explain what's good and what's bad about it equally. In science, you learn a kind of standard integrity and honesty.

In other fields, such as business, it's different. For example, almost every advertisement you see is obviously designed, in some way or another, to fool the customer: The print that they don't want you to read is small; the statements are written in an obscure way. It is obvious to anybody that the product is not being presented in a scientific and balanced way. Therefore, in the selling business, there's a lack of integrity.

—RICHARD FEYNMAN, 1988[1]

THE SHADOW AAAS:
THE ARMCHAIR ARMY OF
ANTISCIENTISTS

Freedom is the main driver of individualism in its relationship to democracy and to science, and that is the core message scientists need to stay focused on. That makes it ironic that many of the most pitched battles over science are being fought by people who proclaim the strength of their values of individualism and freedom quite loudly.

Consider the story of Michael Webber. One day not long ago, Webber stood looking out his office window over Austin, Texas. His business attire of jeans, cowboy boots, and a button-down shirt was not unusual for Texas, and neither was what he had just done. Webber and a group of other parents had just started a homeschooling cooperative, opting out of the public schools or, in Webber's group's case, augmenting them, with material that he believes is vital for his children to know. This was Texas, after all, where religion holds a special place in the public dialogue.

But the cooperative wasn't formed to teach kids about religion—it was formed to teach them about sex. "We're turning that steady march of progress on its head by instead explicitly choosing the widespread indoctrination of stupidity," Webber said in a 2009 opinion piece in Austin's *Statesman*.[2] The cooperative was formed because Texas lawmakers had passed laws cutting sex ed from two six-month courses to a single unit of "abstinence only" education, Webber says.[3]

According to a 2009 report, 94 percent of Texas schools, which at the time were educating more than 3.7 million students, were giving no sex ed whatsoever beyond "abstinence only," a curriculum that includes emphasizing that birth control doesn't work. "You think I want my daughters learning *that?*" Webber asks. Another 2.3 percent of schools had no sex education at all, and 3.6 percent had "abstinence-plus."[4]

Instead of providing fact-based information about birth control and pregnancy prevention, the programs heavily emphasize the risks of sexually transmitted diseases leading to cervical cancer, radical hysterectomy, and death—which in reality are very, very small—together with Christian morality. The report highlighted one Texas public school district's handout entitled "Things to Look for in a Mate."

How they relate to God

A. Is Jesus their first love?

B. Trying to impress people or serve God?

Another public school district had a series of handouts that referenced the Bible as their basis to promote abstinence from sexual activity.

> **Question:** "What does the Bible say about sex before marriage/pre-marital sex?"
>
> **Answer:** Along with all other kinds of sexual immorality, sex before marriage/premarital sex is repeatedly condemned in Scripture (Acts 15:20; Romans 1:29; 1 Corinthians 5:1; 6:13,18; 7:2; 10:8; 2 Corinthians 12:21; Galatians 5:19; Ephesians 5:3; Colossians 3:5; 1 Thessalonians 4:3; Jude 7).

The results? Teen pregnancy in Texas *went up*. It was higher than it had been before the "abstinence only" movement,[5] and, the 2006 Texas teen birth rate, at 63.1 live births per 1,000 teenagers, was more than 50 percent higher than the national average of 41.9 live births per 1,000 teens. The National Center for Health Statistics data for 2008, the most recent available, are nearly identical to these numbers.[6] Even more troubling than this dismal failure, Webber says, was that *repeat* teen pregnancy also went up. It turns out that Texas kids thought that "if birth control doesn't work, why use it?"[7]

One can't argue with their logic. The religious activist proponents of the abstinence-only approach convinced state education and school officials that, contrary to scientific findings, further information about birth control and sex should be taught by parents, according to their morality, in the privacy of their own homes. The thought of teachers talking with their kids about sex made them queasy. But the problem, Webber says, is that those same parents are often too bashful to bring the topic up with their kids at all, and if they do, they often use euphemisms. One girl Webber knew of didn't understand how she had gotten pregnant because she and the boy hadn't actually *slept* together after having sex. It's also extremely tough for teenagers to get contraceptives in Texas. "If you are a kid, even in college, if it's state-funded you have to have parental consent," said Susan Tortolero, director of the Prevention Research Center at the University of Texas in Houston. [8]

"Abstinence works," said Texas governor Rick Perry during an October 15, 2010, televised interview with *Texas Tribune* reporter Evan Smith. The audience laughed and Smith pointed out the state's abysmal teen pregnancy rate. "It works," insisted Perry. "Maybe it's the way it's being taught, or the way it's being applied out there, but the fact of the matter is it is the best form of—uh—to teach our children." Smith asked for a statistic to suggest it works, and Perry replied, "I'm just going to tell you from my own personal life, abstinence works."[9]

Webber and his wife chose to protect their daughters by forming the mother–daughter sex-ed cooperative. They brought in experts and armed the girls with what they needed to know. "But not everybody can afford to do that," he says. "In fact, I'll bet even of those who can, most won't. What I really don't understand is why these people are so insistent on instilling stupidity in the next generation. If it's not sex ed, it's climate change or evolution or vaccinations."[10]

The larger question posed by the reversal of parents homeschooling to teach science-based sex ed instead of the religion-based abstinence mainstreamed in the public schools is, What is the relative value of science versus ideology in policy making and education? That the question must be asked at all is a testament to the success of a movement that was largely fought by an armchair army of antiscientists—what we could call "the Shadow AAAS"—that is loosely organized through evangelical churches and industry-funded nonprofits. These activists cherry-pick science and misconstrue studies to create rhetorical arguments that sound plausible to lawmakers who either don't understand that what is being presented to them is not science or, more likely, don't care because their natural desire is to grease the squeakiest wheel.

The same methods are being used by opponents of evolution, climate change, vaccines, birth control, stem cell research, and other issues. Across the nation, thousands of laypeople are delving into geology, biology, immunology, paleontology, statistics, climatology, meteorology, geophysics, and oceanography, with the support of churches and industry-funded nonprofits who, like the Cornwall Alliance, preach a new gospel of biblical fundamentalism mixed with a heavy dose of Ayn Rand, free market economics, science denial, and no new taxes. This Shadow AAAS is aided and abetted by trained and working scientists and professors like Willie Soon, David Legates, and Michael Behe who are willing to opine and make it look like science, supplying a steady stream of pseudoscience that can be used to sway the public debate.

BAD REASON

These partisans generally make two arguments that sound plausible to average lawmakers or school board members because they themselves use rhetorical arguments to navigate their daily lives. The arguments are the same whether the subject is climate change, evolution, vaccination, or sex education.

The first argument is, *Lacking certainty, we should do nothing.* On the face of it, this sounds prudent. Why take a risk unless you are certain of the need? We cannot be absolutely certain that climate change is occurring; therefore we should do nothing. We cannot be absolutely certain that the theory of evolution is true; therefore we should not teach it. But science has never proposed absolute certainty, only the certainty offered by the preponderance of the evidence. And that is what has made it so uniquely powerful.

The second argument is, *Since the conclusion is not certain, we should get a balanced perspective from both sides.* This also sounds reasonable. If you can't say evolution or climate change is an absolute fact, then it seems reasonable to hear from people whose views differ, to weigh the pros and cons fairly and form a balanced opinion, like a judge would.

This is, once again, applying the rules of *rhetorical* argument, or opinion, to a *knowledge* issue. Imagine that you have been to one hundred doctors. Ninety-seven of them have told you that you have cancer and that it is treatable if they operate now. But that scares you and it's expensive and you'll have to take off from work. Three doctors say there's some doubt. One of them is a homeopath and says he can cure it with water; another says you need more faith in the healing power of Jesus Christ; and the third is a hyperskeptic who says it might be an aberration in the MRI machine, it might be a cyst, it might be benign, it might be malignant but so slow growing that it's not worth the dangers of surgery, we just don't know, and so on. Do you listen to the ninety-seven and get treatment or the three and wait? You're torn, so you pick one doctor from the ninety-seven and one from the three and, like a judge in robes, you say, "Convince me." But you are already the fool.

This is, of course, the current situation with climate change. "For what a man had rather were true he more readily believes," Bacon argued in *Novum Organum*.

> Therefore he rejects difficult things from impatience of research; sober things, because they narrow hope; the deeper things of nature, from superstition; the light of experience, from arrogance and pride, lest his mind should seem to be occupied with things mean and transitory; things not commonly believed, out of deference to the opinion of the vulgar. Numberless in short are the ways, and sometimes imperceptible, in which the affections color and infect the understanding.[11]

SCIENCE IS ABSOLUTE ABOUT FALSITY, BUT NOT ABOUT TRUTH

These two arguments advanced by the Shadow AAAS are driven by opinion and faith, but not knowledge. They capitalize on the fact that legitimate scientific conclusions, because they are empirical—that is to say, inductive and based on observation—will *rarely if ever make absolute statements* without allowing for the possibility of error. Lack of absolute certainty is not a weakness in science, it's the only way you know it *is* science. As the great science philosopher Karl Popper put it, only if a theory allows for the possibility of disproof can it be said to be science. This is because it insists on sticking to only what can be determined from observations. Since we cannot observe the whole universe at once, we can make only provisional statements.

This quality that Popper identified is referred to as a theory's falsifiability. While science can rarely absolutely say that something is true, it *can* absolutely say that something is false. It is absolutely false that the world is flat; we can prove it by going around it. All of our knowledge derived using science stands exposed to the same test of falsifiability: A simple experiment could prove it false. That none has so far gives us pretty high confidence that a given proposition, like the theory of evolution, is probably true. But it is not an article of faith—just the opposite. It's an article of what is left after doubt and scrutiny.

This is how we "bottom" the argument, as Locke suggested, in natural law and physical reality. Because we lack absolute certainty, we are required to assume responsibility for ourselves and for our decisions, which is sometimes uncomfortable, but is what democracy is all about.

Espousing the antiscience argument, that the lack of absolute certainty that something is true is grounds for inaction or doubt, or that we should hear out all perspectives equally, is evidence of either a lack of understanding of reason or motivation by an unreasonable agenda. Such people should not be entrusted with making decisions that will have serious impacts upon others.

The Roots of Partisanship

The adoption by schools of abstinence-only sex ed despite proof of its dismal results highlights a central question about American values: Which is more important in education—adherence to our ideological perspectives as parents or the outcomes it achieves for our children? The same question hangs over the ominous consequences of our decisions about creationism and climate change. Is it morally defensible to insist on an ideologically satisfying position at the cost that doing so will exact upon our children's future?

These questions are critically important. But this moral conflict can only occur if one accepts the possibility of doubt. Evidence suggests that this is a difficult thing to do if one has become wrapped up in the team-sports mentality of a politicized ideology. Partisanship thus works against enacting public policy based on sound science. How can we get past it?

First, let's look at what we know about why it's so powerful. There are several factors we're beginning to understand, and the first we'll call the rhetorical thinking hypothesis. When presented with evidence that confirms our beliefs and conclusions, we tend to accept it uncritically. When presented with evidence that contradicts those same conclusions, however, we subject it to withering scrutiny, ignore it, argue with it, or try to intimidate its proponents, much like the opposing counsel at a trial does. Rhetorical thinking is fundamentally nonscientific. It's not about finding truth or creating knowledge; it's about winning an argument.

The foundation work in this field is a 1979 paper by Charles Lord, Lee Ross, and Mark Lepper. The Stanford University psychologists recruited a group of subjects, half of whom said they supported the death penalty and half who opposed it. They showed each participant one research study that supported and another that contradicted the idea that the death penalty deters violent crime.[12]

Each participant identified extensive methodological problems with the evidence that contradicted his or her preexisting opinion, but didn't critically examine the evidence that supported it.

Eugenie Scott, head of the National Center for Science Education, says she encounters the same thinking when discussing evolution, as does Michael Mann when discussing climate change. "But people are not black or white in their views," says Scott. "In reality, they distribute along a continuum," with the vast majority of people falling in the center somewhere.[13] It's only when they are forced to choose sides that the continuum collapses into partisanship.

THE PARTISAN BRAIN

The second factor has to do with the way our brains process science information. Neuroscientist Sam Harris and colleagues published a neuropsychology study that showed that people use the same brain region in the ventromedial prefrontal cortex for belief as they do for ordinary facts,[14] seemingly confirming the frame conflict Matt Nisbet talks about.

Danish cognitive neuroscientist Uffe Schjødt and his colleagues took that a step further by exploring how the brain responds to authority. Schjødt played recorded prayers to study participants and compared the responses of participants who were charismatic Christians—who believe in speaking in tongues and healing by prayer—to those of non-believers. Both groups were asked to listen to three different groups of recordings:

• prayers read by a non-Christian

• prayers read by an ordinary Christian

• prayers read by a Christian known for his healing powers

But in actuality, all the prayers were read by ordinary Christians.[15]

The researchers used functional magnetic resonance imaging (fMRI) to scan participants' brains as they listened to the readings, and the scans showed that when the Christian subjects were listening to recordings they thought were made by healers, who have special religious authority, they turned off parts of their medial and dorsolateral prefrontal cortices, which play key roles in critical thinking and skepticism. Nonbelievers, in contrast, did not shut down this brain system.

The authors said this may explain why certain individuals who are perceived to have authority can exert influence over others. They suggested that the effect could extend to other interpersonal relationships, such as parents and children, doctors and patients, teachers and students, producers and consumers, and leaders and followers. Perhaps it also assists con men in business suits in fleecing their marks.

Were the Christian subjects idiots or gullible? No. What the fMRIs showed is that the *preconceived authority* of the speaker, regardless of his or her actual expertise, starts a brain process—a rhetorical frame—that makes us less critical of the speaker's commentary. Thus, partisanship.

The brain uses it as a strategy for negotiating the environment and living in social groups.

BAD THINGS IN A JUST WORLD

Beyond mistaken reasoning, rhetorical thinking, and a predisposition not to question authority, Americans as a whole have a high level of what social psychologists call the just world belief,[16] which helps explain why Americans have more trouble accepting scientific evidence of environmental problems than people in most other countries do.

People tend to believe that the world is inherently just: The wicked are eventually punished, the good are rewarded, and problems are corrected. In other words, they believe that people get what they deserve.[17] The relative strength of the belief correlates with differences between those who believe that self-discipline, hard work, and personal responsibility lead to success and those who believe that people are more at the mercy of luck, birth, or other factors beyond their control. The former view is a treasured part of the American ethos, and Americans as a whole have a much stronger belief in a just world than, say, Europeans, who tend to be less idealistic, more cynical, and more likely to believe that good or bad luck rather than individual merit or lack thereof plays a significant role in a person's circumstances.[18]

Climate change and other environmental problems seem to violate the just world belief. The idea that despite your best efforts your fate is influenced by luck or the collective actions of others is antithetical to the classic American story that we have self-determination and that with hard work and responsibility *anyone* can grow up to be president.

Research shows that this conflict makes it more difficult for Americans to accurately assess personal responsibility. For example, the tendency to blame the victim, which is unusually high in Americans,[19] is an effort, psychologists say, to maintain the just world belief that people get what they deserve. A woman who was raped in her apartment by a stranger who sneaked in while she was taking out the trash described how friends and colleagues suggested that she was partly to blame because of her "'negative attitude' that might have 'attracted' more 'negativity,'" or because she had "'bad karma' from a 'previous life.'" A close friend told her she had "asked for it" by choosing to live in that particular neighborhood.[20] A crime victim must have been in the wrong place; a sick person must have done something to deserve it. If we believe we are responsible for our circumstances, this prejudice makes sense. Of course, degrees of just world belief span the ideological spectrum even in America, and correlate to some degree with political ideology. Strong just world believers tend to be more economically and politically conservative, to see the status quo as desirable, to believe in an active God, to be less cynical, to be less socially and politically active, and, interestingly, to more often adhere to the Protestant ethic.[21] This correlation runs headlong into science when science fails to affirm the ethos of freedom in favor of data suggesting that we have limited resources, that behavior has collective effects, that conditions such as drug addiction may limit control over decision making—and that preaching the self-control of abstinence to hormone-saturated teens may make *parents* feel better, but may not be the most effective way to reduce teen pregnancy.

University of California at Berkeley psychologists Robb Willer and Matthew Feinberg published a study showing that the just world belief influences how people respond specifically to the threat of global warming.[22] The presentation of "doomsday" scenarios, as Sarah Palin characterized them—the demise of the polar bear, the forced emigration of tens of millions of people, flooded coastal cities, massive wildfires, devastating hurricanes, tornadoes and droughts, death and destruction—have become a common trope in the battle to curb greenhouse gases. They are employed because they are dramatic, which, it is assumed, will get the public's attention and provide motivation.

But is it really true that the bigger the stick, the more powerful the motivation? What Willer and Feinberg's study found is that it is not necessarily true. Apocalyptic messaging, particularly messaging that emphasizes that innocent children will be the ones to suffer from the dire effects of global warming, may galvanize *resistance* to the message, increasing skepticism of it and reducing efforts to minimize carbon impacts, because it deeply offends the just world belief in self-determination and control.

Thus we see *increased* use of SUVs and flagrant energy consumption as an "In your face!" rejection of the offending message. "I think the evidence we've seen is pretty clear," says Willer. "The research underscores the importance of keeping in mind that people have a deeply seated belief in the world as just, fair, and stable. Scientists should steer clear of apocalyptic messaging, or if they use it they need to present concrete solutions that suggest ways to control the outcome. Even people high in belief in a just world can handle these extremes if there's a concrete solution. But without that, they find it paralyzing and are motivated to disregard the message."[23]

Speaking Conservatese

When the just world belief is held along with a high level of patriotism, this effect seems to be multiplied, Willer and Feinberg found in a follow-up study.[24] "Conservatives are on average more patriotic," says Willer. "One thing that sets up is a great deal of cognitive dissonance when it comes to global warming. You think America is great, you know it's a greenhouse-gas emitter, and then you're told that greenhouse gases are bad for the world." They found that if you experimentally increase people's patriotism, their belief in global warming tends to go down.

In other experiments, Feinberg and Willer found that liberals moralize environmental issues and conservatives don't.[25] So they wondered, "What if you tried to make conservatives think of global warming as a moral issue? Go to a moral foundation conservatives respect but liberals don't?" They found that conservatives are more likely to talk about moral issues in terms of purity and disgust: They want to keep things pure and sacred and condemn what disgusts them and grosses them out. For example, Willer says, they believe that "homosexuality is wrong because it's gross."

Michele Bachmann offers an example of this messaging style. In a March 2004 radio interview Bachmann urged conservatives to attend a rally to ban gay marriage at the Minnesota state capitol. "We will be beseeching the Lord," Bachmann said. "Our state will change forever if gay marriage goes through. Little children will be forced to learn that homosexuality is normal and natural and perhaps they should try it. It will take away the civil rights of little children to be protected in their innocence, but also the rights of parents to control their kids' education, and threaten their deeply held religious beliefs. . . . This is a very serious matter, because it is our children who are the prize for this [gay] community, they are specifically targeting our children. . . . The sex curriculum will essentially be taught by the local gay community."[26]

To test the purity versus disgust messaging observation, Willer says, he and Feinberg made a fake environmental advertisement that was phrased in terms of purity and disgust, essentially saying, "Now more than ever, it's so important to protect our sacred mountains and our rivers from desecration." The ad also talked about impurities in the environment entering our bodies. It had images of people drinking dirty water. "What really surprised us was that it closed the gap between liberals and conservatives," Willer says.[27]

CELEBRATING LIKE MAOISTS

Adding to these heavily emotional factors that drive antiscience partisanship is the celebration of anti-intellectualism in America that has been pushed largely by the religious right. It is a strategy commonly employed by authoritarians from Hitler on the right to Mao on the left, both of whom demonized intellectuals and had followers who banned and burned books. Mao even went so far as to close most of China's university system during China's Cultural Revolution, when every thought and act was to be interpreted in the ideological light of Mao Zedong Thought as published in the *Little Red Book*. In 1968, for example, the national Study Group of Mao Zedong Thought was organized, which denounced many science theories, starting with Albert Einstein's theory of relativity.[28] Biologist and leading sinologist Joseph Needham wrote a famous 1978 report for *Nature* titled "Reborn in China: Rise and Fall of the Anti-Intellectual 'Gang,'" in which he described a pathology profes-sor who had been forced to "lecture on carcinogenesis to medical stu-dents while they were picking cotton."[29]

People are naturally skeptical, but they are also social. They don't like to think of themselves as antiscience, ignorant, or stupid, but when those qualities are put in conflict with group belonging or social identity messaging and reinforced by authority, they can weaken the stigma and skepticism can be suspended. This bottom-wing, authori-tarian activity happens on the right with religion and the "conserva-tive" identity, but it also happens with identity politics on the left.

The right's celebration of anti-intellectualism—Al Gore was "an intel-lectual" while George Bush was "a guy you'd like to have a beer with"—recalls the attitudes of the Cultural Revolution and casts the argument in terms of the lone, other, cold intellectual versus we social people hav-ing fun together in a group. This shifts it from *thought* to an *emotional and social identity* level that is then reinforced by authority figures who repeat the message in the media like conductors.

This shift has been powerful enough to affect national media coverage of events, creating a feedback loop. An example is the October 16, 2006, coverage NBC News anchor Brian Williams gave to the momentous event of the United States crossing the three hundred million mark in population. "Tomorrow morning at 7:46 a.m. Eastern time—and don't ask us how they estimate it—" Williams said, "the US population will click over to three hundred million." [30] But isn't reporting "how they estimate it" a pretty big part of the story? One would hope Williams wasn't simply pandering to anti-intellectualism, but figuring out "how they estimate it" wasn't really all that hard, as Mark Strassmann showed in CBS's coverage. "Every 11 seconds, America moves one person closer, and number three hundred million could come by birth, by oath as a legal immigrant, or by stealth: someone sneaking into history." [31] Even the United Kingdom's *Guardian* newspaper made it look easy: "Nobody knows the precise second at which the US will cross the 300 million mark, though the time given next Tuesday [9:01 and 48 seconds in the morning] is the literal interpretation of US census projections."[32]

As inventor Dean Kamen says, "We get what we celebrate. If we celebrate actors and celebrity, we get the balloon boy and stupid people acting out to get on reality shows. If we celebrate sports, we get a bunch of kids wearing jerseys, but how many of them will actually become millionaire sports heroes? What if we celebrate science and engineering with that same adoration?"[33]

Kamen's inventions range from the Segway Personal Transporter to robotic prosthetic arms to high-tech portable medical devices, but he regards his biggest accomplishment to be forming FIRST—For Inspiration and Recognition of Science and Technology—a nonprofit organization that sponsors an annual competition in which more than 250,000 kids try to build the best robots. FIRST holds a finalist competition each year that fills a major football stadium. "When you get successful like we have been in this country," Kamen says, "it's easy to get complacent and lazy and preoccupied with sports and entertainment and nonsense while the rest of the world looks at America enjoying the wealth created by our great-grandparents. The rest of the world is moving even faster than ever before. On a comparative basis, we'd still not be keeping up. Super Bowls are amusement; superconductors can change the energy outlook of the planet. I don't need a president I want to have a beer with. I need a president that takes on the crap and fights for our kids to succeed."

The celebration of anti-intellectualism offers social identity belonging that allows people to let go of shame at antiscience ignorance, a sort of group bravado in the face of otherwise public stigma—"don't ask us how they estimate it"—and sets up a political and social atmosphere in which people can be sold ideas without any grounding in facts.

AN ALTERNATE UNIVERSE

The effect is now moving online with the establishment of sites such as Conservapedia.com, a "conservative" takeoff on Wikipedia.org founded by Andy Schlafly, the son of bottom-right-wing Catholic social activist Phyllis Schlafly, because he views Wikipedia as liberally biased. Conservapedia seeks to create an alternate intellectual universe by reinterpreting and challenging facts that don't fit its readers' ideology and providing rhetorical ammunition for elected officials and others to build ideologically driven pseudoscience arguments. For example, harkening back to the ideological problems that Nazis and Maoists had with Einstein's theory of relativity, Conservapedia says: "The theory of relativity is a mathematical system that allows no exceptions. It is heavily promoted by liberals who like its encouragement of relativism and its tendency to mislead people in how they view the world."[34]

The site identifies thirty-four "counterexamples" that purport to show that the theory of relativity is incorrect, including "action at a distance by Jesus, described in John 4:46–54"; "despite wasting millions of taxpayer dollars searching for gravity waves predicted by the theory, none have ever been found. Sound like global warming?"; "it is impossible to perform an experiment to determine whether Einstein's theory of relativity is correct, or the older Lorentz aether theory is correct. Believing one over the other is a matter of faith"; and "In Genesis 1:6-8, we are told that one of God's first creations was a firmament in the heavens. This likely refers to the creation of the luminiferous aether."

The battle between social identity belonging and natural skepticism shows that people never completely give up their respect for science. This requires people with authoritarian or antiscience vested interests to build elaborate explanations to get around it, like postmodernism, climate science denial, and Conservapedia—all of them finely constructed rhetorical arguments, but none of them science.

THE DEFICIT MODEL:
THE ZERO-SUM ECONOMICS
OF SCIENCE

Faced with this constant stream of emotion-laden messaging that apparently uses neural pathways similar to those used by the science frame and crowds it out, one can't blame scientists for simply wanting to throw up their hands. In the 2006 evolution-versus-creationism documentary

Flock of Dodos, evolutionary biologist turned filmmaker Randy Olson asked a group of scientists gathered around a poker table about the intelligent design versus evolution debate. One of them responded by saying, "I think people have to stand up and say, You know, you're an idiot."[35]

That may work to a point because it's emotional, but it's likely not going to be a successful long-term strategy unless the broader culture mirrors that attitude—something that is virtually impossible in the age of the Internet and fractured media sources operating in the marketplace of emotions.

This comment highlights the different ways scientists and the lay public often think about these issues. Scientists are trained to avoid rhetorical arguments, the "vulgar Induction" Bacon warned against, and let the chips of reality fall where they may. They highly prize this intellectual honesty because the stakes for them are very high. They know how value judgments, prejudices, and habits of thought can blind you to the truth you are seeking, which will limit or end your career as a scientist.

The lay public does just the opposite. They form frames of reference, prejudices, and value judgments as guides for navigating life and then make rhetorical arguments to get what they need. What feels good is good. The idea that the ignorant or stupid public just needs to be better educated in order to see the light is called the deficit model—the assumption by scientists that the public thinks the same way they do, and therefore that the public's differences with science are because of a knowledge deficit. If that's the case, it makes perfect sense for scientists to simply try to pour in more knowledge—fill the deficit—to win support and eradicate the willful inculcation of stupidity that Michael Webber bemoans in Texas.

But what if the problem is not that the public lacks knowledge? What if the problem is that the public has the knowledge or at least access to it, but rejects science *anyway* in the same way that Texas governor Rick Perry does? The answers given in response to variously worded questions about evolution on the right and the fact that most vaccine rejecters on the left are well educated are evidence of this possibility. What then?

A 2008 Pew Research Center survey found that that is exactly what is happening. Republicans who are college educated, the survey found, were substantially *less* likely to accept the scientific consensus on climate change than were their less-educated peers—19 percent to 31 percent, respectively. The proportion was the opposite for Democrats (75 percent to 52 percent).[36] The more educated Republicans were, the less they "believed" in climate change. They cannot be said to be suffering from a knowledge deficit. As climate change has become politicized, Republican support has collapsed.

The issue can be seen as a microcosm of the public's relationship with science on a host of looming problems, and it's got nothing to do with not having the knowledge. It has to do with the public regarding arguments as rhetorical, a view aided and encouraged by postmodern education and by scientists not understanding how to effectively bridge that communication gap.

Science Impotence

Towson University psychologist Geoffrey Munro showed in a 2010 study how high the stakes really are in this discussion: He found that people who are pushed into antiscience positions in one area tend to generalize that partisanship to *all of science*. Like Lord, Ross, and Lepper, Munro asked subjects a politically loaded question: "Do you believe that homosexuality is associated with mental illness?" He separated them into one group that said yes and another that said no. He then presented each group with fake scientific studies, half of which presented conclusions that confirmed each group's prejudice, and half that contradicted the belief.[37]

Munro then asked the subjects to evaluate the scientific validity of those studies. Regardless of their position on the stereotype, those who were presented with fake studies that confirmed their belief tended to ignore the studies' weaknesses, while those who were presented with fake studies that contradicted their belief were much more adept at finding the weaknesses. More important, they were also more likely to come to the conclusion that *science was impotent to settle the question,* which Munro called the scientific impotence hypothesis.

Munro next asked the subjects questions about their opinions on other sociopolitical issues that science could offer definitive information on and found that subjects whose belief had been contradicted by science now felt that science was less likely to be able to solve *any issues at all.* They had generalized their scientific impotence discounting to *all of science*—precisely the outcome that National Academy of Sciences president Ralph Cicerone had been concerned could happen during Climategate. They had essentially been "radicalized" and set on the path toward an antiscience, antirationalist worldview.

"When a person holds a belief," Munro says, "especially a strong one that is linked to important values (e.g., some sociopolitical beliefs), information threatening that belief creates inconsistency in the cognitive system that threatens one's self-image as a smart person. This produces an unpleasant emotional state."[38]

This has tremendous public policy implications in an age dominated by science as the media sits on its hands and cedes the definition of the social standard to a "marketplace of ideas" dominated by angry rhetoricians and spoofing satirists. It also implies that if we wish to improve science education among schoolchildren and college students, we need to understand that some misconceptions about science are not the result of a knowledge deficit, but of belief resistance, and learn ways to shortcircuit these processes.

That belief resistance—and this is a critically important point—is largely coming from the adult ideological worldview. This is why education is political in the first place, and why the children of scientists are the most likely next generation of scientists[39]—an effect that is slowly striating society into knowledge haves and have-nots and can only increase partisanship along science lines.

Beyond scientist parents, the other major predictor of a child's performing well in science is having immigrant parents. Fully 70 percent of the finalists in the 2011 Intel Science Talent Search competition were the children of immigrants—a stunning figure considering that immigrants make up only 12 percent of the US population, and a troublesome one considering the restrictions that have sharply curtailed immigration since 2001.[40] "Our parents brought us up with love of science as a value," said David Kenneth Tang-Quan, one of the finalists.[41]

The Obama administration has prioritized placing renewed focus on improving science, technology, engineering, and math education following a somewhat traditional deficit model approach. This is all well and good, but evidence suggests that it may have little to no effect in transforming the science readiness of children or the ability of the country to compete in a science-driven global economy because it doesn't address the larger problems: *It's not about the kids, it's about their parents,* and it's not just about a deficit, it's about politics. *Transform the parents and you transform the system.* This is of course something that antiscience conservatives know very well—witness sex education in Texas.

PARENTAL SCIENCE IMPOTENCE

By May 2010, this was becoming more apparent. The National Science Teachers Association did a survey that found that 53 percent of American parents didn't feel equipped to encourage their kids in science or to discuss science with their children. "Science education has been identified as a national priority, but science teachers can't do the job on their own," said NSTA executive director Francis Eberle. "They need the help and support from key stakeholders, especially parents."[42]

When they were asked what they thought prevented parents from being more involved, 77 percent of the teachers surveyed said they believed that parents don't feel comfortable talking about science with their children. The teachers guessed that part of the problem might be a lack of resources and community involvement, and parents said it would help if communities had science museums. Half of science teachers said that, as it stands now, parents don't have access to materials or community resources that would encourage their children's interest in science. But those solutions are, yet again, taking the responsibility out of parents' hands and putting it in the hands of science museums and other community resources or educational materials. They also ignore the plethora of science information that is freely available on the Internet.

WHY SCIENCE DEBATES ARE IMPORTANT TO AMERICA

The larger issue is that science is walled off from the general population, a subject left to experts, science museums, universities, and the odd science festival. It has become commoditized and the public is merely presented with the conclusions and not exposed to the process. And in its absence, other powers have rushed in to fill the vacuum in the public dialogue, making science into their whipping boy when its conclusions don't support their ideological predilections. This is the problem science debates solve: By putting science in its rightful place as an ongoing part of the *policy* discussion of the nation, parents can become educated *in the context in which they are used to taking in information*—policy discussions that affect their lives. This allows them to become familiar with science and knowledge-based argumentation as opposed to mere rhetoric, to learn or relearn how to distinguish the two, and to use this thought process not only in making electoral decisions but also in discussing things with their kids.

The fact that science debates are supported by leading figures on the political right and left provides an important means of breaking down identity politics and partisanship concerning science. Science is political in that it is a top-wing activity—i.e., it grounds arguments in facts, not in authority or vested interests—but it is not partisan. There are scientists who are Democrats and scientists who are Republicans.

The structure of a science debate gets at the roots of the problem by asking politicians to ground their arguments in knowledge in the same way Thomas Jefferson did and by bringing them together with scientists and the public as they do so. Thus, science is reinjected into the discussion and rhetoric is tethered back to objective reality, "till the mind is brought to the source on which it bottoms," as Locke advised.

Parents and the general public often are not interested in science for its own sake, but rather for how it affects their lives and the people, ideas, and policies they care about. Alexis de Tocqueville noted this almost two centuries ago. Giving Americans science in the context of what they are interested in is not the politicization of science, it is the depoliticization of science, the Americanization of science, because it makes it ubiquitous, a tool instead of rhetoric.

The process of a science debate also frees politicians by setting the bounds of play, thus providing some measure of structure and fairness to the game and effectiveness to the policy outcome. We elect politicians to deal with complex, objective, real world issues, not simply to joust on steeds of opinion. We cannot expect progress on objective issues unless arguments reach beyond the subjective and are grounded in reality. Seeing politicians actually do this would likely increase public appreciation of them and of government. This is not only a noble and worthy goal, but also critical to the ongoing strength of the United States and the values it stands for worldwide.

Kamen points out that nearly half the world's population lives on less than $2 per day. "I think the future will depend entirely on whether the next generation of American kids will be more innovative than the two billion kids elsewhere. We have eighty million. We will either lead or watch the decline of the America that we knew." Which is the loss of the ability to innovate, create, and solve problems that Tocqueville warned of. As that falters, so falters the republic.

CHAPTER 13

RETHINKING OUR
RELIGION

"Never again will I curse the ground because of man, even though every inclination of his heart is evil from childhood and never again will I destroy all living creatures as I have done. . . . "

I believe that's the infallible word of God, and that's the way it's going to be for his creation. . . . The Earth will end only when God declares it's time to be over. Man will not destroy this Earth. This Earth will not be destroyed by a flood. . . . I do believe God's word is infallible, unchanging, perfect. . . . There is a theological debate that this is a carbon-starved planet, not too much carbon.[1]

—US REPRESENTATIVE JOHN SHIMKUS (R-IL), 2009*

WHAT DOES IT MEAN TO BE MORAL?

Given the times and what we now know about our world from the tools of science, and given Congressman Shimkus's national leadership role in the United States Congress, can the religious leader who guided him to that conclusion be called a *moral* leader? Is it possible to be moral while insisting on a literalist interpretation of the Bible in the face of scientific evidence to the contrary, when lives and national policy are at stake? This is a question upon which the future of the nation hangs.

* Shimkus is chair of the House Subcommittee on Environment and the Economy in the 112th Congress. He read from Genesis 8:21 in his comments at a March 25, 2009, hearing of the House Subcommittee on Energy and the Environment.

Former Republican political strategist Kevin Phillips argued in his 2006 book *American Theocracy*[2] that the modern Republican Party has been making a steady march back to the old themes of the Deep South, where fundamentalist religion and oil coexist in a close-knit relationship. But as knowledge is extended, the willful promulgation of ignorance from the pulpit, CEO's soapbox, or politician's office is an increasingly immoral act.

Being people of principle, the religious leaders who are helping businesses to purchase America's modern denial of science must ask themselves if such denial is truly a service to God or humanity or even the future economy, and if it is a more moral stand than battling on principle, especially with the mother of all science problems, climate change. Disagreeing is one thing. Denying and undermining the validity of knowledge itself places something far more cherished at risk: our children. How in good conscience can a religious leader look in the eyes of the small children in the pews, so full of trust, so sure of the righteousness and holiness and gentility of life, so true and pure in the simplicity of their belief in the goodness of their parents, their church, and society, and *take* from them in ways they don't even comprehend, because of politics? Would Jesus honor that? "For what shall it profit a man, if he shall gain the whole world, and lose his own soul?" (Mark 8:36, King James Version) Is there any greater moral imperative, to religion, to politics, to leadership, to economics, to science, to individuals, and especially to children, than the ongoing viability of our children and the planet God has entrusted to his creation? Is there anything more pro-life?

THE LORD COMMUNICATES; ARE BELIEVERS LISTENING?

There is an old Christian joke that was popular in New Orleans after it was devastated by Hurricane Katrina in August 2005. You've probably heard some variation of it. A preacher was caught in a hurricane, and as the waters began to rise he decided he'd better climb onto his roof. As he struggled with the ladder, the fast-rising water threatening to sweep him away, a man in a boat came by and said, "Preacher, hop in!" But the preacher was a man of faith and said, "No, no, my son—Jesus will provide." So the man left. After some struggle the preacher got onto his roof just as the ladder was swept away. One of his neighbors, a scientist, came back in a boat and said, "Preacher, I know you don't want to leave, but the measurements we've got say it's gonna get a lot worse!" And the preacher said, "You can measure all you want, but the Lord will provide," and waved him away. But the waters kept rising. Soon the preacher's legs were engulfed in the flow as it rushed over the peak of his roof, and he had to hang on to his chimney for dear life. A helicopter spotted the poor man and lowered a ladder. A man yelled down and said, "Preacher! Grab the ladder!" But the preacher yelled, "Get away, ye of little faith!" And so the helicopter had no choice but to fly away. The waters rose again, and the preacher was swept away into the rush and drowned. He woke up outside the Pearly Gates. You can imagine how mad he was that the Lord had let him down after a lifetime of faith and devotion. He saw Jesus walking inside and called to him, "Lord, I held firm in my faith, but you betrayed me!" And Jesus said, "What are you talking about? I sent two boats and a helicopter!"

Science produced those boats and designed that helicopter, and turning away from it is turning away from the natural law of the Puritans—our ability to reason. "What we understand scripture and God are saying to us is sometimes *better* understood by knowing what science tells us," says the Reverend Peg Chemberlin. Chemberlin is president of the National Council of Churches, the leading organization for ecumenical cooperation among Christians in the United States and the go-to organization when it comes to broadly engaging the Christian community, which includes about forty-five million people in more than one hundred thousand congregations across the nation.

Chemberlin says she understands facts as things to be interpreted by scripture, not ignored by scripture. "It's fairly arrogant to think that we know all the ins and outs of scripture," says Chemberlin. Religious tradition is alive and is always in engagement with the world around it. One world-view can never explain or encompass the nature of the universe, much less the full nature of God. "So to hang on to a worldview and not allow science to engage that worldview keeps us from opening up to that."[3]

NATURAL LAW: A COMMON GROUND?

Chemberlin sees that as a value in both the Protestant and the Catholic churches, and says it harks back to the Puritan belief in natural law and the idea that understanding the physical world is an important part of understanding scripture and therefore of our relationships with both science and God.

"I often ask parishioners," she says, "'How many creation stories are there?' They usually say 'Two,' and I say 'Name them' and they tell me the first two stories in Genesis. But there are more than two: There's also a creation story in Proverbs; there's a creation story in Job; there's a creation story in Corinthians; there's a creation story in the Gospel of John. All of them are different stories. If we're just understanding them on a factual level, they conflict. So we have to do some kind of interpre-tations of those facts—how do we understand them in a broader way—to understand all those scripture stories. And I think that's very much the same kind of process we have to bring to understanding conflicts between scripture and science. With all that science throws at us, it's very exciting to have that invitation in front of us."

Chemberlin says people of faith have endless and exciting opportuni-ties to see their faith from fresh new perspectives as science advances and creates new knowledge about the way the physical world—God's creation—exists and functions. In contrast to the views more often heard from extreme fundamentalists, this approach allows for a more con-structive and responsible exchange on many issues that are critical to the future of the country and the planet. In fact, it was the view that gave birth to science in the first place.

A Place for Ethical Reflection

Chemberlin says that few churches consider religion a naturally conservative force or feel that it is moral to deny scientific findings when they conflict with religion. They see the church as a place for ethical reflection in life, and thus on political and policy questions that sometimes involve science.

"Religion is a basic anchor there, but that doesn't mean conservative. Religious thinking has changed as science has helped open up our understanding of what is true. It's been a long time since anybody in the church believed that the Earth was flat," she says. "It's been a long time since anybody in the church believed that the Earth was at the center of the universe, and it hasn't been so long since the scripture was used as a basis to reject the equality of women with men or the equality of the races. Even in the more evangelical communities that's no longer the case, with a very few extreme exceptions. So in scripture-based negations of science, in all of those situations it changed. There are a couple of places where parts of the church are still saying that scripture negates evolution, but that's a pretty small part of us."

Who Moderates the Debate?

It may be a small part of the church, but it is a vocal part, and in some ways—as shown by opinion dynamics—it is controlling the debate. Recall from Chapter 9 the frustration of Scott Westphal, the head pastor of a Lutheran church in a conservative area, with the harshness of certain AM talk radio personalities, or the unwillingness of certain members of his flock to listen to the more traditional Lutheran teachings that asking questions lies at the foundation of Lutheranism. "Harshness, mockery, and intolerance," he says, "are not a natural part of being either conservative or Christian."[4]

And yet there are the prominent televangelist supporters of the Cornwall Alliance who argue that we should reduce collective ownership of natural resources, increase our appreciation of private property, and that there is no such thing as anthropogenic global warming. Are these theological arguments, or business arguments masquerading as theology?

It's an important issue that is having a profoundly negative impact on the country in a time when half of the incoming freshman GOP legislators in 2010 were vocal and unashamed deniers of the findings of climate science and angry demagogues dominate the airwaves in unending wars of opinion that, without the antiauthoritarian arbiter of knowledge, can only resort to authoritarian brutality to settle their arguments.

"It's very much the moderate world that has to speak up on this," says Chemberlin. "I recently got an award from the Islamic Resource Group [a US-based educational group seeking to further understanding of Islam]. And I told them, 'Your extremists and our extremists want to fight. And then we're all in trouble.' That's true with faith and science too. But in a democracy, because I believe that my God, my transcendent value, shares a value about democracy, then I have to live in the democracy and that means compromises on a daily basis. I'm not going to get everything that I believe is right all the time. If we think our public life should be 'everybody else has to understand it my way,' that's a theocracy, not a democracy. Once you decide God favors a democracy, then you have to act in democratic ways, find common ground with people whom you fundamentally disagree with. A Christian who disagrees with a scientist still has to come into the process and say, 'I disagree, but where do we have a common ground?' Sometimes that's not easy, but if you believe that God values democracy, then that is what you are called to do. Don't work with the crazy part of the other side. There are extremists in science too, who want to say, 'My view is the only view that's right.' Work with the moderate part of the other side and find some common ground. In the faith community there are many more of us who are in the moderate place. We're not as vocal, but don't disregard us. My dad's a doctor; he couldn't be a member of a church that didn't accept evolution. On the other hand, he couldn't just be a scientist."

No matter what one's opinions or beliefs are, there can be no doubt that the knowledge we now have warns of the magnitude of the problems we are facing. And the problems are now not so much problems created by science as they are problems created by the failure of policy makers operating without the moral leadership of the moderate faith community.

We know what climate science is telling us. We know what the declining international science and math rankings of American students are telling us. We know what evolution and biology are telling us. What environmental science is telling us. What studies on economic competitiveness are telling us. What the oceans are telling us. What our runaway population is telling us. But instead of arguing through the particulars of that knowledge and deciding on policies based on reality, we are denying the science because we don't want to hear it. Because it would be hard. Because solutions would require work, and leadership, and saying and doing uncomfortable things. And these are things that religious moderates don't like to do. Lutherans don't like to impose. So they leave it to the angry extremists. But if ever there was an example of the metaphorical ostrich sticking its head in the sand, it is the content of the dominant religious and policy-making discussion about science. And if there ever was a time for moral courage, and for caring, loving, religious moderates to get involved, it is right now.

A Time for Moral Leadership

Some of this is occurring because the science is progressing faster than the social institutions that are charged with working through the ethics and morality have been willing or able to keep up with. And yet it is hard to imagine a creative, spirited preacher who would not be continually excited and grateful for the ongoing flow of opportunities for ethical reflection afforded by science. We have—stupidly—handed over our biggest microphones to the smallest and most vocal minority, largely on one side of the issues, and abandoned arbitration, relying instead on an assumption that economist Adam Smith's "invisible hand" will moderate for us and there will occur a fair "marketplace of ideas." But human history has shown this never to be the case.

This is another term for a lack of leadership. The marketplace of *opinions* is real, but that of *ideas* is a myth. It is little more than relativism masquerading as economics, where all "ways of knowing" have equal merit and only might makes right. Without the primacy of knowledge, there is no shared social contract, no common truth or value upon which all Americans can agree. Market economics like these *destroy* the world of ideas, which, without a process in place to protect the knowledge we have gained, is invariably left to the conscienceless. Thus, bullies now reign.

The public bullying is risking a collapse into barbarism. Opposing it is precisely what led to the foundation of the United States in the first place. To protect against "might makes right" in politics, we have created a system of democracy that is founded on ideas of freedom and knowledge and expressly *not* "but faith, or opinion," the twin components that now dominate our airwaves, where ideologues and pundits argue endlessly over whose reality is the true one.

This is a time for churches to reach out to scientists and to speak about science and politics, because these discussions are so important to the future. We are in a moral crisis, and it matters little whether a preacher is conservative or progressive if he or she is incorporating knowledge into moral reflections. That's challenging! It's a wonderful contribution. We need the church, as Peg Chamberlin put it, to be a place of ethical reflection, but pastors and priests who don't engage with reality are falling down on the job. Then, the opportunity to help congregations parse new knowledge is lost and an important, reflective part of society breaks down and retreats into unthinking bottom-wing authoritarianism and partisanship preached from the pulpit.

There was a time when religious leaders were on the cutting edge of scientific thinking. They should be again. There is every reason for every preacher in America to be a member of the AAAS and to have subscriptions to *Science, Nature,* and *Scientific American,* and to spend time reading the science blogs at Blogs.DiscoverMagazine.com, ScienceBlogs. com, Wired.com, and other sites. How can you provide meaningful reflection if you are not part of the discussion? It is time for a new Protestantism and for the spirit of Luther's questioning to be reborn.

Consider the example of the geological clock. In this clock, one hour represents about 400 million years. The clock starts at 00:00:00 at the formation of the earth. The end of the dinosaurs (65 million years ago) is at about 11:50:15. The earliest known hominids (about 4 million years ago) are at 11:59:24. The start of human civilization (about 30,000 years ago) is at about 11:59:59.7, or only during the last third of a second. That's a wondrous thing to contemplate, especially when you consider the impact we've had on the world in the last 200 years—it's akin to an explosion when viewed in geologic time. There is much there for ethical reflection on the part of churches.

There is considerable evidence that many scientists would welcome this engagement. They do not have the interest or capacity to form "churches of science" to help society ethically reflect, and they are not particularly good at it, by and large. That is not where their skills lie— even though, more and more, they are stepping forward and taking on the demagogues because so many religious leaders have been cowardly in that regard. These scientists are trying to find effective ways to reach out, to help people think through the implications of this expanding knowledge, to understand the truth about climate change, ocean health, the biosciences, the economy and the environment. To anticipate the questions that will be raised by emerging advances in genetics, nanotechnology, synthetic biology, technology, neuroscience, and stem cell research. Religious leaders who love their flocks and care about God's Earth and the stewardship of creation have a moral duty to seize scientists' open hands with open hands of their own, to reach out to scientists in renewed partnership to minister to the whole person's full intelligence, not just to his or her anger, fear, resentment, and memories of the rose-tinted days of yore.

Peg Chemberlin, religious leader, daughter of a physician and scientist, shakes her head in wonder at the possibilities. "The data in the last decade about how much genetic material we share with the whole of creation, for the creationists, they would probably say, 'Bah, humbug, this isn't true.' For the rest of us, we would say, 'Wow that's an extraordinary statement about our connectedness.' There's a line in scripture: 'The earth is the Lord's and the fullness thereof.' He's got the whole world in his hands. I think that's the way in which science can really open up our sense of the relatedness that is the oneness of the Earth and that gets me outside of my own little place on the planet. The Earth is much fuller than that and much more connected than that."

Amen.

CHAPTER 14

ON TRUTH AND BEAUTY

Believe in yourself. You gain strength, courage and confidence by every experience in which you stop to look fear in the face. . . . You must do that which you think you cannot do. . . . The future belongs to those who believe in the beauty of their dreams.

—ELEANOR ROOSEVELT[1]

THE UNREACHABLE STAR

David Sanders is a Purdue University biologist who has run for Congress three times as a Democrat in Indiana's Republican-leaning Fourth District. He realizes that the district tilts heavily against Democrats, but he runs anyway.[2]

Sanders says he feels it is part of his civic duty as a scientist. "So many of the issues that we face as a society have a science or technological basis," he says. "We've simply got to talk about them. I've had National Institutes of Health grants and National Science Foundation grants. By taking that public money, I think it's my responsibility as a scientist to do good science, but also to participate in the public sphere." Win or lose, Sanders puts a face on science and elevates it in the discussion. Sometimes this process can take years to have an impact. But eventually, it does.

"I went to the county fairs and I'd meet lots of people," relates Sanders. "A not infrequent response was 'Oh, since you're a scientist you probably believe in global warming.' This was from people who didn't. So there was this cognitive dissonance. They know that my being a scientist would make me think that global warming was real, and they both didn't believe it themselves and thought that there was some dispute about it. And they would maintain the position that it's an open scientific question. But in person at least, I didn't feel like they were hostile to me. It's hard to know how sustained you have to be.

"At Indiana county fairs there are a lot of church booths. I started introducing myself. Some were like 'It's wonderful to have somebody running for office.' They were the good side of religion. Welcoming, open, wanting what is best. Other ones, the first question was 'Are you a Christian?' I said no. Then they said, 'Well, are you pro-life?' I asked them, 'What do you mean? Do you mean stem cells? Or abortion? Or abortion for women who have been raped?' But they couldn't tell me. So I probed, and asked, 'If it's murder, do you want abortion to be banned, or do you want physicians to go to jail, or women who get abortions to go to jail?' I'd never get a clear answer. They had some sort of concept of what they thought was right and wrong, but you couldn't push them to the logical consequences. It wasn't possible."

Sanders says he ran into a similar thing when he taught microbiology to a class of 250. These students had already had two semesters of biology. Somewhere between 15 and 20 percent of these students, in an advanced biology class, didn't believe in evolution. They were offended by it. They gave him the usual arguments—the flagellum, for example. Flagella are long spiral propellers protruding from many bacteria that are connected to tiny biological motors that drive their rotation. They are one way bacteria move around. Creationists argue that the flagellum couldn't possibly arise through evolution because if you take one part away it doesn't work anymore. It's irreducibly complex and could only arise from design. So Sanders would show them there are intermediates. Ones that turn just one direction, one with two rings, one with three rings. He explained that that's the point of evolution; if you had an extra part, it would go away. He very patiently worked to educate them out of their "vulgar Induction."

The students complained that he worked evolution into every lecture, he says. But biology at this level is essentially evolution in the context of chemistry and physics. Even in these classes, he concluded, there is a certain segment of the population that simply cannot be reached. Sanders says he would have the occasional student come to him upset and say, "Before I believed, and now I am questioning it and I don't know what to do." He'd tell them, "I think that you can continue to practice your religion and still accept evolution. I don't see any conflicts there. If you go with 'What's the message?' I don't think there is any contradiction."

Sanders's approach is not unique. From Vern Ehlers to Bill Foster to Rush Holt, scientists are beginning to run for Congress. But they don't all have to; they just have to have conversations in the community, so someone is speaking for reality. Sanders says he began to realize this when he was teaching. The students who couldn't accept evolution, he realized, were over and over again repeating the arguments coming from a vast and well-organized propaganda network. The frame they would ultimately come to, Sanders says, was "I can accept microevolution, but not macroevolution." So he'd say, "Give me your best shot, but if that proves not to be true, then you're going to agree to change your position." But they wouldn't, even though that's the rational approach, so deep was the conflict—the belief resistance—between what they wanted to or felt a holy duty to believe and what reason was telling them.

If it is this difficult for an educated person in an advanced biology class to accept evidence that contradicts a preconceived belief, if the fear of being cast out by an unforgiving God or an unforgiving family or an unforgiving political party is so profound, imagine how difficult it would be for someone without that education who has little economic stake in a career in science.

"In the end," says Sanders, "some people simply have a different approach to discussion of rational thought. You hope that as a scientist you look at the facts and make a conclusion. That's what we're trained to do." The problem arises when people look for the facts that agree with their preconceptions. In the public and political spheres, that's what happens, instead of people trusting in the evidence of their own senses. As an example, Sanders cites discussion about a common argument of climate science deniers popularized by Christopher Monckton: "'The troposphere isn't warming.'[3] I bet you most people who said that couldn't tell me what the troposphere was. It turns out those data weren't correct, and that was the outlier in a very large body of evidence saying otherwise. But once you dispel that, they go on to some other argument."

This is, of course, because we make rhetorical arguments in order to succeed in life—arguments that we think will convince other people of something even if we don't believe it ourselves. Instead of arguing on the basis of data, we argue to win. At some point, we've made that argument so often or so publicly that it becomes impossible to retreat from it.

Despite the electoral losses and the people he can't reach, Sanders remains upbeat. "I am hopeful as an educator. I have to believe that what I do is worthwhile. I think that if people of goodwill with skills—scientists—don't speak out, then we're totally sunk, then the cynicism has seeped into us as well. We need as a society to take pride not only in our rock stars and sports heroes, but also in the beauty of the science we accomplish."

DEFENDING THE BEAUTY OF A DREAM

In the end this is what matters most: *the beauty*. It's why scientists do science: to apprehend the great beauty of nature. They are Puritans, four hundred years hence, with more data, but still searching for that direct communion with wonder and the aesthetic. In this way science is not unlike art or religion. In fact, art often anticipates and reflects the forms science discovers, as does religion. Science is much bigger than just solving challenges, as important as they are. It is about who we are as human beings, about our ability to love, to wonder, to imagine, to heal, to care for one another, to create a better future, to dream of things unseen.

The English poet Alfred Noyes was on hand on the night of November 1, 1917, as George Ellery Hale's only invited guest for the dedication of the Hooker one-hundred-inch telescope on the top of Mount Wilson. [4] "Your Milton's 'optic tube' has grown in power since Galileo," Hale had written his friend. Hale ordered the giant telescope to be trained, like

Galileo's had been, on Jupiter and its moons. The experience of being among the first to ever see those moons so clearly inspired Noyes to write his epic poem *Watchers of the Sky*, which charts the long emergence of astronomy from religion, and yet their close kinship still.[5] The poet took a break shortly after midnight and wandered out onto the mountaintop in a state of near-rapture. Almost all of the tiny sparkling lights of the distant Los Angeles had vanished by then, and "the whole dark mountain seemed to have lost its earth and to be sailing like a ship through heaven." His poem concluded:

> When I consider the heavens, the work of Thy fingers,
>
> The sun and the moon and the stars which Thou hast ordained,
>
> Though man be as dust I know Thou art mindful of him;
>
> And, through Thy law, Thy light still visiteth him.

The evening would not have been possible without the financial support of Andrew Carnegie, who had given Hale a check for $10 million, making the investment in the unknown and building what was in its time the greatest scientific laboratory on Earth—without having any idea of what it might find. Using the Mount Wilson Observatory, Harlow Shapley showed that the sun was not at the center of the Milky Way. Using it, Edwin Hubble showed that there were thousands of other galaxies besides the Milky Way. Using it, Hubble also defined the expansion of the universe, clarifying Einstein's work and birthing the field of cosmology. Those discoveries did not bear any direct financial returns. They did not add to the national defense. But they did something far more important: They changed our lives and the way we think about ourselves and the universe. They are among the most profound discoveries of science, and they had no financial justification whatsoever. They were seeking truth and beauty.

In 1967, the physicist and sculptor Robert Wilson was hired to build a similar dream of truth and beauty: his "fantasy of a utopian laboratory,"[6] a particle accelerator that would smash subatomic particles together at close to the speed of light, perhaps yielding new understandings about the underlying nature of the universe. But no one knew what practical applications might come of it, and this time science was being funded not by visionary philanthropists, but by the less-than-visionary senators and representatives of the federal government.

Because of this, Wilson's goal was a hard sell. The nation was already spending unprecedented amounts of money on the risky Apollo program. Science was unpopular with the general public, which saw the spending as wasteful when so many pressing social needs were going unmet. Science was despoiling the environment, and nuclear physics in particular had embroiled us in an arms race that would perhaps end us all in mutually assured destruction.

By 1969 the nation had tumbled into yet another recession, the emerging baby boom was loudly questioning America's priorities, and Congress was looking for ways to cut spending. There was no political capital to be gained by funding a massive new science investment like a particle accelerator unless there was some greater justification. Wilson was called up before the congressional Joint Committee on Atomic Energy and asked to supply just that.

Using a cold war rationale, at least there was some chance of still getting approval, and Wilson was no fool. Fear of Russia was running at a high pitch. Science and the military were joined at the hip. Wilson had been a key player in the Manhattan Project and knew these politics well. So when Senator John Pastore (D-RI) asked him to justify allocating $200 million to build the world's largest particle accelerator, the room was stunned by Wilson's reply.

> Pastore: Is there anything here that projects us in a position of being competitive with the Russians, with regard to this race?
>
> Wilson: Only from a long-range point of view, of a developing technology. Otherwise, it has to do with: Are we good painters, good sculptors, great poets? I mean all the things that we really venerate and honor in our country and are patriotic about. In that sense, this new knowledge has all to do with honor and country but it has nothing to do directly with defending our country—except to make it worth defending.[7]

Quite breathtakingly, Wilson had reached not for defense, but for the profound—not fear, but wonder. Miraculously, the funding was approved. Fermilab, as the facility that housed the accelerator came to be called, was built despite offering no quantifiable return to defense or industry, and when it began operating in 1972, it was the most powerful particle accelerator in the world. Reflecting Wilson's hope, the design of its main building, Wilson Hall, recalled the great French cathedrals of Chartres and Beauvais, bucking the attitude in Washington that Atomic Energy Commission buildings "did not have to be cheap, they just had to look cheap." [8] His sense of the important connection between science and the aesthetic went beyond architecture. "I have always felt," he later wrote, "that science, technology, and art are importantly connected, indeed, science and technology seem to many scholars to have grown out of art. In any case, in designing an accelerator I proceed very much as I do in making a sculpture. I felt that just as a theory is beautiful, so, too, is a scientific instrument—or that it should be." The futuristic superconducting technologies Wilson pushed for helped keep the accelerator vital under the leadership of his successor, the physicist, Nobel laureate, and great science humanitarian Leon Lederman, and it remained the world's most powerful until 2006, when the Large Hadron Collider opened at CERN (European Organization for Nuclear Research) in Geneva.

To Awaken, Perchance to Wonder

As Wilson noted, science, like art, is a cultural expression that makes a nation worth defending. Like great art and great music, its true value lies in exploring the unknown. Today, the opposite argument, the commoditization of science, is virtually the only one heard. It has metastasized from the smaller-minded appeals of the cold war to all of human learning and higher education. Education and knowledge are no longer values of truth and beauty that make life worth living, they are means to the ends of greater pay and more consumption, which somehow are supposed to make life worth living.

Legal and humanities scholar Stanley Fish wrote of this triumph of small-mindedness in 2010 in an eloquent criticism of *Securing a Sustainable Future for Higher Education,* a set of recommendations made by an independent panel to the British government. It advocated "student choice" in funding higher education. [9] Among the report's palliatives: "Our proposals put students at the heart of the system." "Our recommendations . . . are based on giving students the ability to make an informed choice of where and what to study." "Students are best placed to make the judgment about what they want to get from participating in higher education." [10] The idea is that the money follows the students. Courses that compete successfully for student attendance survive and prosper; those that do not wither and die. The assumptions of market economics have triumphed: Ideas are now considered commodities.

The problem is that this approach requires no judgment on the part of administrators. In fact, there is no need for them. Left in the hands of the students, there is no curating of an education to be done.

But as Fish pointed out, students are in fact *not* "best placed to make the judgment about what they want to get from participating in higher education." That is the whole *point* of education—to show students what they *don't already know.* In other words, to *help them build a telescope* so they can discover their own talents and interests as scholars and human beings.

This is yet another place where the pernicious myth of the "marketplace of ideas" has corrupted Western thought with authoritarianism. That it was Fish who made this impassioned defense is all the more meaningful since it was Fish who had most prominently criticized the Sokal hoax and defended postmodernism some fifteen years before.

The same triumph of commoditization is plaguing much of American culture, where things that cannot be easily quantified in the marketplace or contained within preexisting software are assumed to have no value and a brutality of warring opinions has come to reign, shorn free from facts. Applying such a commodity approach to education, just as in applying it to art or science, or classics or history, or poetry or math or love or joy, defeats the whole point of living and learning and turns universities into trade schools whose sole purpose is to supply the skills that enable one to get a job to earn money to buy things. Our primary asset is not our money. It is the quality of our time on earth. It is a vast misunderstanding by a generation that has lost touch with—or perhaps never really knew—what education should do: open us up to *wonder* and the great meaning and aesthetic beauty of life.

THE SHINING CITY UPON A HILL

It is *wonder* that we lost in our move to a commoditized, national defense model of science funding, which revolved around dispelling fear. It was perhaps important and perhaps helpful, but it went too far, and by forgetting the real reasons we do science it was ultimately a colossal error. As Sanders says, "Science is no longer about science. It's about marketing. It's either 'Gee whiz' or 'What is it going to do for me?'"[11]

The commoditization can be seen in how we oversell the practical benefits of science. The war on cancer, for example. We say, "We're going to cure cancer in fifteen years," and then when we don't, that's memorable. It's less memorable but more helpful to talk about the basic research that led, eventually, to the discovery of monoclonal antibodies, which actually have application now in fighting cancer, among other things. To value the *process*.

Literature professor Michael Bérubé, once also a leading critic of Sokal and his hoax, now says, "These days, when I talk to my scientist friends, I offer them a deal. I say: I'll admit that you were right about the potential for science studies to go horribly wrong and give fuel to deeply ignorant and/or reactionary people. And in return, you'll admit that I was right about the culture wars, and right that the natural sciences would not be held harmless from the right-wing noise machine. And if you'll go further, and acknowledge that some circumspect, well-informed critiques of actually existing science have merit (such as the criticism that the postwar medicalization of pregnancy and childbirth had some ill effects), I'll go further too, and acknowledge that many humanists' critiques of science and reason are neither circumspect nor well-informed, . . . leading both humanists and scientists to realize that the shared enemies of their enterprises are the religious fundamentalists who reject all knowledge that challenges their faith and the free market fundamentalists whose policies will surely scorch the earth."[12]

Sarah Palin may rip on "fruit fly research in Paris, France. I kid you not!" but if it weren't for people pursuing truth and beauty by working on weird mouse and bird viruses, we'd be nowhere in treating HIV infection. We weren't even aware of human retroviruses until 1979. Because we were willing to let those scientists explore unexpected corners, we have wrestled more freedom for ourselves, turning infection with HIV from a death sentence into a manageable chronic disease, pulling back years from the gaping maw of death and giving patients the freedom of life, and society the benefits of their contributions. Not because we set a goal to do it and did it, but because we valued the truth and beauty of science. The one thing we do know about science, the one thing that is predictable about it, is that if we don't value it, if we become inhospitable to the tolerance, freedom, and open exchange of ideas that stimulates it, if we wall it off and call it a separate culture instead of something we all should do, if we cease funding it, if we try to be overly directive of it, if we elevate ideology over science in our public policies, we will stifle creativity and science will go away. We won't get the big breakthroughs. We won't get the economic boons. We won't get the national defense advantages. We won't get the clean environment or healthy children. Germany already proved this can happen with its precipitous fall from arguably the most powerful science nation on Earth to a nation bereft of scientific enterprise in a single decade of Nazi ideological intolerance. China proved it under Mao. The Soviets proved it with Lysenko. The Vatican proved it with Galileo.

We can't move toward the world we could create if we don't value tolerance, freedom, and the beauty of *wonder*. Ronald Reagan described such a world in his hopeful farewell address, which was markedly different from the fearful farewell delivered by Eisenhower some three decades before.

> And that's about all I have to say tonight, except for one thing. The past few days when I've been at that window upstairs, I've thought a bit of the "shining city upon a hill." The phrase comes from John Winthrop, who wrote it to describe the America he imagined. What he imagined was important because he was an early Pilgrim, an

early freedom man. He journeyed here on what today we'd call a little wooden boat; and like the other Pilgrims, he was looking for a home that would be free. I've spoken of the shining city all my political life, but I don't know if I ever quite communicated what I saw when I said it. But in my mind it was a tall, proud city built on rocks stronger than oceans, windswept, God blessed, and teeming with people of all kinds living in harmony and peace; a city with free ports that hummed with commerce and creativity. And if there had to be city walls, the walls had doors and the doors were open to anyone with the will and the heart to get here. That's how I saw it, and see it still.[13]

"My colleagues think our science is beautiful," says Sanders, "but we don't put any effort into communicating that to the public. We don't think the public is interested."[14]

THE CHOICE

This was the great danger of democracy that French political scholar Alexis de Tocqueville most feared for America when he warned that we might one day come to value utility and the authority of tradition over meditation and the antiauthoritarian newness of creativity. He cautioned against losing the ability to think for ourselves, for we would then go the way of ancient China, where the stream flowed on, but the inhabitants had become unable to increase the flow or alter the course.

In the end, politics is about story. Robert McKee, Hollywood's master of storytelling, views the world from the top of America's other great cultural export—its movies.

"I think that the American ethos is not science-friendly and never has been," he says. "The American model is Thomas Edison and Henry Ford. Guys who never went to college and who were geniuses and invented things, and people like them. The inventor versus the scientist. Somebody who can go west, discover gold mines, and create a lot of money without an education. Unlike Europe or Japan or India, or even China these days. In those cultures they admire and compete to be a really well educated person in some field. That is not the American dream. The American dream is Hollywood, sitting in a drugstore and somebody says, 'You ought to be a movie star.' It's an attitude that life is a game and that what you gotta learn is to play that game well, but it's not on a gridiron where you actually have to practice, it's a game of manipulation and most of that game is somehow bullshit."[15]

If McKee and Tocqueville are right, the influx of European scientists starting in the 1930s and continuing through World War II may have changed the United States and led to a temporary boom. America may be coasting to the end of that momentum now. The importing of such talent declined sharply in the wake of 9/11. We have made the country more difficult and less hospitable for immigrants, closing those doors Ronald Reagan saw standing open to anyone with the will and the heart to get here. And now science, like our corporations, is going global, and the last great engine of the American economy—our ability to innovate—may be slipping from our grasp.

So we are faced with a choice. Will we go the way of the ancient Chinese, nosing our heads comfortably into the warm sand, obedient, productive, agreeably alike in thought, but rigid, paralyzed, no longer able to improve?

Or will we take up the mantle of freedom and leadership that science gave us—the commitment to knowledge over the assertions of "but faith, or opinion" that led to the disquieting idea of equality that is the foundation of democracy? Will we be skeptical of claims that seek to crowd out the space for knowledge in the public dialogue? Will we continue to embrace the antiauthoritarian power of wonder, tolerance, and imagination to create a new future—a shining city upon a hill? Will we reject ideological conformity and reward a facts-based press and science education? Will we set aside the left-right skirmishes of identity politics and focus as our founders did on the top-bottom battle for freedom? Will we protect and fund the conditions that encourage diversity, creativity, and prosperity in art and science, not because of what they do for our pocketbooks but because of what they mean to our values as Americans? Will the people, in short, remain well enough informed to be trusted with their own government?

In a century dominated by the awesome powers and dangers of science, there is no greater moral, economic, or political question.

Appendix

The American Science Pledge

Why a Pledge

We live in a time when the majority of the unsolved policy challenges facing the United States revolve around science. These challenges have continued to accumulate to the point that many of them are threatening the economic well-being of the nation and the ongoing health and vitality of its citizens and their environment.

We elect our representatives, senators, governors, and president to tackle tough issues. Many in this generation of elected officials have retreated from that level of tough leadership. They have failed to solve the accumulating science challenges, preferring to punt them into the future or, increasingly, deny they even exist.

With the continued delay this failure of leadership imposes, these problems get worse, and solutions become even more difficult and expensive. With each step away from reason and into denial, the country moves toward a state of tyranny, in which public policy decisions come to be based not on knowledge, but on the most loudly voiced opinions.

We need candidates of both major parties who will lead on these tough science questions, and who will reassert the primacy of knowledge and science as the best basis for informed, effective, and fair public policies in a diverse nation.

In order to reflect candidates' commitments to basing public policy decisions on knowledge, versus opinion or belief, we need a vehicle. The Contract from America, the Taxpayer Protection Pledge, and the No Climate Tax Pledge all seek to restrict reasoned debate. We need a pledge *for* reason, equality, transparency, and freedom. A pledge that expands reasoned debate. We need an American Science Pledge.

The American Science Pledge asks candidates to commit to the kind of civic-minded leadership citizens are owed in a democratic republic. It seeks to separate freedom lovers from authoritarians, data-based decision makers from those governed by "but faith, or opinion," and independent thinkers from ideologues.

Candidates who care about America are asked to sign this pledge to show their commitment to its five core principles and to agree to debate the fourteen top science questions in public forums.

THE AMERICAN SCIENCE PLEDGE

A renewed commitment to civic leadership based on the principles of freedom, science, and knowledge in America.

America has accomplished great things by setting freedom, knowledge and science as the bases for sound, effective, and equitable public policy. We are freer now than we have ever been. Our lives are longer, our children are healthier, and we have more options open to us than any previous generation. Our houses of worship are free from persecution. But science, reason, and knowledge have increasingly come under attack in our national dialogue and in the highest offices of our republic, causing a growing risk to our freedom. Policy challenges have been allowed to accumulate unresolved because our elected representatives have increasingly come to substitute rhetoric for facts in policy discussions. Because of this failure, the people are being deprived of the full measure of freedom and wealth we might otherwise claim.

Preserving freedom and solving our accumulating unresolved policy challenges requires elected leaders to make a renewed commitment to the following five core principles.

THE FIVE CORE PRINCIPLES

Seeking to be entrusted with the power of the people of the United States, and to represent their interests in the elected offices of the nation, I do hereby pledge to the citizens of America that:

1. PUBLIC DECISIONS MUST BE BASED ON KNOWLEDGE
   I will support public policy decisions based on the knowledge produced by science, which may be informed by economic interests and my values but never superseded by personal opinions or political objectives.

2. KNOWLEDGE IS SUPREME AND MUST NOT BE SUPPRESSED
   I will protect and defend the precious basis of America's freedom, which is the scientific consensus of knowledge, against political forces that seek to deny it, suppress it, or substitute for it rhetoric or opinion.

3. SCIENTIFIC INTEGRITY AND TRANSPARENCY MUST BE PROTECTED
   I will oppose all efforts to reduce freedom by holding back or altering scientific reports because they conflict with personal opinions, economic interests, or political objectives.

4. FREEDOM OF INQUIRY MUST BE ENCOURAGED
   I will oppose acts that reduce freedom by attacking, intimidating, interrogating, prosecuting, disparaging, or silencing scientists and academics whose research or scientific reports conflict with personal opinions, economic interests, or political objectives.

5. THE MAJOR SCIENCE POLICY ISSUES MUST BE OPENLY DEBATED
   To demonstrate my commitment to these principles and to moving America forward in solving these challenges, I will participate in one or more substantive, nonpartisan, public, televised, independently moderated debates on the top science challenges facing America.

I hereby make this pledge on this _____ day of _____, 20___.

_____

Candidate's Signature

_____

Candidate's Name

_____

Office Sought or Held

_____

Street Address

_____

City, State, Zip Code

_____   _____

Telephone   E-mail

Please return this signed pledge to
American Science Pledge
pledge@sciencedebate.org

A revised pledge may be available online at sciencedebate.org.

THE CALL FOR SCIENCE DEBATES

Given the many urgent scientific and technological challenges facing America and the rest of the world, the increasing need for accurate scientific information in political decision making, and the vital role scientific innovation plays in spurring economic growth and competitiveness, we call for candidates for elected office to attend at least one public debate dedicated to the issues of the environment, health and medicine, climate change, the economy, and science and technology policy.

THE 14 TOP SCIENCE QUESTIONS FACING AMERICA

Several of America's leading scientists, science organizations, corporate CEOs, and universities have identified the fourteen top science challenges facing America. The list is periodically reviewed and updated. In Science Debate 2008, they were:

1. Innovation. Science and technology have been responsible for half of the growth of the American economy since World War II. But several recent reports question America's continued leadership in these vital areas. What policies will you support to ensure that America remains the world leader in innovation?

2. Climate Change. Earth's climate is changing and there is concern about the potentially adverse effects of these changes on life on the planet. Please set out what your positions are on the following measures that have been proposed to address global climate change: a cap-and-trade system, a carbon tax, increased fuel-economy standards, firm carbon emissions targets, and/or research? What other policies would you support?

3. Energy. Many policy makers and scientists say energy security and sustainability are major problems facing the United States

during this century. What policies would you support to meet demand for energy while ensuring an economically and environmentally sustainable future?

4. Education. A comparison of fifteen-year-olds in thirty wealthy nations found that average science scores among US students ranked seventeenth, while average US math scores ranked twenty-fourth. What role do you think the federal government should play in preparing K-12 students for the science- and technology-driven twenty-first century?

5. National Security. Science and technology are at the core of national security like never before. What is your view of how science and technology can best be used to ensure national security, and where should we put our focus?

6. Pandemics and Biosecurity. Some estimates suggest that an emerging pandemic could kill more than three hundred million people. In an era of constant and rapid international travel, what steps should the United States take to protect our population from global pandemics and deliberate biological attacks?

7. Genetics Research. The field of genetics has the potential to improve human health and nutrition, but many people are concerned about the effects of genetic modification both in humans and in agriculture. What is the right policy balance between the benefits of genetic advances and their potential risks?

8. Stem Cells. Stem cell research advocates say it may successfully lead to treatments for many chronic diseases and injuries, saving lives, but opponents argue that using embryos as a source for stem cells destroys human life. What are your positions on government regulation and funding of stem cell research?

9. Ocean Health. Scientists estimate that some 75 percent of the world's fisheries are in serious decline and habitats around the world like coral reefs are seriously threatened. What steps, if any, should the United States take during your term to protect ocean health?

10. Water. Thirty-nine states expect some level of water shortage over the next decade, and scientific studies suggest that a majority of our water resources are at risk. What policies would you support to meet demand for water resources?

11. Space. The study of Earth from space can yield important information about climate change; focus on the cosmos can advance our understanding of the universe; and manned space travel can help us inspire new generations of youth to go into science. Can we afford all of them? How would you prioritize space in your administration?

12. Scientific Integrity. Many government scientists have reported political interference in their work. Is it acceptable for elected officials to hold back or alter scientific reports if they conflict with their own views, and how will you balance scientific information with politics and personal beliefs in your decision making?

13. Research. For many years, Congress has recognized the importance of science and engineering research to realizing our national goals. Given that the next Congress will likely face spending constraints, what priority would you give to investment in basic research in upcoming budgets?

14. Health. Americans are increasingly concerned about the cost, quality, and availability of health care. How do you see science, research, and technology contributing to improved health and quality of life, and what do you believe is the solution to America's "health care crisis"?

Leaving these challenges unaddressed takes from Americans the full measure of health, wonder, freedom, and prosperity we might otherwise secure with the help of science and exposes our lands, waters, and bodies to disease and destruction. Tackling them restores the promise of democracy and provides hope for our children.

Notes

Chapter 1. Let's Have a Science Debate

1. Monckton, C. Lord Christopher Monckton at April 15, 2010 Tax Day Tea Party, Washington, DC. YouTube.com, April 15, 2010. www.youtube.com/watch?v=OO-BWhfPqGQ. [audiovisual footage] Monckton spoke at the Tax Day Tea Party Rally on April 15, 2010, in Washington, DC. The rally was sponsored by the oil-industry-supported conservative organization FreedomWorks. FreedomWorks was cofounded in 1984 as Citizens for a Sound Economy by lobbyist and former congressman Dick Armey (R-TX) (Continetti, M. The Paranoid Style in Liberal Politics. *Weekly Standard* magazine 2011;16(28). www.weeklystandard.com/articles/paranoid-style-liberal-politics_555525.html) and David and Charles Koch of Koch Industries, which Forbes ranks as the second-largest private company in America and the largest privately held energy company. Armey is currently the foundation's chairman, and David Koch served as chairman of the board of Citizens for a Sound Economy.

2. Jefferson, T. Letter to Richard Price, January 8, 1789. Thomas Jefferson: Creating a Virginia Republic, Library of Congress, August 3, 2010. www.loc.gov/exhibits/jefferson/jeffrep.html.

3. Smith, A. *An Inquiry Into the Nature and Causes of the Wealth of Nations*. Dublin: Whitestone, 1776.

4. Wilson, E. O. *Consilience: The Unity of Knowledge*. New York: Knopf, 1998.

5. Manning, J. E. *Membership of the 112th Congress: A Profile*. Washington, DC: Congressional Research Service, March 29, 2011. www.fas.org/sgp/crs/misc/R41647.pdf.

6. Otto, S. L., & Kirshenbaum, S. Science on the Campaign Trail. *Issues in Science and Technology* 2009;25(2). www.issues.org/25.2/p_otto.html.

7. ScienceDebate.org. The Top 14 Science Questions Facing America. ScienceDebate.org, n.d. www.sciencedebate.org/questions.html.

8. Kirshenbaum, S. R., et al. Science and the Candidates. *Science* 2008;320(5873):182.

9. Dean, C. No Democratic Science Debate, Yet. The Caucus, NYTimes.com, April 8, 2008. http://thecaucus.blogs.nytimes.com/2008/04/08/no-democratic-science-debate-yet. [blog]

10. Keim, B. Clinton and Obama Talk Religion, Not Science. Wired Science, Wired.com, April 8, 2008. www.wired.com/wiredscience/2008/04/clinton-and-oba. [blog]

11. Committee on Prospering in the Global Economy of the 21st Century: An Agenda for American Science and Technology, National Academy of Sciences, National Academy of Engineering, and Institute of Medicine. *Rising Above the Gathering Storm: Energizing and Employing America for a Brighter Economic Future*. Washington, DC: National Academies Press, 2007.

12. Greenwald, G. David Gregory Shows Why He's the Perfect Replacement for Tim Russert. Salon.com, December 29, 2008. www.salon.com/news/opinion/glenn_greenwald/2008/12/29/gregory.

13. Russell, C. *Covering Controversial Science: Improving Reporting on Science and Public Policy*. Working Paper #2006-4. Cambridge, MA: Joan Shorenstein Center on the Press, Politics and Public Policy, 2006.

14. Freedman, A. NBC Fires Weather Channel Environmental Unit. *Capital Weather Gang,* WashingtonPost.com, November 21, 2008. http://voices.washingtonpost.com/capitalweathergang/2008/11/nbc_fires_twc_environmental_un.html. [blog]

15. Brainard, C. CNN Cuts Entire Science, Tech Team. *The Observatory,* CJR.org, December 4, 2008. www.cjr.org/the_observatory/cnn_cuts_entire_science_tech_t.php. [blog]

16. Russell, C. Globe Kills Health/Science Section, Keeps Staff. *The Observatory,* CJR.org, March 4, 2009. www.cjr.org/the_observatory/globe_kills_healthscience_sect.php. [blog]

17. León, B. Science Related Information in European Television: A Study of Prime-Time News. *Public Understanding of Science* 2008;17(4):443–460.

18. Irwin, A. Science Journalism "Flourishing" in Developing World. Science and Development Network, February 18, 2009. www.scidev.net/en/news/science-journalism-flourishing-in-developing-world.html.

19. Dean, C. Physicists in Congress Calculate Their Influence. *New York Times,* June 10, 2008.

20. Colson, C., & Pearcey, N. *How Now Shall We Live?* Wheaton, IL: Tyndale House, 1999.

21. Cooperman, A. DeLay Criticized for "Only Christianity" Remarks. *Washington Post,* April 20, 2002.

22. Perl, P. Absolute Truth. *Washington Post,* May 13, 2001. www.washingtonpost.com/wp-dyn/content/article/2006/11/28/AR2006112800700.html.

23. Nissimov, R. DeLay's College Advice: Don't Send Your Kids to Baylor or A&M. *Houston Chronicle,* April 18, 2002.

24. Boehner, J. A., & Chabot, S. Letter to Jennifer L. Sheets and Cyrus B. Richardson Jr., Ohio State Board Education, March 15, 2002. www.discovery.org/articleFiles/PDFs/Boehner-OhioLetter.PDF.

25. Bush, G. H. W. Remarks to the National Academy of Sciences, Washington, DC, April 23, 1990. *American Presidency Project,* n.d. www.presidency.ucsb.edu/ws/?pid=18393.

26. Marquis, C. Bush Misuses Science Data, Report Says. *New York Times,* August 8, 2003. www.nytimes.com/2003/08/08/politics/08REPO.html.

27. Revkin, A. C. Climate Expert Says NASA Tried to Silence Him. *New York Times,* January 29, 2006. www.nytimes.com/2006/01/29/science/earth/29climate.html.

28. Revkin, A. Official Played Down Emissions' Links to Global Warming. *New York Times,* June 7, 2005. www.nytimes.com/2005/06/07/science/07cnd-climate.html.

29. Eilperin, J. Censorship Is Alleged at NOAA. *Washington Post,* February 11, 2006. www.washingtonpost.com/wp-dyn/content/article/2006/02/10/AR2006021001766.html. For a more detailed account, see Union of Concerned Scientists. Agencies Control Scientists' Contacts with the Media. n.d. www.ucsusa.org/scientific_integrity/abuses_of_science/agencies-control-scientists.html.

30. US House of Representatives Committee on Oversight and Government Reform, Committee Holds Hearings on Political Influence on Government Climate Change Scientists. n.d. http://democrats.oversight.house.gov/index.php?option=com_content&task=view&id=2607&Itemid=2.

31. Union of Concerned Scientists, Scientific Integrity in Policy Making (March 2004). n.d. www.ucsusa.org/scientific_integrity/abuses_of_science/reports-scientific-integrity.html.

32. Editors. Abortion and Breast Cancer. *New York Times,* January 6, 2003. www.nytimes.com/2003/01/06/opinion/06MON1Y.html. [editorial] For a more detailed examination of this issue, see Malek, K. The Abortion-Breast Cancer Link: How Politics Trumped Science and Informed Consent. *Journal of American Physicians and Surgeons* 2003;8(2):41–45. http://abortionno.org/pdf/breastcancer.pdf.

33. Population Council. Emergency Contraception's Mode of Action Clarified. Population Briefs 2005;11(2):3. www.popcouncil.org/pdfs/popbriefs/pbmay05.pdf. See also Müller, A. L., et al. Postcoital Treatment with Levonorgestrel Does Not Disrupt Postfertilization Events in the Rat. *Contraception* 2003;67(5):415–419.

34. US Food and Drug Administration Center for Drug Evaluation and Research. Transcript of Nonprescription Drugs Advisory Committee in Joint Session with the Advisory Committee for Reproductive Health Drugs. December 16, 2003. www.fda.gov/ohrms/dockets/ac/03/transcripts/4015T1.DOC. The briefing document collection is posted at www.fda.gov/ohrms/dockets/ac/03/briefing/4015b1.htm.

35. Harris, G. F.D.A. Failing in Drug Safety, Official Asserts. *New York Times,* November 19, 2004. www.nytimes.com/2004/11/19/business/19fda.html.

36. Graham, D. Testimony of David J. Graham, MD, MPH, November 18, 2004. ConsumersUnion.org, n.d. www.consumersunion.org/pub/campaignprescriptionforchange/001651.html.

37. Loudon, M. The FDA Exposed: An Interview with Dr. David Graham, the Vioxx Whistleblower.NaturalNews.com, August 30, 2005. www.naturalnews.com/011401.html. For an overview and timeline of the Vioxx catastrophe, see Rubin, R. How did Vioxx Debacle Happen? *USA Today,* October 11, 2004. www.usatoday.com/news/health/2004-10-11-vioxx-main_x.htm.

38. Keim, B. McCain's VP Wants Creationism Taught in School. Wired Science, Wired.com, August 29, 2008. www.wired.com/wiredscience/2008/08/mccains-vp-want. [blog]

39. Jaschik, S. Defending the Fruit Flies from Sarah Palin. *Inside Higher Ed,* October 28, 2008. www.insidehighered.com/news/2008/10/28/palin.

40. Olsen, E. R. John McCain . . . Tough on Pork, Hard on Grizzlies. Environment Blog, Scienceline.org, September 30, 2008. http://scienceline.org/2008/09/blog-olson-grizzlymccain. [blog]

41. Politifact.com. Bear Study Funding Actually Undersold. Politifact.com, n.d. www.politifact
.com/truth-o-meter/statements/2008/sep/26/john-mccain/bear-study-funding-actually-undersold.

42. Henig, J. Did Obama Request a $3 Million "Overhead Projector," as McCain Claimed? Factcheck.org,
October 14, 2008. www.factcheck.org/askfactcheck/did_obama_request_a_3_million_overhead.html.

43. Adler Planetarium. Statement about Senator John McCain's Comments at the Presidential Debate.
October 8, 2008. www.boston.com/bostonglobe/ideas/brainiac/AdlerStatement_aboutdebate.pdf.
[news release]

CHAPTER 2. IS SCIENCE POLITICAL?

1. Washington, G. First Annual Message to Congress, Washington, DC, January 8, 1790. Miller Center of
Public Affairs, University of Virginia, n.d. http://millercenter.org/scripps/archive/speeches/detail/3448.

2. Bacon, F. *Novum Organum.* London: Joannem Billium, 1620.

3. Halsall, P. The Crime of Galileo: Indictment and Abjuration of 1633. Internet Modern History
Sourcebook, January 1999. www.fordham.edu/halsall/mod/1630galileo.html.

4. Hobbes, T. *Leviathan; or, The Matter, Form, and Power of a Commonwealth Ecclesiastical and Civil,*
3rd ed. London: Goerge Routledge and Sons, 1887.

5. Newman, W. L. Age of the Earth. In *Geologic Time.* Reston, VA: United States Geological Survey,
July 9, 2007. http://pubs.usgs.gov/gip/geotime/age.html.

6. Newman, W. L. Radiometric Time Scale. In *Geologic Time.* Reston, VA: United States Geological
Survey, June 13, 2001. http://pubs.usgs.gov/gip/geotime/radiometric.html.

7. Gold, R. B. The Implications of Defining When a Woman Is Pregnant. *Guttmacher Report on Public
Policy* 2005;8(2):7–10. www.guttmacher.org/pubs/tgr/08/2/gr080207.html.

8. J. Craig Venter Institute. First Self-Replicating Synthetic Bacterial Cell. May 20, 2010. www.jcvi
.org/cms/press/press-releases/full-text/article/first-self-replicating-synthetic-bacterial-cell-constructed-
by-j-craig-venter-institute-researcher. [news release]

9. Ferris, T. *The Science of Liberty.* New York: Harper, 2010.

10. Jefferson, T. Letter to James Madison, December 20, 1787. *The Writings of Thomas Jefferson:
Correspondence.* Ed. Henry Augustine Washington. New York: Derby and Jackson, 1859. p. 332.

CHAPTER 3. RELIGION, MEET SCIENCE

1. Jefferson, T. Memorial on the Book Duty. In *Thomas Jefferson: Writings.* Ed. M. D. Peterson. New
York: Library of America, 1984.

2. Ferris, T. *The Science of Liberty.* New York: Harper, 2010.

3. Cohen, I. B., ed. *Puritanism and the Rise of Modern Science: The Merton Thesis.* New Brunswick, NJ:
Rutgers University Press, 1990.

4. Becker, G. Pietism's Confrontation with Enlightenment Rationalism: An Examination of the Relation
Between Ascetic Protestantism and Science. *Journal for the Scientific Study of Religion* 1991;30(2):
139–158.

5. Raven, C. E. *John Ray, Naturalist: His Life and Works.* Cambridge, UK: Cambridge University Press,
1942.

6. St. Germain, C. *The Doctor and Student.* Ed. W. Muchall. Cincinnati: R. Clarke, 1874.

7. Boyer, A. D. *Sir Edward Coke and the Elizabethan Age.* Stanford, CA: Stanford University Press, 2003.
p. 65.

8. Coke, S. E. Calvin's Case, or the Case of the Postnati. In *The Selected Writings and Speeches of Sir Edward
Coke.* Ed. S. E. Coke and S. Sheppard. Indianapolis: Liberty Fund, 2003.

9. Coke, S. E. The First Part of the Institutes of the Lawes of England; Or, a Commentary upon Littleton,
Not the Name of the Author Only, but of the Law It Selfe. In *The Selected Writings and Speeches of Sir
Edward Coke.* Ed. S. E. Coke and S. Sheppard. Indianapolis: Liberty Fund, 2003.

10. Vaughan, R. The Life of John Milton. In *Paradise Lost,* by J. Milton. London: Cassell, 1894.

11. Jacob, M. C. *Scientific Culture and the Making of the Industrial West.* New York: Oxford University
Press, 1997. p. 28.

12. Hillerbrand, H. J. *The Protestant Reformation,* revised edition. New York: HarperPerennial, 2009.

13. Descartes, R. *Discourse on the Method of Rightly Conducting One's Reason and of Seeking Truth in the Sciences.* 1637. www.gutenberg.org/ebooks/59.

14. Aristotle. *Organon.* Rhodes: Andronicus, 50 BC.

15. Bacon, F. *Novum Organum.* London: Joannem Billium, 1620.

16. Ross, S. Scientist: The Story of a Word. *Annals of Science* 1962;18(2):65–85. www.scribd.com/doc/42338381/Ross-1964-Scientist-the-Story-of-a-Word.

17. Jefferson, T. Letter to Peter Carr, September 7, 1814. *68 Letters to and from Jefferson, 1805–1817.* University of Virginia Library, n.d. http://etext.virginia.edu/etcbin/toccer-new2?id=Jef1Gri.sgm&images=images/modeng&data=/texts/english/modeng/parsed&tag=public&part=5&division=div1.

18. Merton, R. K. *Science, Technology and Society in Seventeenth-Century England.* New York: Howard Fertig, 1970.

19. Cohen, I. B. *Puritanism and the Rise of Modern Science: The Merton Thesis.* New Brunswick, NJ: Rutgers University Press, 1990.

20. White, M. *Isaac Newton: The Last Sorcerer.* Reading, MA: Addison-Wesley, 1997.

21. Anonymous. The World Will End in 2060, According to Newton. *London Evening Standard,* June 19, 2007. www.thisislondon.co.uk/news/article-23401099-the-world-will-end-in-2060-according-to-newton.do.

22. White. *Isaac Newton: The Last Sorcerer.*

23. Ibid.

24. Newton, I. *Isaac Newton's Philosophiae Naturalis Principia Mathematica,* 3rd edition. Ed. I. B. Cohen and A. Koyré. London: Cambridge University Press, 1972.

25. Ferris. *The Science of Liberty.*

26. Bedini, S. A. *Jefferson and Science.* Charlottesville, VA: Thomas Jefferson Foundation, 2002.

27. Hayes, K. J. *The Road to Monticello: The Life and Mind of Thomas Jefferson.* Oxford, UK: Oxford University Press, 2008.

28. Jefferson, T. Letter to Meriwether Lewis, June 20, 1803. Monticello.org, n.d. www.monticello.org/site/jefferson/jeffersons-instructions-to-meriwether-lewis.

29. Bedini, S. A. *Thomas Jefferson, Statesman of Science.* New York: Macmillan, 1990.

30. Jefferson, T. Letter to Pierre Samuel du Pont de Nemours, 1809. In *Correspondence between Thomas Jefferson and Pierre Samuel du Pont de Nemours 1798–1817.* Ed. D. Malone. Cambridge, MA: Da Capo Press, 1970.

31. Jefferson, T. Letter to James Madison, February 17, 1826. In *The Writings of Thomas Jefferson, Volume XVI.* Ed. A. E. Bergh. Washington, DC: Thomas Jefferson Memorial Association of the United States, 1907. p. 155.

32. Jefferson, T. *Thomas Jefferson: Writings.* Ed. M. D. Peterson. New York: Library of America, 1984.

33. Jefferson, T. Letter to Benjamin Rush, January 16, 1811. In *Thomas Jefferson, 1743-1826: Letters: Relations with Adams.* University of Virginia Library, n.d. http://etext.virginia.edu/etcbin/toccer-new2?id=JefLett.sgm&images=images/modeng&data=/texts/english/modeng/parsed&tag=public&part=205&division=div1.

34. Locke, J. *An Essay Concerning Human Understanding.* Oxford: Clarendon Press, 1964.

35. Ibid.

36. Ibid.

37. Franklin, B. *The Writings.* Vol. 10. Ed. A. H. Smyth. New York: Macmillan, 1907. p. 148.

38. Jefferson, T., et al. Declaration of Independence Rough Draft with Edits by Franklin and Adams. Declaring Independence: Drafting the Documents, Library of Congress, July 23, 2010. www.loc.gov/exhibits/declara/images/draft1.jpg.

39. Locke. *An Essay Concerning Human Understanding.*

40. Hume, D. An Enquiry Concerning Human Understanding. In *English Philosophers of the Seventeenth and Eighteenth Centuries: Locke, Berkeley, Hume, with Introductions, Notes and Illustrations.* New York: P. F. Collier and Son, 1910.

41. Newton, I. *Sir Isaac Newton: Theological Manuscripts*. Ed. H. McLachlan. Liverpool, UK: Liverpool University Press, 1950.

42. Hume. An Enquiry Concerning Human Understanding.

CHAPTER 4. SCIENCE, MEET FREEDOM

1. Levin, S. Interview, November 21, 2010, by S. Otto.

2. Coke, S. E. Calvin's Case, or the Case of the Postnati. In *The Selected Writings and Speeches of Sir Edward Coke*. Ed. S. E. Coke and S. Sheppard. Indianapolis: Liberty Fund, 2003.

3. Ferris, T. *The Science of Liberty*. New York: HarperCollins, 2010.

4. Tocqueville, A. *Democracy in America*. Ed. F. Bowen. Trans. H. Reeve. Cambridge, MA: Sever and Francis, 1864.

5. Keeter, S. Public Praises Science; Scientists Fault Public, Media. Pew Research Center for the People and the Press, July 9, 2009. http://people-press.org/files/legacy-pdf/528.pdf. [news release]

6. Millikan, R. A. Science and Society. *Science* 1923;58(1503):293–298.

7. Hitchcock, A. S. Remarks on the Scientific Attitude. *Science* 1924;59(1535):476–477.

8. Christianson, G. E. *Edwin Hubble: Mariner of the Nebulae*. Chicago: University of Chicago Press, 1995.

9. Ibid.

10. Nasaw, D. *Andrew Carnegie*. New York: Penguin Press, 2006.

11. Chernow, R. *Titan: The Life of John D. Rockefeller, Sr.* New York: Vintage Books, 1998.

12. Christianson. *Edwin Hubble*.

13. Isaacson, W. *Einstein: His Life and Universe*. New York: Simon & Schuster, 2007.

14. Anonymous. Einstein to Leave Berlin: Is Aroused by Unfair Attacks on Relativity Theory. *New York Times*, August 29, 1920.

15. Van Dongen, J. On Einstein's Opponents, and Other Crackpots. *Studies in History and Philosophy of Science Part B: Studies in History and Philosophy of Modern Physics* 2010;41(1):78–80.

16. Isaacson. *Einstein*.

17. McWilliams, C. Sunlight in My Soul. In *The Aspirin Age, 1919–1941*. Ed. I. Leighton. New York: Simon & Schuster, 1963.

18. Editor. Editor's Addendum. Mr. Bryan on Evolution. n.d. www.icr.org/article/mr-bryan-evolution.

19. Darwin, C. R. Letter 12041—Darwin, C. R. to Fordyce, John, 7 May 1879. Darwin Correspondence Project, n.d. www.darwinproject.ac.uk/entry-12041.

20. Tennessee State Legislature. *Tennessee Anti-Evolution Statutes. Public Acts of the State of Tennessee Passed by the 64th General Assembly 1925. Chapter No. 27. House Bill No. 185*. March 21, 1925. www.law .umkc.edu/faculty/projects/ftrials/scopes/tennstat.htm.

21. Sutton, M. *Aimee Semple McPherson and the Resurrection of Christian America*. Cambridge, MA: Harvard University Press, 2009.

22. Ibid.

23. Miller, K. Interview, August 18, 2010, by S. Otto.

24. Christianson. *Edwin Hubble*.

25. Johnson, G. *Miss Leavitt's Stars: The Untold Story of the Woman Who Discovered How to Measure the Universe*. New York: W.W. Norton, 2005.

26. Shapley, H. *Through Rugged Ways to the Stars*. New York: Scribner, 1969.

27. Christianson. *Edwin Hubble*.

28. Pickover, C. *Archimedes to Hawking: Laws of Science and the Great Minds Behind Them*. New York: Oxford University Press, 2008

29. Christianson. *Edwin Hubble*.

30. Sandage, A. Interview, August 3, 2004, by S. Otto.

31. Ibid.

32. Ibid.

33. Ibid.

34. Davis, E. Letter to Edwin Hubble, November 23, 1951. Huntington Library Collection, HUB Box 10, f. 238. San Marino, CA: Henry Huntington Library. http://cdn.calisphere.org/data/13030/rd/tf7b69n8rd/files/tf7b69n8rd.pdf.

CHAPTER 5. GIMME SHELTER

1. Bird, K., & Sherwin, M. J. *American Prometheus: The Triumph and Tragedy of J. Robert Oppenheimer.* New York: Vintage Books, 2006.

2. Hoffmann, H., & Berchtold, J. *Hitler über Deutschland.* Munich: Franz Eher Nachfolger GmbH, 1932.

3. Norton-Taylor, R, ed. *Nuremberg: The War Crimes Trial: Transcript.* London: Nick Hern Books, 1997. p 55.

4. Wiesner, J. B. *Vannevar Bush: 1890–1974.* Washington, DC: National Academy of Sciences, 1979. www.nap.edu/html/biomems/vbush.pdf.

5 Bethe, H. A. *J. Robert Oppenheimer: 1904–1967.* Washington, DC: National Academy of Sciences, 1997.

6. Laurence, W. Atomic Bombing of Nagasaki Told by Flight Member. *New York Times*, September 9, 1945.

7. Laurence, W. U.S. Atom Bomb Site Belies Tokyo Tales. *New York Times*, September 12, 1945.

8. Laurence, W. Drama of the Atomic Bomb Found Climax in July 16 Test. *New York Times*, September 26, 1945.

9. Freed, F., producer. The Decision to Drop the Bomb. *NBC White Paper,* 1965. [television documentary]

10. Bird. *American Prometheus.*

11. Anonymous. Atomic Education Urged by Einstein; Scientist in Plea for $200,000 to Promote New Type of Essential Thinking. *New York Times*, May 25, 1946.

12. Chargaff, E. *Heraclitean Fire: Sketches from a Life Before Nature.* New York: Rockefeller University Press, 1978.

13. Bradley, O. Armistice Day Speech, Boston, Massachusetts, November 10, 1948. *The Collected Writings of General Omar N. Bradley.* Volume 1. Washington, DC: United States Government Printing Office, 1977. pp. 584–589.

14. Roosevelt, F. D. President Roosevelt's Letter to Vannevar Bush, November 17, 1944. www.nsf.gov/od/lpa/nsf50/vbush1945.htm#letter.

15. Bush, V. *Science, the Endless Frontier.* Washington, DC: US Government Printing Office, 1945. www.nsf.gov/od/lpa/nsf50/vbush1945.htm.

16. Lapp, R. An Interview with Governor Val Peterson. *Bulletin of the Atomic Scientists* 1953;9(7):237–242.

17. Monson, D. Is Dispersal Obsolete? *Bulletin of the Atomic Scientists* 1954;10(10):378–383.

18. Winkler, A. M. *Life Under a Cloud: American Anxiety about the Atom.* Urbana: University of Illinois Press, 1999. p. 117.

19. Schwartz, S., et al. Excerpts from Atomic Audit. *Bulletin of the Atomic Scientists* 1998;54(5):36–43.

20. Lapp, R. An Interview with Governor Val Peterson. *Bulletin of the Atomic Scientists* 1954;10(10):375–377.

21. Brown, J. "A Is for Atom, B Is for Bomb": Civil Defense in American Public Education, 1948–1963. *Journal of American History* 1998;75(1):68–90. www.jstor.org/pss/1889655.

22. Finney, J. Sputnik Acts as a Spur to U.S. Science and Research. *New York Times,* November 2, 1957.

23. Pell, D. *If the Bomb Falls: A Recorded Guide to Survival.* Los Angeles: Tops Records, 1961. [record album]

24. Nicolay, J. G., & Hay, J. *Abraham Lincoln: A History.* Volume 2. New York: Cosimo Classics, 1917.

25. Snow, C. P. *The Two Cultures.* Cambridge, UK: Cambridge University Press, 1960.

26. Editors. The Hundred Most Influential Books Since the War. *Times Literary Supplement,* December 30, 2008. http://entertainment.timesonline.co.uk/tol/arts_and_entertainment/the_tls/article5418361.ece.

27. Eisenhower, D. D. Farewell Address to the American People. *Eisenhower Presidential Library and Museum: Dwight D. Eisenhower Speeches.* n.d. www.eisenhower.archives.gov/all_about_ike/Speeches/WAV%20files/farewell%20address.mp3.

CHAPTER 6. SCIENCE, DRUGS, AND ROCK 'N' ROLL

1. Author's translation, Horace. Poem 11. *The Odes.*

2. Playboy. Bertrand Russell: Playboy Interview. *Playboy*, March 1963. www.playboy.com/articles/bertrand-russell-interview/index.html.

3. Kennedy, J. F., & Webb, J. E. Tape Recording of a Meeting between President John F. Kennedy and NASA Administrator James E. Webb, November 21, 1962. White House Meeting Tape 63. John Fitzgerald Kennedy Library, n.d.

4. Kennedy, J. F. Excerpt from an Address Before a Joint Session of Congress, 25 May 1961. John F. Kennedy Presidential Library and Museum, n.d. www.jfklibrary.org/Asset-Viewer/xzw1gaeeTES6khED14P1Iw.aspx.

5. WAMU. *Washington Goes to the Moon.* WAMU.org, May 24, 2001. http://wamu.org/programs/special/01/washington_goes_to_the_moon.php. [radio broadcast]

6. Dallek, R. *An Unfinished Life: John F. Kennedy, 1917–1963.* Boston: Little, Brown, 2003.

7. Launius, R. Interviewed in Howard McCurdy and Roger Launius on Opposition to Apollo. Background interview for *Washington Goes to the Moon.* WAMU.org, May 24, 2001. http://wamu.org/d/programs/special/moon/mccurdy-launius_opp.txt.

8. McCurdy, H. Interview in *Washington Goes to the Moon.* WAMU.org, May 24, 2001. http://wamu.org/programs/special/01/washington_goes_to_the_moon.php. [radio broadcast]

9. Launius. Interview in *Washington Goes to the Moon.*

10. Launius, R. Exploring the Myth of Popular Support for Project Apollo. Roger Launius's Blog, August 16, 2010. http://launiusr.wordpress.com/2010/8/16/exploding-the-myth-of-popular-support-for-project-apollo. [blog]

11. Augustine, N., et al. Seeking a Human Spaceflight Program Worthy of a Great Nation. Review of US Human Spaceflight Plans Committee, 2009. www.nasa.gov/offices/hsf/meetings/10_22_pressconference.html.

12. Franklin, J. *The New Priesthood—The Scientific Elite and the Uses of Power. Engineering and Science* 1965;28(9):4. http://calteches.library.caltech.edu/2383/1/books.pdf.

13. Lapp, R, The New Priesthood: The Scientific Elite and the Uses of Power. New York: Harper and Row, 1965.

14. Reagan, R. Address by Governor Ronald Reagan, Installation of President Robert Hill, Chico State College, May 20, 1967. Ronald Reagan Presidential Library, n.d. www.reagan.utexas.edu//archives/speeches/govspeech/05201967a.htm.

15. Gleason, R. Bob Dylan: Poet to a Generation. *Jazz and Pop,* December 1968. pp. 36–37. www.loc.gov/folklife/guides/BibDylan.html.

16. Vonnegut, K. American Notes: Vonnegut's Gospel. *Time,* June 29, 1970.

17. Heppenheimer, T. A. *The Space Shuttle Decision: NASA's Search for a Reusable Space Vehicle.* Washington, DC: National Aeronautics and Space Administration, 1999.

18. Launius, R. Managing the Unmanageable: Apollo, Space Age Management and American Social Problems. *Space Policy* 2008;24(3):158–165. http://si-pddr.si.edu/jspui/bitstream/10088/8213/1/Launius_2008_Managing_the_unmanageable.pdf.

19. Laursen, L. @ApolloPlus40—A Colossal Perversion. In the Field, Nature.com, July 7, 2009. http://blogs.nature.com/inthefield/2009/07/apolloplus40_a_colossal_perver.html. [blog]

20. Heppenheimer, T. A. *Winter of Discontent. The Space Shuttle Decision.* NASA History Series SP-4221. Washington, DC: National Aeronautics and Space Administration, 1999. http://history.nasa.gov/SP-4221/ch4.htm.

21. Newport, F. Landing a Man on the Moon: The Public's View. Gallup News Service, July 20, 1999. www.gallup.com/poll/3712/landing-man-moon-publics-view.aspx.

22. Pion, G., & Lipsey, M. Public Attitudes Toward Science and Technology: What Have the Surveys Told Us? *Public Opinion Quarterly* 1981;45(3):303–316.

23. Sagan, C. *The Demon-Haunted World: Science as a Candle in the Dark*. New York: Random House, 1995.

24. Ibid.

25. Mooney, C., & Kirshenbaum, S. *Unscientific America: How Scientific Illiteracy Threatens Our Future*. New York: Basic Books, 2009.

26. Flam, F. What Should It Take to Join Science's Most Exclusive Club? *Science* 1992;256(5059):960–961.

27. Luey, B. Are Fame and Fortune the Kiss of Death? *Publishing Research Quarterly* 2007;19(3):35–44.

28. Hartz, J., & Chappell, R. Worlds Apart: *How the Distance Between Science and Journalism Threatens America's Future*. Nashville, TN: Freedom Forum First Amendment Center, 1997.

29. Shermer, M. Stephen Jay Gould as Historian of Science and Scientific Historian, Popular Scientist and Scientific Popularizer. *Social Studies of Science* 2002;32(4):489–525.

30. Jensen, P., et al. Scientists Who Engage with Society Perform Better Academically. *Science and Public Policy* 2008;35(7):527–541.

31. Keeter, S. Public Praises Science; Scientists Fault Public, Media. The Pew Research Center for the People & the Press, July 9, 2009. p. 2. http://people-press.org/2009/07/09/public-praises-science-scientists-fault-public-media.

CHAPTER 7. AMERICAN ANTISCIENCE

1. Kant, I. An Answer to the Question: What Is Enlightenment? In *Kant: Political Writings*. Ed. H. Reiss, trans. H. Nisbet. Cambridge: Cambridge University Press, 1970.

2. Rasmussen, C. Billy Graham's Star Was Born at His 1949 Revival in Los Angeles. *Los Angeles Times*, September 2, 2007. http://articles.latimes.com/2007/sep/02/local/me-then2.

3. Ibid.

4. Graham, B. Why a Revival? *Billy Graham 1949 Christ for Greater Los Angeles Campaign*. Wheaton, IL: Billy Graham Center, Wheaton College, 1949. http://espace.wheaton.edu/bgc/audio/cn026t5702a.mp3. [audio recording]

5. Kant. An Answer to the Question.

6. Graham, B. The Solution to Modern Problems. *Billy Graham New York Crusade, Madison Square Garden*, June 1, 1957. Billy Graham Center. www.wheaton.edu/bgc/archives/exhibits/NYC57/12sample65.htm. [audiovisual footage]

7. Graham, B. Heart Trouble. *Billy Graham New York Crusade, Madison Square Garden*, May 23, 1957. Billy Graham Center. www.wheaton.edu/bgc/archives/exhibits/NYC57/18sample97-1.htm. [audiovisual footage] Video excerpt at www.youtube.com/watch?v=7i95RXDyY70.

8. Saad, L. Barack Obama, Hillary Clinton Are 2010's Most Admired. Gallup.com, December 27, 2010. www.gallup.com/poll/145394/barack-obama-hillary-clinton-2010-admired.aspx.

9. Hadden, J., & Swann, C. *Prime Time Preachers: The Rising Power of Televangelism*. Reading, MA: Addison-Wesley, 1981. pp. 47–55. http://etext.virginia.edu/etcbin/toccer-new2?id=HadPrim.sgm&images=images/modeng&data=/texts/english/modeng/parsed&tag=public&part=all.

10. Diamond, S. *Not by Politics Alone: The Enduring Influence of the Christian Right*. New York: Guilford Press, 2000.

11. Ibid.

12. Barinaga, M. California Backs Evolution Education. *Science* 1989;246(4932):881.

13. Miller, A. The New Catastrophism and Its Defender. *Science* 1922;55(1435):701–703.

14. Nietzsche, F. *Thus Spake Zarathustra*. London: Macmillan, 1896.

15. Russell, B. *A History of Western Philosophy*. New York: Routledge, 2004.

16. Kuhn, T. *The Structure of Scientific Revolutions*. Chicago: University of Chicago Press, 1962.

17. Planck, M. *Scientific Autobiography and Other Papers*. New York: Philosophical Library, 1949. pp. 33–34.

18. Bacon, F. *Novum Organum*. London: Joannem Billium, 1620.

19. Churchland, P. Interview, August 24, 2010, by S. Otto.

20. Kuhn. *The Structure of Scientific Revolutions.*

21. Feyerabend, P. *Killing Time: The Autobiography of Paul Feyerabend.* Chicago: University of Chicago Press, 1995. pp. 142–143.

22. Green, L. The Messy Relationship between Religion and Science: Revisiting Galileo's Inquisition. Fox News.com, January 16, 2008. www.foxnews.com/story/0,2933,323327,00.html.

23. Editors. Science Wars and the Need for Respect and Rigour. *Nature* 1997;385(6615):373. [editorial]

24. Dart, F. E., & Pradham, P. L. Cross-Cultural Teaching of Science. *Science* 1967;155(3763):649–656. www.wmich.edu/slcsp/SLCSP102/slcsp102.pdf.

25. Wilson, B. The Cultural Contexts of Science and Mathematics Education: Preparation of a Bibliographic Guide. *Studies in Science Education* 1981;8:27–44.

26. Cobern, W. W. Alternative Constructions of Science and Science Education: A Plenary Presentation for the Second Annual Southern African Association for Mathematics and Science Education Research, University of Durban-Westville, Durban, South Africa. *Scientific Literacy and Cultural Studies Project* 1994;122.

27. Richardson, V., ed. *Constructivist Teacher Education: Building a World of New Understandings.* New York: Routledge, 1997. p. 8.

28. Beck, C. Postmodernism, Pedagogy, and Philosophy of Education. *Philosophy of Education Yearbook,* 1993. www.ed.uiuc.edu/eps/PES-Yearbook/93_docs/BECK.HTM.

29. Cobern, W. W., & Loving, C. Defining "Science" in a Multicultural World: Implications for Science Education. *Science Education* 2001;85:50–67.

30. Bloom, A. *The Closing of the American Mind: How Higher Education Has Failed Democracy and Impoverished the Souls of Today's Students.* Chicago: University of Chicago Press, 1987.

31. Bloom. *The Closing of the American Mind.*

32. Gross, P., & Levitt, N. *Higher Superstition: The Academic Left and Its Quarrels with Science.* Baltimore: Johns Hopkins University Press, 1994.

33. Ross, A., ed. *Science Wars.* Durham, NC: Duke University Press, 1996. p. 152.

34. Sokal, A. Transgressing the Boundaries: Towards a Transformative Hermeneutics of Quantum Gravity. *Social Text* 1996;(46/47):217–252. www.physics.nyu.edu/sokal/transgress_v2/transgress_v2_singlefile.html.

35. Sokal, A. A Physicist Experiments with Cultural Studies. *Lingua Franca,* May/June 1996. http://linguafranca.mirror.theinfo.org/9605/sokal.html.

36. Limbaugh, R. Unofficial Summary of the *Rush Limbaugh Show,* May 22, 1996. Ed. J. Switzer. http://jwalsh.net/projects/sokal/articles/rlimbaugh.html.

37. Scott, J. Postmodern Gravity Deconstructed, Slyly. *New York Times,* May 18, 1996. www.nytimes.com/1996/05/18/nyregion/postmodern-gravity-deconstructed-slyly.html.

38. Rosen, R. A Physics Prof Drops a Bomb on the Faux Left. *Los Angeles Times,* May 23, 1996. p. A11. www.physics.nyu.edu/faculty/sokal/rosen.html.

39. Bérubé, M. The Science Wars Redux. *Democracy* 2011(19). www.democracyjournal.org/19/6789.php.

40. Pollitt, K. Pomolotov Cocktail. *Nation,* June 10, 1996.

41. Krauss, L. Equal Time for Nonsense. *New York Times,* July 30, 1996. [opinion]

42. Schick, T., & Vaughn, L. *How to Think About Weird Things: Critical Thinking for a New Age.* Mountain View, CA: Mayfield, 1995.

43. Bacon. *Novum Organum.*

CHAPTER 8. THE DESCENT OF THOUGHT

1. Bacon, F. *Novum Organum.* London: Joannem Billium, 1620.

2. Brodeur, P. Annals of Radiation. *New Yorker* June 12, 1989–December 7, 1992. www.newyorker.com/magazine/bios/paul_brodeur/search?contributorName=paul%20brodeur.

3. Brodeur, P. *The Great Power-Line Cover-Up: How the Utilities and the Government Are Trying to Hide the Cancer Hazard Posed by Electromagnetic Fields.* Boston: Little, Brown, 1993.

4. Dennis, J. A., et al. Epidemiological Studies of Exposures to Electromagnetic Fields: II. Cancer. *Journal of Radiological Protection* 1991;11(1):13–25.

5. Park, R. L. Cellular Telephones and Cancer: How Should Science Respond? *Journal of the National Cancer Institute* 2001;93(3):166–167. http://jnci.oxfordjournals.org/content/93/3/166.full.

6. Johansen, C., et al. Cellular Telephones and Cancer—A Nationwide Cohort Study in Denmark. *Journal of the National Cancer Institute* 2001;93(3):203–207. http://jnci.oxfordjournals.org/content/93/3/203.full.

7. Guild, P. B., & Garger, S. *Marching to Different Drummers.* Alexandria, VA: Association for Supervision and Curriculum Development, 1985.

8. Yong, E. Arsenic Bacteria—A Post-Mortem, a Review, and Some Navel-Gazing. Not Exactly Rocket Science, Discovermagazine.com, December 10th, 2010. http://blogs.discovermagazine.com/notrocketscience/2010/12/10/arsenic-bacteria-a-post-mortem-a-review-and-some-navel-gazing. [blog]

9. Kadanoff, L. Hard Times. *Physics Today* 1992;45(10):9–11.

10. Halpern, M. Interview by S. Otto, March 28, 2010.

11. Bush, V. As We May Think. *The Atlantic,* July 1945. www.theatlantic.com/magazine/archive/1945/07/as-we-may-think/3881.

12. McNutt, M. Interview by S. Otto, August 27, 2010.

13. McKee, R. Interview by S. Otto, October 7, 2010.

14. Wakefield, AJ, et al. Retracted: Ileal-lymphoid-nodular hyperplasia, non-specific colitis, and pervasive developmental disorder in children. *The Lancet* 1998;351(9103):637–641. http://www.thelancet.com/journals/lancet/article/PIIS0140-6736(97)11096-0/abstract.

15. Godlee, F., et al. Wakefield's Article Linking MMR Vaccine and Autism Was Fraudulent. *BMJ* 2011;342:c7452. www.bmj.com/content/342/bmj.c7452.full.

16. Park, M. Medical Journal Retracts Study Linking Autism to Vaccine. CNN.com, February 2, 2010. http://articles.cnn.com/2010-02-02/health/lancet.retraction.autism_1_andrew-wakefield-mmr-vaccine-and-autism-general-medical-council?_s=PM:HEALTH.

17. US Food and Drug Administration. Thimerosal in Vaccines. March 31, 2010. www.fda.gov/biologicsbloodvaccines/safetyavailability/vaccinesafety/ucm096228.htm.

18. Kennedy, R. F. Deadly Immunity. *Rolling Stone,* July 14, 2005.

19. Smith, P., et al. Children Who Have Received No Vaccines: Who Are They and Where Do They Live? *Pediatrics* 2004;114(1):187–195. http://pediatrics.aappublications.org/cgi/content/full/114/1/187.

20. Karnowski, S. Autism Fears, Measles Spike among Minn. Somalis. Minnesota Public Radio News [Associated Press], April 2, 2011. http://minnesota.publicradio.org/display/web/2011/04/02/somali-autism-vaccines.

21. Porter, J. Interview by S. Otto, August 31, 2009.

22. Anonymous. Summary of Notifiable Diseases, United States, 1994. *Morbidity and Mortality Weekly Report* 1995;43(53):Tables 11-12. www.cdc.gov/mmwr/preview/mmwrhtml/00039679.htm.

23. Porter. Interview.

24. Committee on Prospering in the Global Economy of the 21st Century: An Agenda for American Science and Technology, National Academy of Sciences, National Academy of Engineering, and Institute of Medicine. *Rising Above the Gathering Storm: Energizing and Employing America for a Brighter Economic Future.* National Academies. Washington, DC: National Academies Press, 2007.

25. Medrich, E., & Griffith, J. *International Mathematics and Science Assessment: What Have We Learned?* Washington, DC: National Center for Education Statistics, 1992. p. 80. http://nces.ed.gov/pubs92/92011.pdf.

26. Organisation for Economic Co-Operation and Development Programme for International Student Assessment. *PISA 2006 Science Competencies for Tomorrow's World.* Paris: Organisation for Economic Co-operation and Development, April 12, 2007. www.oecd.org/document/2/0,3343,en_32252351_32236191_39718850_1_1_1_1,00.html.

27. Leshner, A. Interview by S. Otto, December 1, 2010.

28. Keeter, S. Public Praises Science; Scientists Fault Public, Media. The Pew Research Center for the People & the Press, July 9, 2009. http://people-press.org/2009/07/09/public-praises-science-scientists-fault-public-media.

29. Charlton Research. *Your Congress–Your Health Survey.* Washington, DC: Research!America, 2009. p. 31. www.researchamerica.org/uploads/YourCongress2009.pdf.

CHAPTER 9. TEACHING EVOLUTION: THE VALUES BATTLE

1. Spiegel, A. N., et al. Museum Visitors' Understanding of Evolution. *Museums and Social Issues* 2006;1(1):69–86. www-personal.umich.edu/~evansem/SpiegelEvansGramDiamond.pdf.

2. Tremblay, F. The Eeeevil Evolutionists and the Bumbling Christians. Goosing the Antithesis, August 2, 2008. http://goosetheantithesis.blogspot.com/2008/08/eeeevil-evolutionists-and-bumbling.html. [blog]

3. Huff, G. Schools Should Not Limit Origins-of-Life Discussions to Evolution. *Stillwater Gazette,* September 27, 2005. www.stillwatergazette.com/articles/2003/10/02/export160.txt.

4. Anonymous. *Can You Tell the Difference between Evolution and Natural Selection?* Powder Springs, GA: Creation Ministries International, n.d. www.creation.com/images/pdfs/flyers/can-you-tell-the-difference-between-evolution-and-natural-selection-p.pdf. [pamphlet]

5. Huff. Schools Should Not Limit.

6. Behe, M. *Darwin's Black Box: The Biochemical Challenge to Evolution.* New York: Touchstone, 1996.

7. Lehigh University Department of Biological Sciences. Department Position on Evolution and "Intelligent Design." n.d. www.lehigh.edu/~inbios/news/evolution.htm.

8. Locke, J. *An Essay Concerning Human Understanding.* Oxford, UK: Clarendon Press, 1964.

9. Huff. Schools Should Not Limit.

10. National Science Board. Chapter 7 in *Science and Engineering Indicators—2002.* Arlington, VA: National Science Foundation, 2001. www.nsf.gov/statistics/seind02/c7/c7s1.htm#c7s1l4a.

11. National Science Board. Chapter 7 in *Science and Engineering Indicators—2008.* Arlington, VA: National Science Foundation, 2008. www.nsf.gov/statistics/seind08/c7/c7s2.htm.

12. Kachka, B. Are You There, God? It's Me, Hitchens. *New York,* April 26, 2007. http://nymag.com/arts/books/features/31244.

13. Kirkpatrick, D. Bush Allies Till Fertile Soil, Among Baptists, for Votes. *New York Times,* June 18, 2004. www.nytimes.com/2004/06/18/politics/campaign/18baptists.html.

14. Kirkpatrick, D. Churches See an Election Role and Spread the Word on Bush. *New York Times,* August 9, 2004. www.nytimes.com/2004/08/09/us/churches-see-an-election-role-and-spread-the-word-on-bush.html.

15. Mayer, W., ed. *The Swing Voter in American Politics.* Washington, DC: Brookings Institution, 2008.

16. Brownstein, R. The Hidden History of the American Electorate. *National Journal,* October 18, 2008. www.nationaljournal.com/magazine/the-hidden-history-of-the-american-electorate-20081018.

17. Reed, R. Growing Grassroots. *NewsHour with Jim Lehrer.* Washington, DC: MacNeil/Lehrer Productions, March 2, 2004. www.pbs.org/newshour/bb/media/jan-june04/grassroots_03-02.html. [panel discussion]

18. Kirkpatrick. Churches See an Election Role.

19. Kirkpatrick. Bush Allies Till Fertile Soil.

20. Tackett, M. Laying Claim to the Nation. *Chicago Tribune,* November 7, 2004. www.chicagotribune.com/news/opinion/chi-0411070166nov07,0,595361.story.

21. Kirkpatrick. Bush Allies Till Fertile Soil.

22. Jones, J. *Memorandum and Order, Kitzmiller, et al. v. Dover School District, et al.* United States District Court, Middle District of Pennsylvania, December 20, 2005. www.pamd.uscourts.gov/kitzmiller/decision.htm.

23. Chapman, M. *40 Days and 40 Nights: Darwin, Intelligent Design, God, OxyContin and Other Oddities on Trial in Pennsylvania.* New York: Collins, 2007.

24. Berkman, M. B., et al. Evolution and Creationism in America's Classrooms: A National Portrait. *PLoS Biology* 2008:6(5):e124. www.plosbiology.org/article/info:doi/10.1371/journal.pbio.0060124.

25. Flam, F. The Difference between Science and Religion. *Philadelphia Inquirer,* April 18, 2011. http://articles.philly.com/2011-04-18/news/29443540.

26. Jones. *Memorandum and Order.*

27. Lebo, L. Judge in Dover Case Reports Hostile E-Mails: Jones and His Family Were Under Marshals' Protection in December. *York Daily Record,* March 24, 2006.

28. Jones, J. The Myth of "Activist Judges." *College News,* November 16, 2006. http://collegenews.org/editorials/2006/the-myth-of-activist-judges.html.

29. Karp, D. IRS Warns Churches: No Politics Allowed. *St. Petersburg Times,* September 15, 2004. www.sptimes.com/2004/09/15/State/IRS_warns_churches__n.shtml.

30. Seeyle, K. Moral Values Cited as a Defining Issue of the Election. *New York Times,* November 4, 2004. www.nytimes.com/2004/11/04/politics/campaign/04poll.html.

31. Priests for Life. Voter's Guide for Serious Catholics. PriestsforLife.org, 2004. www.priestsforlife.org/elections/voterguide.htm.

32. Berkowitz, B. The Christian Right's Compassion Deficit. *Dissident Voice,* December 30, 2004. http://dissidentvoice.org/Dec2004/Berkowitz1230.htm.

33. Hunter, M. Religious Watchdog Group Warns Clergy Against Passing Out Voter Guides. *Crosswalk,* October 20, 2004. www.crosswalk.com/news/religion-today/religious-watchdog-group-warns-clergy-against-passing-out-voter-guides-1291673.html.

34. Newport, F. Third of Americans Say Evidence Has Supported Darwin's Evolution Theory. Gallup.com, November 19, 2004. www.gallup.com/poll/14107/Third-Americans-Say-Evidence-Has-Supported-Darwins-Evolution-Theory.aspx.

35. Seeyle. Moral Values Cited as a Defining Issue.

36. Cosgrove-Mather, B. Poll: Creationism Trumps Evolution. CBSNews.com, November 22, 2004. www.cbsnews.com/stories/2004/11/22/opinion/polls/main657083.shtml

37. Harris, S., et al. The Neural Correlates of Religious and Nonreligious Belief. *PLoS One* 2009;4(10):e7272. www.plosone.org/article/info:doi%2F10.1371%2Fjournal.pone.0007272.

38. Bhattacharjee, Y. Scientific Literacy: NSF Board Draws Flak for Dropping Evolution from Indicators. *Science* 2010;328(5975):150–151.

39. National Science Board. Chapter 7 in *Science and Engineering Indicators—2006.* Arlington, VA: National Science Foundation, 2007. p. 19. www.nsf.gov/statistics/seind06/pdf/c07.pdf.

40. Nisbet, M. Interview by S. Otto, April 22, 2010.

41. Feynman, R. *"What Do You Care What Other People Think?" Further Adventures of a Curious Character.* New York: W. W. Norton, 1988.

42. Bacon, F. *Novum Organum.* London: Joannem Billium, 1620.

43. McKee, R. Interview by S. Otto, October 7, 2010.

44. Kaufman, L. Darwin Foes Add Warming to Targets. *New York Times,* March 3, 2010. www.nytimes.com/2010/03/04/science/earth/04climate.html.

45. Palfreman, J. The Vaccine War. *Frontline,* April 27, 2010. www.pbs.org/wgbh/pages/frontline/vaccines/view.

46. Scott, E. Interview by S. Otto, August 25, 2009.

47. Myers, P. Z. Ken Ham Brags About His Websites. Pharyngula, March 22, 2011. http://scienceblogs.com/pharyngula/2011/03/ken_ham_brags_about_his_websit.php. [blog]

48. Chyi, H. I., & Lewis, S. Use of Online Newspaper Sites Lags Behind Print Editions. *Newspaper Research Journal* 2009:30(4):38–52. http://umn.academia.edu/SethLewis/Papers/114952/Use_of_Online_Newspaper_Sites_Lags_Behind_Print_Editions.

49. Myers, P. Z. Interview by S. Otto, August 24, 2009.

50. Levin, S. Interview by S. Otto, November 21, 2010.

51. Nisbet. Interview.

52. Myers. Interview.

53. Leshner, A. I. We Need to Reward Those Who Nurture a Diversity of Ideas in Science. *Chronicle of Higher Education,* March 6, 2011. www.aaas.org/programs/centers/pe/news_svc/media/2011/0306che_leshner_reward_diversifiers.pdf.

54. Westphal, S. P. Interview by S. Otto, April 14, 2010.

55. Scott. Interview.

CHAPTER 10. CLIMATE CHANGE: THE MONEY BATTLE

1. Limbaugh, R. UN Climate Change Plan Fits with Obama's Anti-Capitalism Scheme. *Rush Limbaugh Show,* March 27, 2009. www.rushlimbaugh.com/home/daily/site_032709/content/01125111.guest.html.

2. Beck, G. "AMBER ALERT: Al Gore MIA during blizzards." *Glenn Beck: The Fusion of Entertainment and Enlightenment.* February 10, 2010. http://www.glennbeck.com/content/articles/article/198/36153.

3. Arrhenius, S. On the Influence of Carbonic Acid in the Air upon the Temperature of the Ground. *London, Edinburgh and Dublin Philosophical Magazine and Journal of Science* 1896;41(251):237–276. http://nsdl.org/archives/onramp/classic_articles/issue1_global_warming/n4.Arrhenius1896.pdf.

4. Committee on the Science of Climate Change, National Research Council. *Climate Change Science: An Analysis of Some Key Questions.* Washington, DC: National Academies Press, 2001.

5. Bush, G. W. President Bush Discusses Global Climate Change. June 11, 2001. http://georgewbush-whitehouse.archives.gov/news/releases/2001/06/20010611-2.html. [news release]

6. Cicerone, R. *Climate Change: Evidence and Future Projections.* Statement before the Oversight and Investigations Subcommittee, Committee on Energy and Commerce, US House of Representatives. Washington, DC: National Academies Press, 2006. www7.nationalacademies.org/ocga/testimony/Climate_Change_Evidence_and_Future_Projections.asp.

7. Committee on Radiative Forcing Effects on Climate, Climate Research Committee, National Research Council. *Radiative Forcing of Climate Change: Expanding the Concept and Addressing Uncertainties.* Washington, DC: National Academies Press, 2005.

8. Morrison, J. The Incredible Shrinking Polar Bears. *National Wildlife,* February 1, 2004.

9. Siegel, K., & Cummings, B. *Petition to List the Polar Bear* (Ursus maritimus) *as a Threatened Species under the Endangered Species Act.* Center for Biological Diversity, February 16, 2005. www.biologicaldiversity.org/species/mammals/polar_bear/pdfs/15976_7338.pdf.

10. Siegel, K. Conservation Group Petitions the United States Government to List the Polar Bear as a Threatened Species under the Endangered Species Act. Center for Biological Diversity, February 16, 2005. www.biologicaldiversity.org/news/press_releases/polarbear2-16-05.html. [news release]

11. Siegel, K. Interview by S. Otto, June 13, 2011.

12. Shnayerson, M. The Edge of Extinction. *Vanity Fair,* May 2008. www.vanityfair.com/politics/features/2008/05/polarbear200805.

13. Stirling, I., et al. Polar Bear Population Status in the Northern Beaufort Sea. Reston, VA: US Geological Survey, 2007. www.usgs.gov/newsroom/special/polar_bears.

14. Burnett, H. S. Are Polar Bears Dying? *Environment and Climate News,* May 1, 2006. www.heartland.org/policybot/results/18971/Are_Polar_Bears_Dying.html

15. Legates, D. *Climate Science: Climate Change and Its Impacts.* Dallas: National Center for Policy Analysis, 2006. www.ncpa.org/pub/st285?pg=7.

16. Burnett. Are Polar Bears Dying?

17. DeMelle, B. Disinformation Database: David Legates. *DeSmogBlog,* n.d. www.desmogblog.com/node/2830. [blog]

18. Taylor, M. Last Stand of Our Wild Polar Bears. *Toronto Star,* May 1, 2006. http://ff.org/centers/csspp/library/co2weekly/20060505/20060505_17.html. [opinion]

19. Burnett, H. S. ESA Listing Not Needed for Polar Bears. *Environment and Climate News,* March 1, 2007. www.heartland.org/policybot/results/20631/ESA_Listing_Not_Needed_for_Polar_Bears.html.

20. Editorial Board. Polar Bear Politics. *Wall Street Journal,* January 3, 2007. [editorial]

21. Stirling, I. Polar Bears and Seals in the Eastern Beaufort Sea and Amundsen Gulf: A Synthesis of Population Trends and Ecological Relationships over Three Decades. *Arctic* 2002;55(Suppl 1):59–76. http://arctic.synergiesprairies.ca/arctic/index.php/arctic/article/download/735/761.

22. Government of Nunavut. Minister Accepts Decisions of the Nunavut Wildlife Management Board on Polar Bear Management. January 7, 2005. www.gov.nu.ca/Nunavut/English/news/2005/jan/jan7.pdf. [news release]

23. George, J. Global Warming Won't Hurt Polar Bears, GN Says. *Nunatsiaq News,* May 26, 2006. www.nunatsiaqonline.ca/archives/60526/news/climate/60526_01.html.

24. Thompson, J. Polar Bear Die-Off Unlikely: GN Official. *Nunatsiaq News,* September 14, 2007. www.nunatsiaqonline.ca/archives/2007/709/70914/news/nunavut/70914_498.html.

25. Soon, W., et al. Reconstructing Climate and Environmental Changes of the Past 1000 Years: A Reappraisal. *Energy and Environment* 2003;14(2–3):233–296. www.marshall.org/pdf/materials/132.pdf.

26. Dyck, M. G., et al. Polar Bears of Western Hudson Bay and Climate Change: Are Warming Spring Air Temperatures the "Ultimate" Survival Control Factor? *Ecological Complexity* 2007;4(3):73–84.

27. Shaw, P. *The Philosophical Works of Francis Bacon, Baron of Verulam, Viscount St. Albans, and Lord High-Chancellor of England: Methodized, and Made English, from the Originals. Volume II*. London: J. J. and P. Knapton, et al., 1733.

28. Sanchez, I. Warming Study Draws Fire: Harvard Scientists Accused of Politicizing Research. *Harvard Crimson,* September 12, 2003. www.thecrimson.com/article/2003/9/12/warming-study-draws-fire-a-study.

29. Editorial Staff. Guide for Authors: Conflict of Interest. Elsevier.com, n.d. www.elsevier.com/wps/find/journaldescription.cws_home/701873/authorinstructions#7000.

30. DeMelle, B. Disinformation Database: An Extensive Database of Individuals Involved in the Global Warming Denial Industry. *DeSmogBlog,* n.d. www.desmogblog.com/global-warming-denier-database.

31. Wahl-Jorgensen, K., ed. *The Handbook of Journalism Studies.* New York: Taylor and Francis, 2009.

32. Limbaugh, R. Global Warming Hoax: Polar Bears Are Just Fine! *Rush Limbaugh Show,* March 8, 2007. www.rushlimbaugh.com/home/daily/site_030807/content/01125108.LogIn.html.

33. Green, K. P. Is the Polar Bear Endangered, or Just Conveniently Charismatic? Environmental Policy Outlook #2, American Enterprise Institute for Public Policy Research, May 2008. www.aei.org/outlook/27918.

34. Carroll, C. Morning Bell: Blame Canada. *The Foundry,* May 1, 2008. http://blog.heritage.org/2008/05/01/morning-bell-blame-canada. [blog]

35. Lieberman, B. Do Polar Bears Belong on the Endangered Species List? No: Bears Are Thriving; Greens Tread on Thin Ice. McClatchy-Tribune News Service, February 21, 2008. http://seattletimes.nwsource.com/html/opinion/2004192501_polarcon21.html. [syndicated opinion]

36. The George C. Marshall Institute. *Climate Change Science.* Arlington, VA: George C. Marshall Institute. www.marshall.org/subcategory.php?id=49.

37. Editorial Board. Polar Bear Politics. *Wall Street Journal.* January 3, 2007.

38. Inhofe, J. Inhofe Speech on Polar Bears and Global Warming. *Inhofe EPW Press Blog,* January 5, 2007. http://epw.senate.gov/public/index.cfm?FuseAction=PressRoom.Blogs&ContentRecord_id=f339c09a-802a-23ad-4202-611ef8047a6b. [blog]

39. Lydersen, K. Oil Group Joins Alaska in Suing to Overturn Polar Bear Protection. *Washington Post,* August 31, 2008. www.washingtonpost.com/wp-dyn/content/article/2008/08/30/AR2008083001538.html.

40. Palin, S. State Comments on Proposed FWS Polar Bear Rule. Juneau, AK: State of Alaska, April 9, 2007. p. 11. www.adfg.alaska.gov/static/species/specialstatus/pdfs/polarbear_2007_soa_comments_4_9.pdf.

41. Kempthorne, D. "Secretary Kempthorne Announces Decision to Protect Polar Bears under Endangered Species Act: Rule will allow continuation of vital energy production in Alaska," Office of the Secretary, US Department of the Interior, May 14, 2008. http://www.doi.gov/archive/news/08_News_Releases/080514a.html.

42. US House of Representatives, 110th Congress. *On Thin Ice: The Future of the Polar Bear.* Hearing Before the Select Committee on Energy Independence and Global Warming, House of Representatives, January 17, 2008. Washington, DC: US Government Printing Office, 2010. http://globalwarming.house.gov/files/HRG/FullTranscripts/110-22_2008-01-17.pdf.

43. Greenpeace USA. *Koch Industries Secretly Funding the Climate Denial Machine.* March 2010. http://graphics8.nytimes.com/images/blogs/greeninc/koch.pdf

44. Karoli. Koch Industries Denies Funding Tea Parties, but Official Filings Say Otherwise. Crooks and Liars, April 18, 2010. http://crooksandliars.com/karoli/koch-industries-denies-funding-freedomworks. [blog]

45. Weiss, D., et al. Dirty Money: Oil Companies and Special Interests Spend Millions to Oppose Climate Legislation. Center for American Progress Action Fund, September 27, 2010. www.americanprogressaction.org/issues/2010/09/dirty_money.html.

46. Lefton, R., & Nielsen, N. Interactive: Big Polluters' Big Ad Spending. Center for American Progress Action Fund, October 27, 2010. www.americanprogressaction.org/issues/2010/10/bigoilmoney.html.

47. Greenpeace USA. *Koch Industries Secretly Funding.*

48. Soon, W., & Balliunas, S. *Lessons and Limits of Climate History: Was the 20th Century Climate Unusual?* Washington, DC: George C. Marshall Institute, 2003. www.marshall.org/pdf/materials/136.pdf.

49. Pachauri, R. K., & Reisinger, A., eds. *Climate Change 2007.* Geneva: Intergovernmental Panel on Climate Change, 2007.

50. Mann, M., et al. Global-Scale Temperature Patterns and Climate Forcing over the Past Six Centuries. *Nature* 1998;392:779–787.

51. Otto, S. Selected Videos from Our Signers. ScienceDebate.org, 2008. www.sciencedebate.org/videostatements.html.

52. CNN. Obama Visits Jon Stewart and *The Daily Show* in D.C. CNN.com, October 28, 2010. www.cnn.com/2010/POLITICS/10/28/obama.daily.show/index.html.

53. Monckton, C. Lord Monckton Tells Obama Global Warming Is Bullshit. April 15, 2010 Tax Day Tea Party, Washington, DC. YouTube.com, April 17, 2010. www.youtube.com/watch?v=gJdRwZG5ssA. [audiovisual footage]

54. Id, J. Leaked FOIA Files 62 mb of Gold. *Air Vent,* November 19, 2009. http://noconsensus.wordpress.com/2009/11/19/leaked-foia-files-62-mb-of-gold. [blog]

55. Watts, A. Breaking News Story: CRU Has Apparently Been Hacked—Hundreds of Files Released. *Watts Up with That?* November 19, 2009. http://wattsupwiththat.com/2009/11/19/breaking-news-story-hadley-cru-has-apparently-been-hacked-hundreds-of-files-released. [blog]

56. RealClimate Group. The CRU Hack. RealClimate.org, November 20, 2009. www.realclimate.org/index.php/archives/2009/11/the-cru-hack/#more-1853. [blog]

57. Mann, M. Interview by S. Otto, April 27, 2010.

58. Committee on Surface Temperature Reconstructions for the Last 2,000 Years, National Research Council. *Surface Temperature Reconstructions for the Last 2,000 Years.* Washington, DC: National Academies Press, 2006. www.nap.edu/catalog/11676.html.

59. Briffa, K., et al. Influence of Volcanic Eruptions on Northern Hemisphere Summer Temperature over the Past 600 Years. *Nature* 1998;393:450–455.

60. Limbaugh. UN Climate Change Plan Fits.

61. Palin, S. Mr. President: Boycott Copenhagen; Investigate Your Climate Change "Experts." Facebook.com, December 3, 2009. www.facebook.com/notes/sarah-palin/mr-president-boycott-copenhagen-investigate-your-climate-change-experts/188540473434.

62. Mann. Interview.

63. Limbaugh, R. Three Trees Said to Prove Warming! *Rush Limbaugh Show,* November 24, 2009. www.rushlimbaugh.com/home/daily/site_112409/content/01125112.guest.html.

64. Mann. Interview.

65. Archer, D., et al. An Open Letter to Congress from U.S. Scientists on Climate Change and Recently Stolen Emails. Union of Concerned Scientists, December 4, 2009. www.ucsusa.org/assets/documents/global_warming/scientists-statement-on.pdf.

66. Palin. *Mr. President: Boycott Copenhagen.*

67. Nisbet, M. Interview by S. Otto, April 22, 2010.

68. Dimiero, B. Foxleaks: Fox Boss Ordered Staff to Cast Doubt on Climate Science. Media Matters for America, December 15, 2010. http://mediamatters.org/iphone/blog/201012150004.

69. Hoyt, C. Stolen E-Mail, Stoking the Climate Debate. *New York Times,* December 5, 2009. www .nytimes.com/2009/12/06/opinion/06pubed.html?scp=54&sq=climategate&st=nyt.

70. Richert, C. Inhofe Claims That E-Mails "Debunk" Science Behind Climate Change. Politifact.com, December 11, 2009. http://politifact.com/truth-o-meter/statements/2009/dec/11/james-inhofe/ inhofe-claims-cru-e-mails-debunk-science-behind-cl.

71. Henig, J. "Climategate": Hacked E-Mails Show Climate Scientists in a Bad Light but Don't Change Scientific Consensus on Global Warming. FactCheck.org, December 10, 2009. www.factcheck .org/2009/12/climategate.

72. Borenstein, S., et al. Review: E-Mails Show Pettiness, Not Fraud. Associated Press, December 12, 2009. www.msnbc.msn.com/id/34392959/ns/us_news-environment.

73. McClatchy Washington Bureau. Commentary: "Climategate" Is a Lesson in the Politics of Science. McClatchy Newspapers, December 15, 2009. www.mcclatchydc.com/2009/12/15/v-print/80663/ commentary-climategate-is-a-lesson.html. [opinion]

74. Sensenbrenner, F. J. Sensenbrenner Urges IPCC to Exclude Climategate Scientists. Letters from the Select Committee on Energy Independence and Global Warming, December 8, 2009. http://republicans .globalwarming.house.gov/Press/PRArticle.aspx?NewsID=2749.

75. Palin, S. Sarah Palin on the Politicization of the Copenhagen Climate Conference. *Washington Post,* December 9, 2009. www.washingtonpost.com/wp-dyn/content/article/2009/12/08/AR2009120803402.html.

76. Working Group II to the Fourth Assessment Report of the Intergovernmental Panel on Climate Change. Chapter 10.6.2: The Himalayan Glaciers. In *Climate Change 2007: Impacts, Adaptation and Vulnerability.* New York: Cambridge University Press, 2007. www.ipcc.ch/publications_and_data/ar4/ wg2/en/ch10s10-6-2.html.

77. Working Group II to the Fourth Assessment Report of the Intergovernmental Panel on Climate Change. Introduction to the Working Group II Fourth Assessment Report. In *Climate Change 2007: Impacts, Adaptation and Vulnerability.* New York: Cambridge University Press, 2007. p. 4. www.ipcc.ch/ pdf/assessment-report/ar4/wg2/ar4-wg2-intro.pdf.

78. Pearce, F. Flooded Out. *New Scientist* #2189, June 5, 1999.

79. Koylyakov. V. M., ed. Variations of Snow and Ice in the Past and at Present on a Global and Regional Scale. International Hydrological Program, IHP-IV Project H-4.1. Paris: UNESCO, 1996.

80. Bagla, P. Himalayan Glaciers Melting Deadline "a Mistake." *BBC News,* December 5, 2009. http:// news.bbc.co.uk/2/hi/south_asia/8387737.stm.

81. US Senate Committee on Environment and Public Works Minority Staff. *"Consensus" Exposed: The CRU Controversy.* Washington, DC: US Senate, 2010.

82. Laing, A. "Climategate" Professor Phil Jones "Considered Suicide over Email Scandal." *Daily Telegraph,* February 7, 2010. www.telegraph.co.uk/earth/environment/climatechange/7180154/ Climategate-Professor-Phil-Jones-considered-suicide-over-email-scandal.html.

83. Schneider, S. *Science as a Contact Sport.* Washington, DC: National Geographic Society, 2009.

84. Fogarty, D. Climate Debate Gets Ugly as World Moves to Curb CO2. Reuters, April 26, 2010. www .reuters.com/article/2010/04/26/us-climate-abuse-2-feature-idUSTRE63P00A20100426.

85. MacKenzie, D. Battle Over Climate Science Spreads to US Schoolrooms. *New Scientist* #2751, March 11, 2010.

86. South Dakota Legislative Assembly, 85th Session. *House Concurrent Resolution No. 1009.* March 2, 2009. http://legis.state.sd.us/sessions/2010/Bill.aspx?File=HCR1009P.htm.

87. Stuart, C. Oh, Mann: Cuccinelli Targets UVA papers in Climategate Salvo. *The Hook,* April 29, 2010. www.readthehook.com/67811/oh-mann-cuccinelli-targets-uva-papers-climategate-salvo.

88. State of Virginia. Virginia Fraud Against Taxpayers Act: Chapter 842, Article 19.1. April 17, 2002. www.taf.org/virginiafca.htm.

89. Helderman, R. S. State Attorney General Demands Ex-Professor's Files from University of Virginia. *Washington Post,* May 4, 2010. www.washingtonpost.com/wp-dyn/content/article/2010/05/03/ AR2010050304139.html.

90. Cuccinelli, K. Updated Statement Regarding University of Virginia CID and Investigation. Commonwealth of Virginia, Office of the Attorney General, May 19, 2010. www.vaag.com/PRESS_RELEASES/Cuccinelli/51910_VA_Tech.html. [news release]

91. Grifo, F. Interview by S. Otto, May 7, 2010.

92. Editors. Into Ignorance. *Nature* 2011;471(7338):265–266. www.nature.com/nature/journal/v471/n7338/full/471265b.html. [editorial]

93. Whitfield, E. Biography. n.d. http://whitfield.house.gov/about/bio.shtml.

94. Whitfield, E. Energy. n.d. http://whitfield.house.gov/issues/energy.shtml.

95. State of Minnesota, House of Representatives, 87th Legislative Session. House File 1010, May 18, 2011.

96. Moyle, J. Wild Rice in Minnesota. *Journal of Wildlife Management* 1944;8(3):177–184.

97. Harman, O. Cyril Dean Darlington: The Man Who "Invented" the Chromosome. *Nature Reviews Genetics 2005*;6:79–85. www.nature.com/nrg/journal/v6/n1/box/nrg1506_BX3.html.

98. Joravsky, D. *The Lysenko Affair.* Chicago: University of Chicago Press, 1970. p. 242.

99. Schneider, L. *Biology and Revolution in Twentieth-Century China.* Lanham, MD: Rowman & Littlefield, 2005. p. 179.

100. Harms, W. China's Great Leap Forward. *University of Chicago Chronicle* 1996;15(13). http://chronicle.uchicago.edu/960314/china.shtml.

101. Pappas, M. The Election Mandate: The Contract from America. FreedomWorks.org, November 3, 2010. www.freedomworks.org/blog/max/the-mandate-the-contract-from-america. [blog]

102. Americans for Prosperity. Americans for Prosperity Congratulates 165 Climate Tax Pledge Signers on Election Victory. NoClimateTax.com, November 3, 2010. www.noclimatetax.com/2010/11/pledge-signers-sweep-into-office. [news release]

103. Mayer, J. Covert Operations: The Billionaire Brothers Who Are Waging a War Against Obama. *New Yorker,* August 30, 2010.

104. DeCarlo, S., et al. America's Largest Private Companies. *Forbes Magazine,* November 3, 2010. www.forbes.com/lists/2010/21/private-companies-10_land.html.

105. Snyder, J., & Chipman, K. Global Warming Skeptics Ascend in Congress. *Businessweek,* November 24, 2010. www.businessweek.com/magazine/content/10_49/b4206033143446.htm.

106. Otto, S. American Denialism. ShawnOtto.com, May 22, 2011. www.shawnotto.com/blog20110522.html. [blog]

107. Huizenga, B. On the Issues. HuizengaforCongress.com, n.d. http://huizengaforcongress.com/on-the-issues.

108. Huizenga. On the Issues.

109. Galbraith, K. Boehner: Calling Carbon Dioxide Dangerous Is "Almost Comical." Green, April 21, 2009. http://green.blogs.nytimes.com/2009/04/21/boehner-calling-carbon-dioxide-dangerous-is-almost-comical. [blog]

110. Boehlert, S. Science the GOP Can't Wish Away. *Washington Post,* November 19, 2010. www.washingtonpost.com/wp-dyn/content/article/2010/11/18/AR2010111806072.html. [opinion]

111. Valvo, J., & Oberg, C. Exposing the Special Interests Behind Waxman-Markey. Arlington, VA: Americans for Prosperity, September 2009. http://americansforprosperity.org/files/Policy_Paper_0909_0.pdf.

112. US Environmental Protection Agency. Cap and Trade: Acid Rain Program Results. 2003. www.epa.gov/capandtrade/documents/ctresults.pdf.

113. Holtz-Eakin, D. Interview by S. Otto, November 22, 2010.

114. Viser, M. "Romney reaffirms stance that global warming is real." *Boston Globe,* June 4, 2011. http://www.boston.com/news/science/articles/2011/06/04/romney_reaffirms_stance_that_global_warming_is_real/

115. Limbaugh, R. Romney Vaults Ahead of Obama; Declares Climate Change Manmade. *The Rush Limbaugh Show.* Excellence in Broadcasting Network, June 7, 2011. http://www.rushlimbaugh.com/home/daily/site_060711/content/01125108.guest.html

116. Environmental Pollution Panel of the President's Science Advisory Committee. *Restoring the Quality of Our Environment.* Washington, DC: US Government Printing Office, 1965. http://dge .stanford.edu/labs/caldeiralab/Caldeira%20downloads/PSAC,%201965,%20Restoring%20the%20 Quality%20of%20Our%20Environment.pdf.

117. Johnson, L. B. Conservation and Restoration of Natural Beauty. November 6, 1965. www.lbjlib .utexas.edu/johnson/archives.hom/speeches.hom/650208.asp.

118. Robock, A. Interview by S. Otto, November 16, 2010.

119. Diggles, M. *U.S. Geological Survey Fact Sheet 113-97: The Cataclysmic 1991 Eruption of Mount Pinatubo, Philippines.* Vancouver, WA: Cascades Volcano Observatory, US Geological Survey, 2005. http://pubs.usgs.gov/fs/1997/fs113-97.

120. Borenstein, S. Obama Global Warming Plan Involves Cooling Air. Associated Press, April 8, 2009. www.huffingtonpost.com/2009/04/08/obama-global-warming-plan_n_184657.html.

121. Hooper, R. Media Frenzy Prompts White House Clarification on Geoengineering. NewScientist.com, April 9, 2009. www.newscientist.com/blogs/shortsharpscience/2009/04/holdren-clarifies-the-white-ho .html. [blog]

122. Broad, W. How to Cool a Planet (Maybe). *New York Times,* June 27, 2006. www.nytimes .com/2006/06/27/science/earth/27cool.html.

123. Crutzen, P. Albedo Enhancement by Stratospheric Sulfur Injections: A Contribution to Resolve a Policy Dilemma? *Climatic Change* 2006;77(3–4):211–220. www.cogci.dk/news/Crutzen_albedo%20 enhancement_sulfur%20injections.pdf.

124. Robock. Interview.

125. Levitt, S., & Dubner, S. *Superfreakonomics.* New York: HarperCollins, 2009. p. 180.

126. Levitt. *Superfreakonomics.*

127. Pierrehumbert, R. Interview by S. Otto, November 18, 2010.

128. Holtz-Eakin. Interview.

129. Fleming, J. *Fixing the Sky: The Checkered History of Weather and Climate Control.* New York: Columbia University Press, 2010.

130. United Nations. *Convention on the Prohibition of Military or Any Hostile Use of Environmental Modification Techniques.* December 10, 1976. www.icrc.org/ihl.nsf/FULL/460?OpenDocument.

131. Feinberg, M., & Willer, R. Apocalypse Soon? Dire Messages Reduce Belief in Global Warming by Contradicting Just-World Beliefs. *Psychological Science* 2011;22(1):34–38. http://willer.berkeley.edu/ FeinbergWiller2011.pdf.

132. Mann. Interview.

133. Pierrehumbert. Interview.

134. Kaufman, L. In Kansas, Climate Skeptics Embrace Cleaner Energy. *New York Times,* October 18, 2010. www.nytimes.com/2010/10/19/science/earth/19fossil.html.

135. Fuller, M., et al. *Driving Demand for Home Energy Improvements.* Lawrence Berkeley National Laboratory, September 2010. http://drivingdemand.lbl.gov.

136. Otto, R. *Best Practices Review: Reducing Energy Use in Local Governments.* Office of the Minnesota State Auditor, July 2, 2008. www.auditor.state.mn.us/default.aspx?page=20080702.001.

137. Otto, R. Interview by S. Otto, November 22, 2010.

CHAPTER 11. FREEDOM AND THE COMMONS

1. US Department of Defense. *Quadrennial Defense Review Report,* February 2010. www.defense.gov/qdr/ images/QDR_as_of_12Feb10_1000.pdf.

2. Smith, A. *An Inquiry into the Nature and Causes of the Wealth of Nations.* Ed. J. Manis. Electronic Classics Series. Hazleton, PA: Pennsylvania State University, n.d. www2.hn.psu.edu/faculty/jmanis/ adam-smith/Wealth-Nations.pdf.

3. Hardin, G. The Tragedy of the Commons. *Science* 1968;162(3859):1243–1248. www.sciencemag.org/ content/162/3859/1243.full.

4. Rand, A. *The Virtue of Selfishness.* New York: Signet, 1964.

5. Hume, D. *An Enquiry Concerning Human Understanding.* Harvard Classics Vol. 37. New York: P. F. Collier and Son, 1910.

6. McNutt, M. Interview by S. Otto, August 27, 2010.

7. Friedman, M. *Capitalism and Freedom.* Chicago: University of Chicago Press, 1962. p. 34.

8. Lubchenco, J. Interview by S. Otto, March 28, 2010.

9. Levin, S. Interview by S. Otto, November 21, 2010.

10. Hazardous E-Waste Surging in Developing Countries. *ScienceDaily,* February 23, 2010. www
.sciencedaily.com/releases/2010/02/100222081911.htm.

11. Costanza, R., et al. The Value of the World's Ecosystem Services and Natural Capital. *Nature* 1997;387:253–260.

12. Roush, W. Putting a Price Tag on Nature's Bounty. *Science* 1997;276(5315):1029.

13. Pimm, S. Interview by S. Otto, September 7, 2010.

14. The Cornwall Alliance. An Evangelical Declaration on Global Warming. n.d. www.cornwallalliance
.org/articles/read/an-evangelical-declaration-on-global-warming.

15. The Cornwall Alliance. The Cornwall Declaration on Environmental Stewardship. n.d. www
.cornwallalliance.org/articles/read/the-cornwall-declaration-on-environmental-stewardship.

16. May, R. Interview by S. Otto, June 1, 2011.

17. Avon, N. Why More Americans Don't Travel Abroad. CNN.com, February 4, 2011. http://articles
.cnn.com/2011-02-04/travel/americans.travel.domestically_1_western-hemisphere-travel-initiative-
passports-tourism-industries?_s=PM:TRAVEL.

18. Pimm. Interview.

19. Moreno, J. Interview by S. Otto, August 23, 2010.

20. Pimm. Interview.

21. Mills, E., et al. *Insurance in a Climate of Change: Availability and Affordability.* Lawrence Berkeley National Laboratory, US Department of Energy, 2009. http://insurance.lbl.gov/availability-affordability
.html.

22. Feely, R. Interview by S. Otto, November 4, 2010.

23. Bureau of Economic Analysis. Table 1.1.5: Gross Domestic Product. US Department of Commerce, 2011. www.bea.gov/national/txt/dpga.txt.

24. Mills. *Insurance in a Climate of Change.*

25. Munich Reinsurance America. *2010 Natural Catastrophe Year in Review.* January 10, 2011. www
.munichreamerica.com/webinars/2011_01_natcatreview/index.shtm. [webinar]

26. Munich Reinsurance America. *2010 Half-Year Natural Catastrophe Review.* July 7, 2010. www.amre
.com/webinars/2010_07_natcatreview/natcat_webinar_record/player.html. [webinar]

27. Christenson, G. *Edwin Hubble: Mariner of the Nebulae.* Chicago: University of Chicago Press, 1995. p. 23.

28. Costanza, R., et al. The Perfect Spill: Solutions for Averting the Next Deepwater Horizon. *Solutions Online,* June 16, 2010. www.thesolutionsjournal.com/node/629.

29. Weidenbaum, D., et al. *The Annual Report of the Council of Economic Advisers.* Washington, DC: Council of Economic Advisers, February 6, 1982. p. 45. http://fraser.stlouisfed.org/publications/erp/
issue/1385/ . . . /ERP_ARCEA_1982.pdf.

30. Balmford, A., et al. Economic Reasons for Conserving Wild Nature. *Science* 2002;297(5583):950–953.

31. US Environmental Protection Agency. Full Cost Accounting. September 30, 2008. www.epa.gov/osw/
conserve/tools/fca/index.htm.

32. World Bank. Countries and Economies. n.d. http://data.worldbank.org/country.

CHAPTER 12. TALKING ABOUT SCIENCE IN AMERICA

1. Feynman, R. *"What Do You Care What Other People Think?" Further Adventures of a Curious Character.* New York: W. W. Norton, 1988.

2. Webber, M. E. Webber: Don't Dumb Down Texas. Statesman.com, September 15, 2009. www .statesman.com/opinion/content/editorial/stories/2009/09/16/0916webber_edit.html. [opinion]

3. Webber, M. Interview by S. Otto, July 30, 2010.

4. Wiley, D., & Wilson, K. *Just Say Don't Know: Sexuality Education in Texas Public Schools.* Texas Freedom Network Education Fund, 2009. www.tfn.org/site/DocServer/SexEdRort09_web. pdf?docID=981.

5. Wiley. *Just Say Don't Know.*

6. Mathews, T. J., et al. *State Disparities in Teenage Birth Rates in the United States.* Department of Health and Human Services Publication No. 2011-1209. Hyattsville, MD: National Center for Health Statistics, 2011.

7. Webber. Interview.

8. Collins, G. Mrs. Bush, Abstinence and Texas. *New York Times,* February 16, 2011. www.nytimes .com/2011/02/17/opinion/17gailcollins.html.

9. Hu, E. Gov. Rick Perry on Abstinence, Sanctuary Cities. *Texas Tribune,* October 15, 2010. www .texastribune.org/texas-politics/2010-texas-governors-race/gov-rick-perry-on-abstinence-sanctuary-cities/. [audiovisual footage]

10. Webber. Interview.

11. Bacon, F. *Novum Organum.* London: Joannem Billium, 1620.

12. Lord, C. G., et al. Biased Assimilation and Attitude Polarization: The Effects of Prior Theories on Subsequently Considered Evidence. *Journal of Personality and Social Psychology* 1979;37(11):2098–2109. www.psych.umn.edu/courses/spring07/borgidae/psy5202/readings/lord,%20ross%20&%20lepper%20 %281979%29.pdf.

13. Scott, E. Interview by S. Otto, August 25, 2009.

14. Harris, S., et al. The Neural Correlates of Religious and Nonreligious Belief. *PLoS ONE* 2009;4(10):e7272. www.plosone.org/article/info:doi%2F10.1371%2Fjournal.pone.0007272.

15. Schjoedt, U., et al. The Power of Charisma—Perceived Charisma Inhibits the Frontal Executive Network of Believers in Intercessory Prayer. *Social Cognitive and Affective Neuroscience* 2011;6(1):119–127.

16. Lerner, M. *The Belief in a Just World: A Fundamental Delusion.* New York: Plenum Press, 1980.

17. Furnham, A. Belief in a Just World: Research Progress over the Past Decade. *Personality and Individual Differences* 2002;34(5):795–817.

18. Bénabou, R., & Tirole, J. Belief in a Just World and Redistributive Politics. *Quarterly Journal of Economics* 2006;121(2):699–746. http://qje.oxfordjournals.org/content/121/2/699.abstract.

19. Bénabou & Tirole. Belief in a Just World.

20. Raine, N. V. *After Silence: Rape & My Journey Back.* New York: Crown, 1998. p. 91.

21. Bénabou & Tirole. Belief in a Just World.

22. Feinberg, M., & Willer, R. Apocalypse Soon? Dire Messages Reduce Belief in Global Warming by Contradicting Just-World Beliefs. *Psychological Science* 2011;22(1):34–38. http://willer.berkeley.edu/ FeinbergWiller2011.pdf.

23. Willer, R. Interview by S. Otto, December 16, 2010.

24. Ibid.

25. Ibid.

26. Bachmann, M. Interview by J. Markell, March 20, 2004. *Prophetic Views Behind The News,* KKMS, 980-AM Twin Cities Christian Talk Radio, Minneapolis. [radio broadcast]

27. Willer. Interview.

28. Hu, D. The Reception of Relativity in China. *Isis* 2007;98:539–557.

29. Needham, J. Science Reborn in China: Rise and Fall of the Anti-Intellectual "Gang." *Nature* 1978;274:832–834.

30. Williams, B. An American Milestone. *NBC Nightly News.* NBC.com, October 16, 2006. [audiovisual footage]

31. Straussmann, M. Countdown to 300 Million. *CBS News,* October 12, 2006. www.cbsnews.com/video/watch/?id=2086917n&tag=mncol;lst;7. [audiovisual footage]

32. Pilkington, E. 300 Million and Counting . . . US Reaches Population Milestone. *Guardian,* October 13, 2006. www.guardian.co.uk/world/2006/oct/13/usa.topstories3.

33. Kamen, D. Interview by S. Otto, August 28, 2009.

34. Conservapedia. Counterexamples to Relativity. Conservapedia.com, March 5, 2011. www .conservapedia.com/Counterexamples_to_Relativity.

35. *Flock of Dodos: The Evolution-Intelligent Design Circus,* 2006. R. Olson, Director. [motion picture]

36. Pew Research Center for People and the Press. A Deeper Partisan Divide Over Global Warming. May 8, 2008. http://people-press.org/report/417/a-deeper-partisan-divide-over-global-warming.

37. Munro, G. D. The Scientific Impotence Excuse: Discounting Belief-Threatening Scientific Abstracts. *Journal of Applied Social Psychology* 40(3):579–600.

38. Ibid.

39. Berger, J. *The Young Scientists: America's Future and the Winning of the Westinghouse.* New York: Perseus Books, 1993.

40. Anderson, S. New Research Finds 70 Percent of the Nation's Top High School Science Students Are the Children of Immigrants. National Foundation for American Policy, May 23, 2011. [news release]

41. Anderson, S. *The Impact of the Children of Immigrants on Scientific Achievement in America.* National Foundation for American Policy, May 23, 2011. www.nfap.com/pdf/Children_of_Immigrants_in_Science_and_Math_NFAP_Policy_Brief_May_2011.pdf.

42. National Science Teachers Association. New Survey Finds Parents Need Help Encouraging Their Kids in Science. May 10, 2010. www.nsta.org/about/pressroom.aspx?id=57403. [news release]

CHAPTER 13. RETHINKING OUR RELIGION

1. Shimkus, J. Rep. John Shimkus: God Decides When the "Earth Will End." YouTube.com, March 25, 2009. www.youtube.com/watch?feature=player_embedded&v=_7h08RDYA5E.

2. Phillips, D. *American Theocracy: The Peril and Politics of Radical Religion, Oil, and Borrowed Money in the 21st Century.* New York: Viking, 2006.

3. Chemberlin, P. R. Interview by S. Otto, November 23, 2010.

4. Westphal, S. P. Interview by S. Otto, April 14, 2010.

CHAPTER 14. ON TRUTH AND BEAUTY

1. Roosevelt, E. *It Seems to Me: Selected Letters of Eleanor Roosevelt.* Eds. L. C. Schlup & D. W. Whisenhunt. Lexington, KY: University Press of Kentucky, 2001.

2. Sanders, D. Interview by S. Otto, December 14, 2010.

3. Monckton, C. *Greenhouse Warming? What Greenhouse Warming?* Science and Public Policy Institute, August 22, 2007. http://scienceandpublicpolicy.org/monckton/greenhouse_warming_what_greenhouse_warming_.html.

4. Christensen, G. *Edwin Hubble: Mariner of the Nebulae.* Chicago: University of Chicago Press, 1995. p. 113.

5. Noyes, A. *The Torch Bearers: Watchers of the Sky.* New York: Frederick A. Stokes, 1922.

6. Wilson, R. R. *Starting Fermilab.* Batavia, IL: Fermi National Accelerator Laboratory, 1992. http://history.fnal.gov/GoldenBooks/gb_wilson2.html.

7. Perricone. M. Some Words of Wisdom. *Fermi News* 2000;23(2). www.fnal.gov/pub/ferminews/ ferminews00-01-28/p3.html.

8. Wilson. *Starting Fermilab.*

9. Fish, S. The Value of Higher Education Made Literal. *New York Times,* December 13, 2010. http:// opinionator.blogs.nytimes.com/2010/12/13/the-value-of-higher-education-made-literal. [blog]

10. Review Panel. *Securing a Sustainable Future for Higher Education: An Independent Review of Higher Education Funding and Student Finance.* October 12, 2010. www.bis.gov.uk/assets/biscore/corporate/ docs/s/10-1208-securing-sustainable-higher-education-browne-report.pdf.

11. Sanders. Interview.

12. Bérubé, M. The Science Wars Redux. *Democracy* 2011;(19). www.democracyjournal.org/19/6789 .php?page=5.

13. Reagan, R. Farewell Address to the Nation. Miller Center of Public Affairs, January 11, 1989. http:// millercenter.org/scripps/archive/speeches/detail/3418.

14. Sanders. Interview.

15. McKee, R. Interview by S. Otto, October 7, 2010.

BIBLIOGRAPHY

Adler Planetarium. Statement about Senator John McCain's Comments at the Presidential Debate. October 8, 2008. www.boston.com/bostonglobe/ideas/brainiac/AdlerStatement_aboutdebate .pdf. [news release]

Americans for Prosperity. Americans for Prosperity Congratulates 165 Climate Tax Pledge Signers on Election Victory. NoClimateTax.com, November 3, 2010. www.noclimatetax .com/2010/11/pledge-signers-sweep-into-office. [news release]

Anderson, S. New Research Finds 70 Percent of the Nation's Top High School Science Students Are the Children of Immigrants. National Foundation for American Policy, May 23, 2011. www.nfap.com/pressreleases/DAY_OF_RELEASE_Children_of_Immigrants_May%202011. pdf. [news release]

Anderson, S. *The Impact of the Children of Immigrants on Scientific Achievement in America.* National Foundation for American Policy, May 23, 2011. www.nfap.com/pdf/Children_of_ Immigrants_in_Science_and_Math_NFAP_Policy_Brief_May_2011.pdf.

Anonymous. *Can You Tell the Difference between Evolution and Natural Selection?* Powder Springs, GA: Creation Ministries International, n.d.

www.creation.com/images/pdfs/flyers/can-you-tell-the-difference-between-evolution-and-natural-selection-p.pdf. [pamphlet]

Anonymous. Einstein to Leave Berlin: Is Aroused by Unfair Attacks on Relativity Theory. *New York Times*, August 29, 1920.

Anonymous. Atomic Education Urged by Einstein; Scientist in Plea for $200,000 to Promote New Type of Essential Thinking. *New York Times*, May 25, 1946.

Anonymous. Summary of Notifiable Diseases, United States, 1994. *Morbidity and Mortality Weekly Report* 1995;43(53):Tables 11-12. www.cdc.gov/mmwr/preview/mmwrhtml/00039679. htm.

Anonymous. The World Will End in 2060, According to Newton. *London Evening Standard,* June 19, 2007. www.thisislondon.co.uk/news/article-23401099-the-world-will-end-in-2060-according-to-newton.do.

Archer, D., et al. An Open Letter to Congress from U.S. Scientists on Climate Change and Recently Stolen Emails. Union of Concerned Scientists, December 4, 2009. www.ucsusa.org/ assets/documents/global_warming/scientists-statement-on.pdf.

Aristotle. *Organon.* Rhodes: Andronicus, 50 BC.

Arrhenius, S. On the Influence of Carbonic Acid in the Air upon the Temperature of the Ground. *London, Edinburgh and Dublin Philosophical Magazine and Journal of Science* 1896;41(251):237–276. http://nsdl.org/archives/onramp/classic_articles/issue1_global_ warming/n4.Arrhenius1896.pdf.

Augustine, N., et al. Seeking a Human Spaceflight Program Worthy of a Great Nation. Review of US Human Spaceflight Plans Committee, 2009. www.nasa.gov/offices/hsf/meetings/10_22_ pressconference.html.

Avon, N. Why More Americans Don't Travel Abroad. CNN.com, February 4, 2011. http://articles .cnn.com/2011-02-04/travel/americans.travel.domestically_1_western-hemisphere-travel-initiative-passports-tourism-industries?_s=PM:TRAVEL.

Bachmann, M. Interview by J. Markell, March 20, 2004. *Prophetic Views Behind the News,* KKMS, 980-AM Twin Cities Christian Talk Radio, Minneapolis. [radio broadcast]

Bacon, F. *Novum Organum.* London: Joannem Billium, 1620.

Bagla, P. Himalayan Glaciers Melting Deadline "a Mistake." *BBC News,* December 5, 2009. http://news.bbc.co.uk/2/hi/south_asia/8387737.stm.

Balmford, A., et al. Economic Reasons for Conserving Wild Nature. *Science* 2002;297(5583):950–953.

Barinaga, M. California Backs Evolution Education. *Science* 1989;246(4932):881.

Beck, C. Postmodernism, Pedagogy, and Philosophy of Education. *Philosophy of Education Yearbook,* 1993. www.ed.uiuc.edu/eps/PES-Yearbook/93_docs/BECK.HTM.

Beck, G. "Amber Alert: Al Gore MIA During Blizzards." *Glenn Beck: The Fusion of Entertainment and Enlightenment,* February 10, 2010. www.glennbeck.com/content/articles/article/198/36153.

Becker, G. Pietism's Confrontation with Enlightenment Rationalism: An Examination of the Relation Between Ascetic Protestantism and Science. *Journal for the Scientific Study of Religion* 1991;30(2):139–158.

Bedini, S. A. *Thomas Jefferson, Statesman of Science.* New York: Macmillan, 1990.

Bedini, S. A. *Jefferson and Science.* Charlottesville, VA: Thomas Jefferson Foundation, 2002.

Behe, M. *Darwin's Black Box: The Biochemical Challenge to Evolution.* New York: Touchstone, 1996.

Bénabou, R., & Tirole, J. Belief in a Just World and Redistributive Politics. *Quarterly Journal of Economics* 2006;121(2):699–746. http://qje.oxfordjournals.org/content/121/2/699.abstract.

Berger, J. *The Young Scientists: America's Future and the Winning of the Westinghouse.* New York: Perseus Books, 1993.

Berkman, M. B., et al. Evolution and Creationism in America's Classrooms: A National Portrait. *PLoS Biology* 2008:6(5):e124. www.plosbiology.org/article/info:doi/10.1371/journal.pbio.0060124.

Berkowitz, B. The Christian Right's Compassion Deficit. *Dissident Voice,* December 30, 2004. http://dissidentvoice.org/Dec2004/Berkowitz1230.htm.

Bérubé, M. The Science Wars Redux. *Democracy* 2011(19). www.democracyjournal.org/19/6789.php.

Bethe, H. A. *J. Robert Oppenheimer: 1904–1967.* Washington, DC: National Academy of Sciences, 1997.

Bhattacharjee, Y. Scientific Literacy: NSF Board Draws Flak for Dropping Evolution from Indicators. *Science* 2010;328(5975):150–151.

Bird, K., & Sherwin, M. J. *American Prometheus: The Triumph and Tragedy of J. Robert Oppenheimer.* New York: Vintage Books, 2006.

Bloom, A. *The Closing of the American Mind: How Higher Education Has Failed Democracy and Impoverished the Souls of Today's Students.* Chicago: University of Chicago Press, 1987.

Boehlert, S. Science the GOP Can't Wish Away. *Washington Post,* November 19, 2010. www.washingtonpost.com/wp-dyn/content/article/2010/11/18/AR2010111806072.html. [opinion]

Boehner, J. A., and Chabot, S. Letter to Jennifer L. Sheets and Cyrus B. Richardson Jr., Ohio State Board Education, March 15, 2002. www.discovery.org/articleFiles/PDFs/Boehner-OhioLetter.PDF.

Borenstein, S. Obama Global Warming Plan Involves Cooling Air. Associated Press, April 8, 2009. www.huffingtonpost.com/2009/04/08/obama-global-warming-plan_n_184657.html.

Borenstein, S., et al. Review: E-Mails Show Pettiness, Not Fraud. Associated Press, December 12, 2009. www.msnbc.msn.com/id/34392959/ns/us_news-environment.

Boyer, A. D. *Sir Edward Coke and the Elizabethan Age.* Stanford, CA: Stanford University Press, 2003.

Bradley, O. Armistice Day Speech, Boston, Massachusetts, November 10, 1948. *The Collected Writings of General Omar N. Bradley.* Volume 1. Washington, DC: US Government Printing Office, 1977. pp. 584–589.

Brainard, C. CNN Cuts Entire Science, Tech Team. *The Observatory,* CJR.org, December 4, 2008. www.cjr.org/the_observatory/cnn_cuts_entire_science_tech_t.php. [blog]

Briffa, K., et al. Influence of Volcanic Eruptions on Northern Hemisphere Summer Temperature over the Past 600 Years. *Nature* 1998;393:450–455.

Broad, W. How to Cool a Planet (Maybe). *New York Times,* June 27, 2006. www.nytimes .com/2006/06/27/science/earth/27cool.html.

Brodeur, P. Annals of Radiation. *New Yorker* June 12, 1989–December 7, 1992. www.newyorker .com/magazine/bios/paul_brodeur/search?contributorName=paul%20brodeur.

Brodeur, P. *The Great Power-Line Cover-Up: How the Utilities and the Government Are Trying to Hide the Cancer Hazard Posed by Electromagnetic Fields.* Boston: Little, Brown, 1993.

Brown, J. "A Is for Atom, B Is for Bomb": Civil Defense in American Public Education, 1948-1963. *Journal of American History* 1998;75(1):68–90. www.jstor.org/pss/1889655.

Brownstein, R. The Hidden History of the American Electorate. *National Journal,* October 18, 2008. www.nationaljournal.com/magazine/the-hidden-history-of-the-american-electorate-20081018.

Burnett, H. S. Are Polar Bears Dying? *Environment and Climate News,* May 1, 2006. www .heartland.org/policybot/results/18971/Are_Polar_Bears_Dying.html.

Burnett, H. S. ESA Listing Not Needed for Polar Bears. *Environment and Climate News,* March 1, 2007. www.heartland.org/policybot/results/20631/ESA_Listing_Not_Needed_for_Polar_ Bears.html.

Bush, G. H. W. Remarks to the National Academy of Sciences, Washington, DC, April 23, 1990. *American Presidency Project.* www.presidency.ucsb.edu/ws/?pid=18393.

Bush, G. W. President Bush Discusses Global Climate Change. June 11, 2001. http://georgewbush-whitehouse.archives.gov/news/releases/2001/06/20010611-2.html. [news release]

Bush, V. As We May Think. *The Atlantic,* July 1945. www.theatlantic.com/magazine/ archive/1945/07/as-we-may-think/3881.

Bush, V. *Science, the Endless Frontier.* Washington, DC: US Government Printing Office, 1945. www.nsf.gov/od/lpa/nsf50/vbush1945.htm.

Carroll, C. Morning Bell: Blame Canada. *The Foundry,* May 1, 2008. http://blog.heritage .org/2008/05/01/morning-bell-blame-canada. [blog]

Chapman, M. *40 Days and 40 Nights: Darwin, Intelligent Design, God, OxyContin and Other Oddities on Trial in Pennsylvania.* New York: Collins, 2007.

Chargaff, E. *Heraclitean Fire: Sketches from a Life Before Nature.* New York: Rockefeller University Press, 1978.

Charlton Research. *Your Congress–Your Health Survey.* Washington, DC: Research!America, 2009. p. 31. www.researchamerica.org/uploads/YourCongress2009.pdf.

Chemberlin, P. R. Interview by S. Otto, November 23, 2010.

Chernow, R. *Titan: The Life of John D. Rockefeller, Sr.* New York: Vintage Books, 1998.

Christianson, G. E. *Edwin Hubble: Mariner of the Nebulae.* Chicago: University of Chicago Press, 1995.

Churchland, P. Interview, August 24, 2010, by S. Otto.

Chyi, H. I., & Lewis, S. Use of Online Newspaper Sites Lags Behind Print Editions. *Newspaper Research Journal* 2009;30(4):38–52. http://umn.academia.edu/SethLewis/Papers/114952/Use_of_ Online_Newspaper_Sites_Lags_Behind_Print_Editions.

Cicerone, R. *Climate Change: Evidence and Future Projections.* Statement before the Oversight and Investigations Subcommittee, Committee on Energy and Commerce, US House of Representatives. Washington, DC: National Academies Press, 2006. www7.nationalacademies. org/ocga/testimony/Climate_Change_Evidence_and_Future_Projections.asp.

CNN. Obama Visits Jon Stewart and *The Daily Show* in D.C. CNN.com, October 28, 2010. www .cnn.com/2010/POLITICS/10/28/obama.daily.show/index.html.

Cobern, W. W. Alternative Constructions of Science and Science Education: A Plenary Presentation for the Second Annual Southern African Association for Mathematics and Science Education Research, University of Durban-Westville, Durban, South Africa. *Scientific Literacy and Cultural Studies Project* 1994;122.

Cobern, W. W., & Loving, C. Defining "Science" in a Multicultural World: Implications for Science Education. *Science Education* 2001;85:50–67.

Cohen, I. B., ed. *Puritanism and the Rise of Modern Science: The Merton Thesis.* New Brunswick, NJ: Rutgers University Press, 1990.

Coke, S. E. Calvin's Case, or the Case of the Postnati. In *The Selected Writings and Speeches of Sir Edward Coke.* Ed. S. E. Coke and S. Sheppard. Indianapolis: Liberty Fund, 2003.

Coke, S. E. The First Part of the Institutes of the Lawes of England; Or, a Commentary upon Littleton, Not the Name of the Author Only, but of the Law It Selfe. In *The Selected Writings and Speeches of Sir Edward Coke.* Ed. S. E. Coke and S. Sheppard. Indianapolis: Liberty Fund, 2003.Colson, C., & Pearcey, N. *How Now Shall We Live?* Wheaton, IL: Tyndale House, 1999.

Collins, G. Mrs. Bush, Abstinence and Texas. *New York Times,* February 16, 2011. www.nytimes.com/2011/02/17/opinion/17gailcollins.html.

Committee on Prospering in the Global Economy of the 21st Century: An Agenda for American Science and Technology, National Academy of Sciences, National Academy of Engineering, and Institute of Medicine. *Rising Above the Gathering Storm: Energizing and Employing America for a Brighter Economic Future.* National Academies. Washington, DC: National Academies Press, 2007.

Committee on Radiative Forcing Effects on Climate, Climate Research Committee, National Research Council. *Radiative Forcing of Climate Change: Expanding the Concept and Addressing Uncertainties.* Washington, DC: National Academies Press, 2005.

Committee on Surface Temperature Reconstructions for the Last 2,000 Years, National Research Council. *Surface Temperature Reconstructions for the Last 2,000 Years.* Washington, DC: National Academies Press, 2006. www.nap.edu/catalog/11676.html.

Committee on the Science of Climate Change, National Research Council. *Climate Change Science: An Analysis of Some Key Questions.* Washington, DC: National Academies Press, 2001.

Conservapedia. Counterexamples to Relativity. Conservapedia.com, March 5, 2011. www.conservapedia.com/Counterexamples_to_Relativity.

Continetti, M. The Paranoid Style in Liberal Politics. *Weekly Standard* magazine 2011;16(28). www.weeklystandard.com/articles/paranoid-style-liberal-politics_555525.html.

Cooperman, A. DeLay Criticized for "Only Christianity" Remarks. *Washington Post,* April 20, 2002.

Cornwall Alliance. An Evangelical Declaration on Global Warming. n.d. www.cornwallalliance.org/articles/read/an-evangelical-declaration-on-global-warming.

Cornwall Alliance. The Cornwall Declaration on Environmental Stewardship. n.d. www.cornwallalliance.org/articles/read/the-cornwall-declaration-on-environmental-stewardship.

Cosgrove-Mather, B. Poll: Creationism Trumps Evolution. CBSNews.com, November 22, 2004. www.cbsnews.com/stories/2004/11/22/opinion/polls/main657083.shtml.

Costanza, R., et al. The Value of the World's Ecosystem Services and Natural Capital. *Nature* 1997;387:253–260.

Costanza, R., et al. The Perfect Spill: Solutions for Averting the Next Deepwater Horizon. *Solutions Online,* June 16, 2010. www.thesolutionsjournal.com/node/629.

Crutzen, P. Albedo Enhancement by Stratospheric Sulfur Injections: A Contribution to Resolve a Policy Dilemma? *Climatic Change* 2006;77(3–4):211–220. www.cogci.dk/news/Crutzen_albedo%20enhancement_sulfur%20injections.pdf.

Cuccinelli, K. Updated Statement Regarding University of Virginia CID and Investigation. Commonwealth of Virginia, Office of the Attorney General, May 19, 2010. www.vaag.com/PRESS_RELEASES/Cuccinelli/51910_VA_Tech.html. [news release]

Dallek, R. *An Unfinished Life: John F. Kennedy, 1917–1963.* Boston: Little, Brown, 2003.

Dart, F. E., & Pradham, P. L. Cross-Cultural Teaching of Science. *Science* 1967;155(3763):649–656. www.wmich.edu/slcsp/SLCSP102/slcsp102.pdf.

Darwin, C. R. Letter 12041—Darwin, C. R. to Fordyce, John, 7 May 1879. Darwin Correspondence Project, n.d. www.darwinproject.ac.uk/entry-12041.

Davis, E. Letter to Edwin Hubble, November 23, 1951. Huntington Library Collection, HUB Box 10 , f. 238. San Marino, CA: Henry Huntington Library. http://cdn.calisphere.org/data/13030/rd/tf7b69n8rd/files/tf7b69n8rd.pdf.

Dean, C. No Democratic Science Debate, Yet. *The Caucus,* NYTimes.com, April 8, 2008. http://thecaucus.blogs.nytimes.com/2008/04/08/no-democratic-science-debate-yet. [blog]

Dean, C. Physicists in Congress Calculate Their Influence. *New York Times,* June 10, 2008.

DeCarlo, S., et al. America's Largest Private Companies. *Forbes Magazine,* November 3, 2010. www.forbes.com/lists/2010/21/private-companies-10_land.html.

DeMelle, B. Disinformation Database: David Legates. *DeSmogBlog,* n.d. www.desmogblog.com/node/2830. [blog]

DeMelle, B. Disinformation Database: An Extensive Database of Individuals Involved in the Global Warming Denial Industry. *DeSmogBlog,* n.d. www.desmogblog.com/global-warming-denier-database.

Dennis, J. A., et al. Epidemiological Studies of Exposures to Electromagnetic Fields: II. Cancer. *Journal of Radiological Protection* 1991;11(1):13–25.

Descartes, R. *Discourse on the Method of Rightly Conducting One's Reason and of Seeking Truth in the Sciences.* 1637. www.gutenberg.org/ebooks/59.

Diamond, S. *Not By Politics Alone: The Enduring Influence of the Christian Right.* New York: Guilford Press, 2000.

Diggles, M. *U.S. Geological Survey Fact Sheet 113-97: The Cataclysmic 1991 Eruption of Mount Pinatubo, Philippines.* Vancouver, WA: Cascades Volcano Observatory, US Geological Survey, 2005. http://pubs.usgs.gov/fs/1997/fs113-97.

Dimiero, B. Foxleaks: Fox Boss Ordered Staff to Cast Doubt on Climate Science. Media Matters for America, December 15, 2010. http://mediamatters.org/iphone/blog/201012150004.

Dyck, M. G., et al. Polar Bears of Western Hudson Bay and Climate Change: Are Warming Spring Air Temperatures the "Ultimate" Survival Control Factor? *Ecological Complexity* 2007;4(3):73–84.

Editor. Editor's Addendum. Mr. Bryan on Evolution. n.d. www.icr.org/article/mr-bryan-evolution.

Editorial Board. Polar Bear Politics. *Wall Street Journal,* January 3, 2007. [editorial]

Editorial Staff. Guide for Authors: Conflict of Interest. Elsevier.com, n.d. www.elsevier.com/wps/find/journaldescription.cws_home/701873/authorinstructions#7000.

Editors. The Hundred Most Influential Books Since the War. *Times Literary Supplement,* December 30, 2008. http://entertainment.timesonline.co.uk/tol/arts_and_entertainment/the_tls/article5418361.ece.

Editors. Science Wars and the Need for Respect and Rigour. *Nature* 1997;385(6615):373. [editorial]

Editors. Abortion and Breast Cancer. *New York Times,* January 6, 2003. www.nytimes.com/2003/01/06/opinion/06MON1Y.html. [editorial]

Editors. Into Ignorance. *Nature* 2011;471(7338):265–266. www.nature.com/nature/journal/v471/n7338/full/471265b.html. [editorial]

Eilperin, J. Censorship Is Alleged at NOAA. *Washington Post,* February 11, 2006. www.washingtonpost.com/wp-dyn/content/article/2006/02/10/AR2006021001766.html.

Eisenhower, D. D. Farewell Address to the American People. *Eisenhower Presidential Library and Museum: Dwight D. Eisenhower Speeches.* n.d. www.eisenhower.archives.gov/all_about_ike/Speeches/WAV%20files/farewell%20address.mp3.

Environmental Pollution Panel of the President's Science Advisory Committee. *Restoring the Quality of Our Environment.* Washington, DC: US Government Printing Office, 1965. http://dge.stanford.edu/labs/caldeiralab/Caldeira%20downloads/PSAC,%201965,%20Restoring%20the%20Quality%20of%20Our%20Environment.pdf.

Feely, R. Interview by S. Otto, November 4, 2010.

Feinberg, M., & Willer, R. Apocalypse Soon? Dire Messages Reduce Belief in Global Warming by Contradicting Just-World Beliefs. *Psychological Science* 2011;22(1):34–38. http://willer.berkeley.edu/FeinbergWiller2011.pdf.

Ferris, T. *The Science of Liberty.* New York: Harper, 2010.

Feyerabend, P. *Killing Time: The Autobiography of Paul Feyerabend.* Chicago: University of Chicago Press, 1995.

Feynman, R. *"What Do You Care What Other People Think?" Further Adventures of a Curious Character.* New York: W. W. Norton, 1988.

Finney, J. Sputnik Acts as a Spur to U.S. Science and Research. *New York Times,* November 2, 1957.

Fish, S. The Value of Higher Education Made Literal. *New York Times,* December 13, 2010. http://opinionator.blogs.nytimes.com/2010/12/13/the-value-of-higher-education-made-literal. [blog]

Flam, F. What Should It Take to Join Science's Most Exclusive Club? *Science* 1992;256(5059):960–961.

Flam, F. The Difference between Science and Religion. *Philadelphia Inquirer,* April 18, 2011. http://articles.philly.com/2011-04-18/news/29443540.

Fleming, J. *Fixing the Sky: The Checkered History of Weather and Climate Control.* New York: Columbia University Press, 2010.

Flock of Dodos: The Evolution-Intelligent Design Circus, 2006. R. Olson, Director. [motion picture]

Fogarty, D. Climate Debate Gets Ugly as World Moves to Curb CO2. Reuters, April 26, 2010. www.reuters.com/article/2010/04/26/us-climate-abuse-2-feature-idUSTRE63P00A20100426.

Franklin, B. *The Writings.* Vol. 10. Ed. A. H. Smyth. New York: Macmillan, 1907.

Franklin, J. The New Priesthood—The Scientific Elite and the Uses of Power. *Engineering and Science* 1965;28(9):4. http://calteches.library.caltech.edu/2383/1/books.pdf.

Freed, F., producer. The Decision to Drop the Bomb. *NBC White Paper,* 1965. [television documentary]

Freedman, A. NBC Fires Weather Channel Environmental Unit. *Capital Weather Gang,* WashingtonPost.com, November 21, 2008. http://voices.washingtonpost.com/capitalweathergang/2008/11/nbc_fires_twc_environmental_un.html. [blog]Friedman, M. *Capitalism and Freedom.* Chicago: University of Chicago Press, 1962. p. 34.

Fuller, M., et al. *Driving Demand for Home Energy Improvements.* Lawrence Berkeley National Laboratory, September 2010. http://drivingdemand.lbl.gov.

Furnham, A. Belief in a Just World: Research Progress over the Past Decade. *Personality and Individual Differences* 2002;34(5):795–817.

Galbraith, K. Boehner: Calling Carbon Dioxide Dangerous Is "Almost Comical." *Green,* April 21, 2009. http://green.blogs.nytimes.com/2009/04/21/boehner-calling-carbon-dioxide-dangerous-is-almost-comical. [blog]

George, J. Global Warming Won't Hurt Polar Bears, GN Says. *Nunatsiaq News,* May 26, 2006. www.nunatsiaqonline.ca/archives/60526/news/climate/60526_01.html.

George C. Marshall Institute. *Climate Change Science.* Arlington, VA: George C. Marshall Institute. www.marshall.org/subcategory.php?id=49.

Gleason, R. Bob Dylan: Poet to a Generation. *Jazz and Pop,* December 1968. pp. 36–37. www.loc.gov/folklife/guides/BibDylan.html.

Godlee, F., et al. Wakefield's Article Linking MMR Vaccine and Autism Was Fraudulent. *BMJ* 2011;342:c7452. www.bmj.com/content/342/bmj.c7452.full.

Gold, R. B. The Implications of Defining When a Woman Is Pregnant. *Guttmacher Report on Public Policy* 2005;8(2):7–10. www.guttmacher.org/pubs/tgr/08/2/gr080207.html.

Government of Nunavut. Minister Accepts Decisions of the Nunavut Wildlife Management Board on Polar Bear Management. January 7, 2005. www.gov.nu.ca/Nunavut/English/news/2005/jan/jan7.pdf. [news release]

Graham, B. Why a Revival? *Billy Graham 1949 Christ for Greater Los Angeles Campaign.* Wheaton, IL: Billy Graham Center, Wheaton College, 1949. http://espace.wheaton.edu/bgc/audio/cn026t5702a.mp3. [audio recording]

Graham, B. Heart Trouble. *Billy Graham New York Crusade, Madison Square Garden,* May 23, 1957. Billy Graham Center. www.wheaton.edu/bgc/archives/exhibits/NYC57/18sample97-1.htm. [audiovisual footage]

Graham, B. The Solution to Modern Problems. *Billy Graham New York Crusade, Madison Square Garden,* June 1, 1957. Billy Graham Center. www.wheaton.edu/bgc/archives/exhibits/NYC57/12sample65.htm. [audiovisual footage]

Graham, D. Testimony of David J. Graham, MD, MPH, November 18, 2004. ConsumersUnion .org, n.d. www.consumersunion.org/pub/campaignprescriptionforchange/001651.html.

Green, K. P. Is the Polar Bear Endangered, or Just Conveniently Charismatic? Environmental Policy Outlook #2, American Enterprise Institute for Public Policy Research, May 2008. www .aei.org/outlook/27918.

Green, L. The Messy Relationship between Religion and Science: Revisiting Galileo's Inquisition. Fox News.com, January 16, 2008. www.foxnews.com/story/0,2933,323327,00.html.

Greenpeace USA. *Koch Industries Secretly Funding the Climate Denial Machine.* March 2010. http://graphics8.nytimes.com/images/blogs/greeninc/koch.pdf.

Greenwald, G. David Gregory Shows Why He's the Perfect Replacement for Tim Russert. Salon.com, December 29, 2008. www.salon.com/news/opinion/glenn_greenwald/2008/12/29/gregory.

Grifo, F. Interview by S. Otto, May 7, 2010.

Gross, P., & Levitt, N. *Higher Superstition: The Academic Left and Its Quarrels with Science.* Baltimore: Johns Hopkins University Press, 1994.

Guild, P. B., & Garger, S. *Marching to Different Drummers.* Alexandria, VA: Association for Supervision and Curriculum Development, 1985.

Hadden, J., & Swann, C. *Prime Time Preachers: The Rising Power of Televangelism.* Reading, MA: Addison-Wesley, 1981. pp. 47–55. http://etext.virginia.edu/etcbin/toccer-new2?id=HadPrim .sgm&images=images/modeng&data=/texts/english/modeng/parsed&tag=public&part=all.

Halpern, M. Interview by S. Otto, March 28, 2010.

Halsall, P. The Crime of Galileo: Indictment and Abjuration of 1633. Internet Modern History Sourcebook, January 1999. www.fordham.edu/halsall/mod/1630galileo.html.

Hardin, G. The Tragedy of the Commons. *Science* 1968;162(3859):1243–1248. www.sciencemag .org/content/162/3859/1243.full.

Harman, O. Cyril Dean Darlington: The Man Who "Invented" the Chromosome. *Nature Reviews Genetics* 2005;6:79–85. www.nature.com/nrg/journal/v6/n1/box/nrg1506_BX3.html.

Harms, W. China's Great Leap Forward. *University of Chicago Chronicle* 1996;15(13). http://chronicle.uchicago.edu/960314/china.shtml.

Harris, G. F.D.A. Failing in Drug Safety, Official Asserts. *New York Times,* November 19, 2004. www.nytimes.com/2004/11/19/business/19fda.html.

Harris, S., et al. The Neural Correlates of Religious and Nonreligious Belief. *PLoS One* 2009;4(10):e7272. www.plosone.org/article/info:doi%2F10.1371%2Fjournal.pone.0007272.

Hartz, J., & Chappell, R. *Worlds Apart: How the Distance Between Science and Journalism Threatens America's Future.* Nashville, TN: Freedom Forum First Amendment Center, 1997.

Hayes, K. J. *The Road to Monticello: The Life and Mind of Thomas Jefferson.* Oxford, UK: Oxford University Press, 2008.

Helderman, R. S. State Attorney General Demands Ex-Professor's Files from University of Virginia. *Washington Post,* May 4, 2010. www.washingtonpost.com/wp-dyn/content/article/2010/05/03/AR2010050304139.html.

Henig, J. Did Obama Request a $3 Million "Overhead Projector," as McCain Claimed? Factcheck .org, October 14, 2008. www.factcheck.org/askfactcheck/did_obama_request_a_3_million_ overhead.html.

Henig, J. "Climategate": Hacked E-Mails Show Climate Scientists in a Bad Light but Don't Change Scientific Consensus on Global Warming. FactCheck.org, December 10, 2009. www.factcheck .org/2009/12/climategate.

Heppenheimer, T. A. *The Space Shuttle Decision: NASA's Search for a Reusable Space Vehicle.* Washington, DC: National Aeronautics and Space Administration, 1999.

Hillerbrand, H. J. *The Protestant Reformation,* revised edition. New York: HarperPerennial, 2009.

Hitchcock, A. S. Remarks on the Scientific Attitude. *Science* 1924;59(1535):476–477.

Hobbes, T. *Leviathan; or, The Matter, Form, and Power of a Commonwealth Ecclesiastical and Civil,* 3rd ed. London: George Routledge and Sons, 1887.

Hoffmann, H., & Berchtold, J. *Hitler über Deutchland.* Munich: Franz Eher Nachfolger GmbH, 1932.

Holdren, J. Email sent to scientists and science journalists re: geoengineering and White House policy, April 8, 2009.

Holtz-Eakin, D. Interview by S. Otto, November 22, 2010.

Hooper, R. Media Frenzy Prompts White House Clarification on Geoengineering. NewScientist. com, April 9, 2009. www.newscientist.com/blogs/shortsharpscience/2009/04/holdren-clarifies-the-white-ho.html. [blog]

Horace. Poem 11. *The Odes.*

Hoyt, C. Stolen E-Mail, Stoking the Climate Debate. *New York Times,* December 5, 2009. www .nytimes.com/2009/12/06/opinion/06pubed.html?scp=54&sq=climategate&st=nyt.

Hu, D. The Reception of Relativity in China. *Isis* 2007;98:539–557.

Hu, E. Gov. Rick Perry on Abstinence, Sanctuary Cities. *Texas Tribune,* October 15, 2010. www .texastribune.org/texas-politics/2010-texas-governors-race/gov-rick-perry-on-abstinence-sanctuary-cities/. [audiovisual footage]

Huff, G. Schools Should Not Limit Origins-of-Life Discussions to Evolution, Republican Legislators Say. Dem. Rep. Otto Disagrees, Argues That Educators Can't Teach Personal Values. *Stillwater Gazette,* September 27, 2005. www.stillwatergazette.com/articles/2003/10/02/export160.txt. Originally published as: Local Republicans: "Schools Should Teach Creationism." *Stillwater Gazette,* September 29, 2003.

Huizenga, B. On the Issues. HuizengaforCongress.com, n.d. http://huizengaforcongress.com/on-the-issues.

Hume, D. *An Enquiry Concerning Human Understanding.* Harvard Classics Vol. 37. New York: P. F. Collier and Son, 1910.

Hunter, M. Religious Watchdog Group Warns Clergy Against Passing Out Voter Guides. *Crosswalk,* October 20, 2004. www.crosswalk.com/news/religion-today/religious-watchdog-group-warns-clergy-against-passing-out-voter-guides-1291673.html.

Id, J. Leaked FOIA Files 62 mb of Gold. *Air Vent,* November 19, 2009. http://noconsensus .wordpress.com/2009/11/19/leaked-foia-files-62-mb-of-gold. [blog]

Infoplease.com. Life Expectancy by Age, 1850–2004. n.d. www.infoplease.com/ipa/A0005140.html.

Inhofe, J. Inhofe Speech on Polar Bears and Global Warming. *Inhofe EPW Press Blog,* January 5, 2007. http://epw.senate.gov/public/index.cfm?FuseAction=PressRoom .Blogs&ContentRecord_id=f339c09a-802a-23ad-4202-611ef8047a6b. [blog]

Irwin, A. Science Journalism "Flourishing" in Developing World. Science and Development Network, February 18, 2009. www.scidev.net/en/news/science-journalism-flourishing-in-developing-world.html.

Isaacson, W. *Einstein: His Life and Universe.* New York: Simon & Schuster, 2007.

Jacob, M. C. *Scientific Culture and the Making of the Industrial West.* New York: Oxford University Press, 1997.

Jaschik, S. Defending the Fruit Flies from Sarah Palin. *Inside Higher Ed,* October 28, 2008. www .insidehighered.com/news/2008/10/28/palin.

J. Craig Venter Institute. First Self-Replicating Synthetic Bacterial Cell. May 20, 2010. www.jcvi.org/cms/press/press-releases/full-text/article/first-self-replicating-synthetic-bacterial-cell-constructed-by-j-craig-venter-institute-researcher. [news release]

Jefferson, T. *Thomas Jefferson: Writings.* Ed. M. D. Peterson. New York: Library of America, 1984.

Jefferson, T. Letter to Richard Price, January 8, 1789. Thomas Jefferson: Creating a Virginia Republic, Library of Congress, August 3, 2010. www.loc.gov/exhibits/jefferson/jeffrep.html.

Jefferson, T. Letter to James Madison, December 20, 1787. *The Writings of Thomas Jefferson: Correspondence.* Ed. Henry Augustine Washington. New York: Derby and Jackson, 1859.

Jefferson, T. Letter to James Madison, February 17, 1826. In *The Writings of Thomas Jefferson, Volume XVI*. Ed. A. E. Bergh. Washington, DC: Thomas Jefferson Memorial Association of the United States, 1907.

Jefferson, T. Letter to Peter Carr, September 7, 1814. *68 Letters to and from Jefferson, 1805-1817*. University of Virginia Library, n.d. http://etext.virginia.edu/etcbin/toccer-new2?id=Jef1Gri .sgm&images=images/modeng&data=/texts/english/modeng/parsed&tag=public&part=5&d ivision=div1.

Jefferson, T. Letter to Meriwether Lewis, June 20, 1803. Monticello.org, n.d. www.monticello.org/ site/jefferson/jeffersons-instructions-to-meriwether-lewis.

Jefferson, T. Letter to Benjamin Rush, January 16, 1811. In *Thomas Jefferson, 1743–1826: Letters: Relations with Adams*. University of Virginia Library, n.d. http://etext.virginia.edu/etcbin/ toccer-new2?id=JefLett.sgm&images=images/modeng&data=/texts/english/modeng/parsed &tag=public&part=205&division=div1.

Jefferson, T. Letter to Pierre Samuel du Pont de Nemours, 1809. In *Correspondence between Thomas Jefferson and Pierre Samuel du Pont de Nemours 1798–1817*. Ed. D. Malone. Cambridge, MA: Da Capo Press, 1970.

Jefferson, T. Memorial on the Book Duty. In *Thomas Jefferson: Writings*. Ed. M. D. Peterson. New York: Library of America, 1984.

Jefferson, T., et al. Declaration of Independence Rough Draft with Edits by Franklin and Adams. Declaring Independence: Drafting the Documents, Library of Congress, July 23, 2010. www .loc.gov/exhibits/declara/images/draft1.jpg.

Jensen, P., et al. Scientists Who Engage with Society Perform Better Academically. *Science and Public Policy* 2008;35(7):527–541.

Johansen, C., et al. Cellular Telephones and Cancer—A Nationwide Cohort Study in Denmark. *Journal of the National Cancer Institute* 2001;93(3):203–207. http://jnci.oxfordjournals.org/ content/93/3/203.full.

Johnson, G. *Miss Leavitt's Stars: The Untold Story of the Woman Who Discovered How to Measure the Universe*. New York: W. W. Norton, 2005.

Johnson, L. B. Conservation and Restoration of Natural Beauty. November 6, 1965. www.lbjlib .utexas.edu/johnson/archives.hom/speeches.hom/650208.asp.

Jones, J. *Memorandum and Order, Kitzmiller, et al. v. Dover School District, et al*. United States District Court, Middle District of Pennsylvania, December 20, 2005. www.pamd.uscourts .gov/kitzmiller/decision.htm.

Jones, J. The Myth of "Activist Judges." *College News*, November 16, 2006. http://collegenews.org/ editorials/2006/the-myth-of-activist-judges.html.

Joravsky, D. *The Lysenko Affair*. Chicago: University of Chicago Press, 1970.

Kachka, B. Are You There, God? It's Me, Hitchens. *New York*, April 26, 2007. http://nymag.com/ arts/books/features/31244.

Kadanoff, L. Hard Times. *Physics Today* 1992;45(10):9–11.

Kamen, D. Interview by S. Otto, August 28, 2009.

Kant, I. An Answer to the Question: What Is Enlightenment? In *Kant: Political Writings*. Ed. H. Reiss, trans. H. Nisbet. Cambridge: Cambridge University Press, 1970.

Karnowski, S. Autism Fears, Measles Spike among Minn. Somalis. Minnesota Public Radio News [Associated Press], April 2, 2011. http://minnesota.publicradio.org/display/web/2011/04/02/ somali-autism-vaccines.

Karoli Koch Industries Denies Funding Tea Parties, but Official Filings Say Otherwise. *Crooks and Liars*, April 18, 2010. http://crooksandliars.com/karoli/koch-industries-denies-funding- freedomworks. [blog]

Karp, D. IRS Warns Churches: No Politics Allowed. *St. Petersburg Times*, September 15, 2004. www.sptimes.com/2004/09/15/State/IRS_warns_churches__n.shtml.

Kaufman, L. In Kansas, Climate Skeptics Embrace Cleaner Energy. *New York Times*, October 18, 2010. www.nytimes.com/2010/10/19/science/earth/19fossil.html.

Kaufman, L. Darwin Foes Add Warming to Targets. *New York Times,* March 3, 2010. www
.nytimes.com/2010/03/04/science/earth/04climate.html.

Keeter, S. Public Praises Science; Scientists Fault Public, Media. Pew Research Center for the People
and the Press, July 9, 2009. http://people-press.org/files/legacy-pdf/528.pdf. [news release]

Keim, B. Clinton and Obama Talk Religion, Not Science. *Wired Science,* Wired.com, April 8,
2008. www.wired.com/wiredscience/2008/04/clinton-and-oba. [blog]

Keim, B. McCain's VP Wants Creationism Taught in School. *Wired Science,* Wired.com, August
29, 2008. www.wired.com/wiredscience/2008/08/mccains-vp-want. [blog]

Kennedy, J. F. Excerpt from an Address Before a Joint Session of Congress, 25 May 1961. John F.
Kennedy Presidential Library and Museum, n.d. www.jfklibrary.org/Asset-Viewer/
xzw1gaeeTES6khED14P1Iw.aspx.

Kennedy, J. F., & Webb, J. E. Tape Recording of a Meeting between President John F. Kennedy and
NASA Administrator James E. Webb, November 21, 1962. *White House Meeting Tape 63.* John
Fitzgerald Kennedy Library. n.d.

Kennedy, R. F. Deadly Immunity. *Rolling Stone,* July 14, 2005.

Kirkpatrick, D. Bush Allies Till Fertile Soil, Among Baptists, for Votes. *New York Times,* June 18,
2004. www.nytimes.com/2004/06/18/politics/campaign/18baptists.html.

Kirkpatrick, D. Churches See an Election Role and Spread the Word on Bush. *New York Times,*
August 9, 2004. www.nytimes.com/2004/08/09/us/churches-see-an-election-role-and-spread-
the-word-on-bush.html.

Kirshenbaum, S. R., et al. Science and the Candidates. *Science* 2008;320(5873):182.

Koylyakov. V. M., ed. Variations of Snow and Ice in the Past and at Present on a Global and Regional
Scale. International Hydrological Program, IHP-IV Project H-4.1. Paris: UNESCO, 1996.

Krauss, L. Equal Time for Nonsense. *New York Times,* July 30, 1996. [opinion]

Kuhn, T. *The Structure of Scientific Revolutions.* Chicago: University of Chicago Press, 1962.

Laing, A. "Climategate" Professor Phil Jones "Considered Suicide over Email Scandal." *Daily
Telegraph,* February 7, 2010. www.telegraph.co.uk/earth/environment/climatechange/7180154/
Climategate-Professor-Phil-Jones-considered-suicide-over-email-scandal.html.

Lapp, R. An Interview with Governor Val Peterson, *Bulletin of the Atomic Scientists* 1953;9(7):237–242.

Lapp, R. An Interview with Governor Val Peterson. *Bulletin of the Atomic Scientists*
1954;10(10):375–377.

Lapp, R. *The New Priesthood: The Scientific Elite and the Uses of Power.* New York: Harper &
Row, 1965.

Launius, R. Interviewed in Howard McCurdy and Roger Launius on Opposition to Apollo.
Background interview for *Washington Goes to the Moon.* WAMU.org, May 24, 2001. http://
wamu.org/d/programs/special/moon/mccurdy-launius_opp.txt.

Launius, R. Managing the Unmanageable: Apollo, Space Age Management and American Social
Problems. *Space Policy* 2008;24(3):158–165. http://si-pddr.si.edu/jspui/bitstream/10088/8213/1/
Launius_2008_Managing_the_unmanageable.pdf.

Laurence, W. Atomic Bombing of Nagasaki Told by Flight Member. *New York Times,* September
9, 1945.

Laurence, W. U.S. Atom Bomb Site Belies Tokyo Tales. *New York Times,* September 12, 1945.

Laurence, W. Drama of the Atomic Bomb Found Climax in July 16 Test. *New York Times,*
September 26, 1945.

Laursen, L. @ApolloPlus40—A Colossal Perversion. *In the Field,* Nature.com, July 7, 2009. http://
blogs.nature.com/inthefield/2009/07/apolloplus40_a_colossal_perver.html. [blog]

Lebo, L. Judge in Dover Case Reports Hostile E-Mails: Jones and His Family Were Under
Marshals' Protection in December. *York Daily Record,* March 24, 2006.

Lefton, R., & Nielsen, N. Interactive: Big Polluters' Big Ad Spending. Center for American
Progress Action Fund, October 27, 2010. www.americanprogressaction.org/issues/2010/10/
bigoilmoney.html.

Legates, D. *Climate Science: Climate Change and Its Impacts.* Dallas: National Center for Policy Analysis, 2006. www.ncpa.org/pub/st285?pg=7.

Lehigh University Department of Biological Sciences. Department Position on Evolution and "Intelligent Design." n.d. www.lehigh.edu/~inbios/news/evolution.htm.

León, B. Science Related Information in European Television: A Study of Prime-Time News. *Public Understanding of Science* 2008;17(4):443–460.

Lerner, M. *The Belief in a Just World: A Fundamental Delusion.* New York: Plenum Press, 1980.

Leshner, A. Interview by S. Otto, December 1, 2010.

Leshner, A. I. We Need to Reward Those Who Nurture a Diversity of Ideas in Science. *Chronicle of Higher Education,* March 6, 2011. www.aaas.org/programs/centers/pe/news_svc/media/2011/0306che_leshner_reward_diversifiers.pdf.

Levin, S. Interview, November 21, 2010, by S. Otto.

Levitt, S., & Dubner, S. *Superfreakonomics.* New York: Harper Collins, 2009. p. 180.

Lieberman, B. Do Polar Bears Belong on the Endangered Species List? No: Bears Are Thriving; Greens Tread on Thin Ice. McClatchy-Tribune News Service, February 21, 2008. http://seattletimes.nwsource.com/html/opinion/2004192501_polarcon21.html. [syndicated opinion]

Limbaugh, R. Unofficial Summary of the *Rush Limbaugh Show,* May 22, 1996. Ed. J. Switzer. http://jwalsh.net/projects/sokal/articles/rlimbaugh.html.

Limbaugh, R. Global Warming Hoax: Polar Bears Are Just Fine! *Rush Limbaugh Show,* March 8, 2007. www.rushlimbaugh.com/home/daily/site_030807/content/01125108.LogIn.html.

Limbaugh, R. UN Climate Change Plan Fits with Obama's Anti-Capitalism Scheme. *Rush Limbaugh Show,* March 27, 2009. www.rushlimbaugh.com/home/daily/site_032709/content/01125111.guest.html.

Limbaugh, R. Three Trees Said to Prove Warming! *Rush Limbaugh Show,* November 24, 2009. www.rushlimbaugh.com/home/daily/site_112409/content/01125112.guest.html.

Locke, J. *An Essay Concerning Human Understanding.* Oxford: Clarendon Press, 1964.

Lord, C. G., et al. Biased Assimilation and Attitude Polarization: The Effects of Prior Theories on Subsequently Considered Evidence. *Journal of Personality and Social Psychology* 1979;37(11):2098–2109.

Loudon, M. The FDA Exposed: An Interview with Dr. David Graham, the Vioxx Whistleblower. NaturalNews.com, August 30, 2005. www.naturalnews.com/011401.html.

Lubchenco, J. Interview by S. Otto, March 28, 2010.

Luey, B. Are Fame and Fortune the Kiss of Death? *Publishing Research Quarterly* 2007;19(3):35–44.

Lydersen, K. Oil Group Joins Alaska in Suing to Overturn Polar Bear Protection. *Washington Post,* August 31, 2008. www.washingtonpost.com/wp-dyn/content/article/2008/08/30/AR2008083001538.html.

MacKenzie, D. Battle Over Climate Science Spreads to US Schoolrooms. *New Scientist* #2751, March 11, 2010.

Malek, K. The Abortion-Breast Cancer Link: How Politics Trumped Science and Informed Consent. *Journal of American Physicians and Surgeons* 2003;8(2):41–45. http://abortionno.org/pdf/breastcancer.pdf.

Mann, M. Interview by S. Otto, April 27, 2010.

Manning, J. E. *Membership of the 112th Congress: A Profile.* Washington, DC: Congressional Research Service, March 29, 2011. www.fas.org/sgp/crs/misc/R41647.pdf

Mann, M., et al. Global-Scale Temperature Patterns and Climate Forcing over the Past Six Centuries. *Nature* 1998;392:779–787.

Marquis, C. Bush Misuses Science Data, Report Says. *New York Times,* August 8, 2003. www.nytimes.com/2003/08/08/politics/08REPO.html.

Mathews, T. J., et al. *State Disparities in Teenage Birth Rates in the United States.* Department of Health and Human Services Publication No. 2011-1209. Hyattsville, MD: National Center for Health Statistics, 2011.

May, R. Interview by S. Otto, June 1, 2011.

Mayer, J. Covert Operations: The Billionaire Brothers Who Are Waging a War against Obama. *New Yorker,* August 30, 2010.

Mayer, W., ed. *The Swing Voter in American Politics.* Washington, DC: Brookings Institution, 2008.

McClatchy Washington Bureau. Commentary: "Climategate" Is a Lesson in the Politics of Science. McClatchy Newspapers, December 15, 2009. www.mcclatchydc.com/2009/12/15/v-print/80663/commentary-climategate-is-a-lesson.html. [opinion]

McCurdy, H. Interview in *Washington Goes to the Moon.* WAMU.org, May 24, 2001. http://wamu.org/programs/special/01/washington_goes_to_the_moon.php. [radio broadcast]

McKee, R. Interview by S. Otto, October 7, 2010.

McNutt, M. Interview by S. Otto, August 27, 2010.

McWilliams, C. Sunlight in My Soul. In *The Aspirin Age, 1919–1941.* Ed. I. Leighton. New York: Simon & Schuster, 1963.

Medrich, E., & Griffith, J. *International Mathematics and Science Assessment: What Have We Learned?* Washington, DC: National Center for Education Statistics, 1992. p. 80. http://nces.ed.gov/pubs92/92011.pdf.

Merton, R. K. *Science, Technology and Society in Seventeenth-Century England.* New York: Howard Fertig, 1970.

Miller, A. The New Catastrophism and Its Defender. *Science* 1922;55(1435):701–703.

Miller, K. Interview, August 18, 2010, by S. Otto.

Millikan, R. A. Science and Society. *Science* 1923;58(1503):293–298.

Mills, E., et al. *Insurance in a Climate of Change: Availability and Affordability.* Lawrence Berkeley National Laboratory, US Department of Energy, 2009. http://insurance.lbl.gov/availability-affordability.html.

Monckton, C. *Greenhouse Warming? What Greenhouse Warming?* Science and Public Policy Institute, August 22, 2007. http://scienceandpublicpolicy.org/monckton/greenhouse_warming_what_greenhouse_warming_.html.

Monckton, C. Lord Christopher Monckton at April 15, 2010 Tax Day Tea Party Washington DC. YouTube.com, April 15, 2010. www.youtube.com/watch?v=OO-BWhfPqGQ. [audiovisual footage]

Monckton, C. Lord Monckton Tells Obama Global Warming Is Bull Shit. April 15, 2010 Tax Day Tea Party Washington DC. YouTube.com, April 17, 2010. www.youtube.com/watch?v=gJdRwZG5ssA. [audiovisual footage]

Monson, D. Is Dispersal Obsolete? *Bulletin of the Atomic Scientists* 1954;10(10):378–383.

Mooney, C., & Kirshenbaum, S. *Unscientific America: How Scientific Illiteracy Threatens Our Future.* New York: Basic Books, 2009.

Moreno, J. Interview by S. Otto, August 23, 2010.

Morrison, J. The Incredible Shrinking Polar Bears. *National Wildlife,* February 1, 2004. www.biologicaldiversity.org/species/mammals/polar_bear/pdfs/15976_7338.pdf.

Moyle, J. Wild Rice in Minnesota. *Journal of Wildlife Management* 1944;8(3):177–184.

Siegel, K., & Cummings, B. *Petition to List the Polar Bear (Ursus maritimus) as a Threatened Species under the Endangered Species Act.* Center for Biological Diversity, February 16, 2005.

Müller, A. L., et al. Postcoital Treatment with Levonorgestrel Does Not Disrupt Postfertilization Events in the Rat. *Contraception* 2003;67(5):415–419.

Munich Reinsurance America. *2010 Half-Year Natural Catastrophe Review.* July 7, 2010. www.amre.com/webinars/2010_07_natcatreview/natcat_webinar_record/player.html. [webinar]

Munich Reinsurance America. *2010 Natural Catastrophe Year in Review.* January 10, 2011. www.munichreamerica.com/webinars/2011_01_natcatreview/index.shtm. [webinar]

Munro, G. D. The Scientific Impotence Excuse: Discounting Belief-Threatening Scientific Abstracts. *Journal of Applied Social Psychology* 40(3):579–600.

Myers, P. Z. Interview by S. Otto, August 24, 2009.

Myers, P. Z. Ken Ham Brags About His Websites. Pharyngula, March 22, 2011. http://scienceblogs
.com/pharyngula/2011/03/ken_ham_brags_about_his_websit.php. [blog]

Nasaw, D. *Andrew Carnegie*. New York: Penguin Press, 2006.

National Science Board. Chapter 7 in *Science and Engineering Indicators—2002*. Arlington, VA:
National Science Foundation, 2001. www.nsf.gov/statistics/seind02/c7/c7s1.htm#c7s1l4a.

National Science Board. Chapter 7 in *Science and Engineering Indicators—2006*. Arlington, VA:
National Science Foundation, 2007. p. 19. www.nsf.gov/statistics/seind06/pdf/c07.pdf.

National Science Board. Chapter 7 in *Science and Engineering Indicators—2008*. Arlington, VA:
National Science Foundation, 2008. www.nsf.gov/statistics/seind08/c7/c7s2.htm.

National Science Teachers Association. New Survey Finds Parents Need Help Encouraging Their
Kids in Science. May 10, 2010. www.nsta.org/about/pressroom.aspx?id=57403. [news release]

Needham, J. Science Reborn in China: Rise and Fall of the Anti-Intellectual "Gang." *Nature*
1978;274:832–834.

Newman, W. L. Radiometric Time Scale. In *Geologic Time*. Reston, VA: United States Geological
Survey, June 13, 2001. http://pubs.usgs.gov/gip/geotime/radiometric.html.

Newman, W. L. Age of the Earth. In *Geologic Time*. Reston, VA: United States Geological Survey,
July 9, 2007. http://pubs.usgs.gov/gip/geotime/age.html.

Newport, F. Landing a Man on the Moon: The Public's View. Gallup News Service, July 20, 1999.
www.gallup.com/poll/3712/landing-man-moon-publics-view.aspx.

Newport, F. Third of Americans Say Evidence Has Supported Darwin's Evolution Theory. Gallup.
com, November 19, 2004. www.gallup.com/poll/14107/Third-Americans-Say-Evidence-Has-
Supported-Darwins-Evolution-Theory.aspx.

Newton, I. *Isaac Newton's Philosophiae Naturalis Principia Mathematica*, 3rd edition. Ed. I. B.
Cohen and A. Koyré. London: Cambridge University Press, 1972.

Newton, I. *Sir Isaac Newton: Theological Manuscripts*. Ed. H. McLachlan. Liverpool, UK:
Liverpool University Press, 1950.

Nicolay, J. G., & Hay, J. *Abraham Lincoln: A History*. Volume 2. New York: Cosimo Classics, 1917.

Nietzsche, F. *Thus Spake Zarathustra*. London: Macmillan, 1896.

Nisbet, M. Interview by S. Otto, April 22, 2010.

Nissimov, R. DeLay's College Advice: Don't Send Your Kids to Baylor or A&M. *Houston Chronicle*,
April 18, 2002.

Norton-Taylor, R, ed. *Nuremberg: The War Crimes Trial: Transcript*. London: Nick Hern Books,
1997. p 55.

Noyes, A. *The Torch Bearers: Watchers of the Sky*. New York: Frederick A. Stokes, 1922.

Olsen, E. R. John McCain . . . Tough on Pork, Hard on Grizzlies. *Environment Blog*, Scienceline
.org, September 30, 2008. http://scienceline.org/2008/09/blog-olson-grizzlymccain. [blog]

Organisation for Economic Co-Operation and Development Programme for International Student
Assessment. *PISA 2006 Science Competencies for Tomorrow's World*. Paris: Organisation for
Economic Co-Operation and Development, April 12, 2007. www.oecd.org/document/2/0,3343
,en_32252351_32236191_39718850_1_1_1,00.html.

Otto, R. *Best Practices Review: Reducing Energy Use in Local Governments*. Office of the Minnesota
State Auditor, July 2, 2008. www.auditor.state.mn.us/default.aspx?page=20080702.001.

Otto, R. Interview by S. Otto, November 22, 2010.

Otto, S. Selected Videos from Our Signers. ScienceDebate.org, 2008. www.sciencedebate.org/
videostatements.html.

Otto, S. American Denialism. ShawnOtto.com, May 22, 2011. www.shawnotto.com/blog20110522
.html. [blog]

Otto, S.L., & Kirshenbaum, S. Science on the Campaign Trail. *Issues in Science and Technology*
2009;25(2). www.issues.org/25.2/p_otto.html.

Pachauri, R. K., & Reisinger, A., eds. *Climate Change 2007.* Geneva: Intergovernmental Panel on Climate Change, 2007.

Palfreman, J. "The Vaccine War." *Frontline,* April 27, 2010. www.pbs.org/wgbh/pages/frontline/vaccines/view.

Palin, S. State Comments on Proposed FWS Polar Bear Rule. Juneau, AK: State of Alaska, April 9, 2007. p. 11. www.adfg.alaska.gov/static/species/specialstatus/pdfs/polarbear_2007_soa_comments_4_9.pdf.

Palin, S. Mr. President: Boycott Copenhagen; Investigate Your Climate Change "Experts." Facebook.com, December 3, 2009. www.facebook.com/notes/sarah-palin/mr-president-boycott-copenhagen-investigate-your-climate-change-experts/188540473434.

Palin, S. Sarah Palin on the Politicization of the Copenhagen Climate Conference. *Washington Post,* December 9, 2009. www.washingtonpost.com/wp-dyn/content/article/2009/12/08/AR2009120803402.html.

Pappas, M. The Election Mandate: The Contract from America. FreedomWorks.org, November 3, 2010. www.freedomworks.org/blog/max/the-mandate-the-contract-from-america. [blog]

Park, M. Medical Journal Retracts Study Linking Autism to Vaccine. CNN.com, February 2, 2010. http://articles.cnn.com/2010-02-02/health/lancet.retraction.autism_1_andrew-wakefield-mmr-vaccine-and-autism-general-medical-council?_s=PM:HEALTH.

Park, R. L. Cellular Telephones and Cancer: How Should Science Respond? *Journal of the National Cancer Institute* 2001;93(3):166–167. http://jnci.oxfordjournals.org/content/93/3/166.full.

Pearce, F. Flooded Out. *New Scientist* #2189, June 5, 1999.

Pearcey Report. About. n.d. www.pearceyreport.com/about.php.

Pell, D. *If the Bomb Falls: A Recorded Guide to Survival.* Los Angeles: Tops Records, 1961. [record album]

Perl, P. Absolute Truth. *Washington Post,* May 13, 2001. www.washingtonpost.com/wp-dyn/content/article/2006/11/28/AR2006112800700.html.

Perricone. M. Some Words of Wisdom. *Fermi News* 2000;23(2). www.fnal.gov/pub/ferminews/ferminews00-01-28/p3.html.

Pew Research Center for People and the Press. A Deeper Partisan Divide Over Global Warming. May 8, 2008. http://people-press.org/report/417/a-deeper-partisan-divide-over-global-warming.

Phillips, D. *American Theocracy: The Peril and Politics of Radical Religion, Oil, and Borrowed Money in the 21st Century.* New York: Viking, 2006.

Pickover, C. *Archimedes to Hawking: Laws of Science and the Great Minds Behind Them.* New York: Oxford University Press, 2008.

Pierrehumbert, R. Interview by S. Otto, November 18, 2010.

Pilkington, E. 300 Million and Counting . . . US Reaches Population Milestone. *Guardian,* October 13, 2006. www.guardian.co.uk/world/2006/oct/13/usa.topstories3.

Pimm, S. Interview by S. Otto, September 7, 2010.

Pion, G., & Lipsey, M. Public Attitudes Toward Science and Technology: What Have the Surveys Told Us? *Public Opinion Quarterly* 1981;45(3):303–316.

Planck, M. *Scientific Autobiography and Other Papers.* New York: Philosophical Library, 1949.

Playboy. Bertrand Russell: *Playboy* Interview. *Playboy,* March 1963. www.playboy.com/articles/bertrand-russell-interview/index.html.

Politifact.com. Bear Study Funding Actually Undersold. Politifact.com, n.d. www.politifact.com/truth-o-meter/statements/2008/sep/26/john-mccain/bear-study-funding-actually-undersold.

Pollitt, K. Pomolotov Cocktail. *Nation,* June 10, 1996.

Population Council. Emergency Contraception's Mode of Action Clarified. *Population Briefs* 2005;11(2):3. www.popcouncil.org/pdfs/popbriefs/pbmay05.pdf.

Porter, J. Interview by S. Otto, August 31, 2009.

Priests for Life. Voter's Guide for Serious Catholics. PriestsforLife.org, 2004. www.priestsforlife
.org/elections/voterguide.htm.

Raine, N. V. *After Silence: Rape and My Journey Back*. New York: Crown, 1998.

Rand, A. *The Virtue of Selfishness*. New York: Signet, 1964.

Rasmussen, C. Billy Graham's Star Was Born at His 1949 Revival in Los Angeles. *Los Angeles Times*, September 2, 2007. http://articles.latimes.com/2007/sep/02/local/me-then2.

Raven, C. E. *John Ray, Naturalist: His Life and Works*. Cambridge, UK: Cambridge University Press, 1942.

Reagan, R. Address by Governor Ronald Reagan, Installation of President Robert Hill, Chico State College, May 20, 1967. Ronald Reagan Presidential Library, n.d. www.reagan.utexas.edu// archives/speeches/govspeech/05201967a.htm.

Reagan, R. Farewell Address to the Nation. Miller Center of Public Affairs, January 11, 1989. http://millercenter.org/scripps/archive/speeches/detail/3418.

RealClimate Group. The CRU Hack. RealClimate.org, November 20, 2009. www.realclimate.org/ index.php/archives/2009/11/the-cru-hack/#more-1853. [blog]

Reed, R. Growing Grassroots. *NewsHour with Jim Lehrer*. Washington, DC: MacNeil/Lehrer Productions, March 2, 2004. www.pbs.org/newshour/bb/media/jan-june04/grassroots_03-02. html. [panel discussion]

Review Panel. *Securing a Sustainable Future for Higher Education: An Independent Review of Higher Education Funding and Student Finance*. October 12, 2010. www.bis.gov.uk/assets/ biscore/corporate/docs/s/10-1208-securing-sustainable-higher-education-browne-report. pdf.

Revkin, A. Official Played Down Emissions' Links to Global Warming. *New York Times*, June 7, 2005. www.nytimes.com/2005/06/07/science/07cnd-climate.html.

Revkin, A. C. Climate Expert Says NASA Tried to Silence Him. *New York Times*, January 29, 2006. www.nytimes.com/2006/01/29/science/earth/29climate.html.

Richardson, V., ed. *Constructivist Teacher Education: Building a World of New Understandings*. New York: Routledge, 1997.

Richert, C. Inhofe Claims That E-Mails "Debunk" Science Behind Climate Change. Politifact.com, December 11, 2009. http://politifact.com/truth-o-meter/statements/2009/dec/11/james-inhofe/ inhofe-claims-cru-e-mails-debunk-science-behind-cl.

Robock, A. Interview by S. Otto, November 16, 2010.

Roosevelt, E. *It Seems to Me: Selected Letters of Eleanor Roosevelt*. Eds. L. C. Schlup & D. W. Whisenhunt. Lexington, KY: University Press of Kentucky, 2001.

Roosevelt, F. D. President Roosevelt's Letter to Vannevar Bush, November 17, 1944. www.nsf.gov/ od/lpansf50/vbush1945.htm#letter.

Rosen, R. A Physics Prof Drops a Bomb on the Faux Left. *Los Angeles Times*, May 23, 1996. p. A11. www.physics.nyu.edu/faculty/sokal/rosen.html.

Ross, A., ed. *Science Wars*. Durham, NC: Duke University Press, 1996. p. 152.

Ross, S. Scientist: The Story of a Word. *Annals of Science* 1962;18(2):65–85. www.scribd.com/ doc/42338381/Ross-1964-Scientist-the-Story-of-a-Word.

Roush, W. Putting a Price Tag on Nature's Bounty. *Science* 1997;276(5315):1029.

Rubin, R. How did Vioxx Debacle Happen? *USA Today*, October 11, 2004. www.usatoday.com/ news/health/2004-10-11-vioxx-main_x.htm.

Russell, B. *A History of Western Philosophy*. New York: Routledge, 2004.

Russell, C. *Covering Controversial Science: Improving Reporting on Science and Public Policy*. Working Paper #2006-4. Cambridge, MA: Joan Shorenstein Center on the Press, Politics and Public Policy, 2006.

Russell, C. Globe Kills Health/Science Section, Keeps Staff. *The Observatory*, CJR.org, March 4, 2009. www.cjr.org/the_observatory/globe_kills_healthscience_sect.php. [blog]

Saad, L. Barack Obama, Hillary Clinton Are 2010's Most Admired. Gallup.com, December 27, 2010. www.gallup.com/poll/145394/barack-obama-hillary-clinton-2010-admired.aspx.

Sagan, C. *The Demon-Haunted World: Science as a Candle in the Dark.* New York: Random House, 1995.

St. Germain, C. *The Doctor and Student.* Ed. W. Muchall. Cincinnati: R. Clarke, 1874.

Sanchez, I. Warming Study Draws Fire: Harvard Scientists Accused of Politicizing Research. *Harvard Crimson,* September 12, 2003. www.thecrimson.com/article/2003/9/12/warming-study-draws-fire-a-study.

Sandage, A. Interview by S. Otto, August 3, 2004.

Sanders, D. Interview by S. Otto, December 14, 2010.

Schick, T., & Vaughn, L. *How to Think About Weird Things: Critical Thinking for a New Age.* Mountain View, CA: Mayfield, 1995.

Schjoedt, U., et al. The Power of Charisma—Perceived Charisma Inhibits the Frontal Executive Network of Believers in Intercessory Prayer. *Social Cognitive and Affective Neuroscience* 2011;6(1):119–127.

Schneider, L. *Biology and Revolution in Twentieth-Century China.* Lanham, MD: Rowman & Littlefield, 2005. p. 179.

Schneider, S. *Science as a Contact Sport.* Washington, DC: National Geographic Society, 2009.

Schwartz, S., et al. Excerpts from Atomic Audit. *Bulletin of the Atomic Scientists* 1998;54(5):36–43.

ScienceDaily. Hazardous E-Waste Surging in Developing Countries. *ScienceDaily,* February 23, 2010. www.sciencedaily.com/releases/2010/02/100222081911.htm.

ScienceDebate.org. The Top 14 Science Questions Facing America. ScienceDebate.org, n.d. www.sciencedebate.org/questions.html.

Scott, E. Interview by S. Otto, August 25, 2009.

Scott, J. Postmodern Gravity Deconstructed, Slyly. *New York Times,* May 18, 1996. www.nytimes.com/1996/05/18/nyregion/postmodern-gravity-deconstructed-slyly.html.

Seeyle, K. Moral Values Cited as a Defining Issue of the Election. *New York Times,* November 4, 2004. www.nytimes.com/2004/11/04/politics/campaign/04poll.html.

Sensenbrenner, F. J. Sensenbrenner Urges IPCC to Exclude Climategate Scientists. Letters from the Select Committee on Energy Independence and Global Warming, December 8, 2009. http://republicans.globalwarming.house.gov/Press/PRArticle.aspx?NewsID=2749.

Shapley, H. *Through Rugged Ways to the Stars.* New York: Scribner, 1969.

Shaw, P. *The Philosophical Works of Francis Bacon, Baron of Verulam, Viscount St. Albans, and Lord High-Chancellor of England: Methodized, and Made English, from the Originals.* Volume II. London: J. J. and P. Knapton, et al., 1733.

Shermer, M. Stephen Jay Gould as Historian of Science and Scientific Historian, Popular Scientist and Scientific Popularizer. *Social Studies of Science* 2002;32(4):489–525.

Shimkus, J. Rep. John Shimkus: God Decides When the "Earth Will End." YouTube.com, March 25, 2009. www.youtube.com/watch?feature=player_embedded&v=_7h08RDYA5E.

Shnayerson, M. The Edge of Extinction. *Vanity Fair,* May 2008. www.vanityfair.com/politics/features/2008/05/polarbear200805.

Siegel, K. Conservation Group Petitions the United States Government to List the Polar Bear as a Threatened Species Under the Endangered Species Act. Center for Biological Diversity, February 16, 2005. www.biologicaldiversity.org/news/press_releases/polarbear2-16-05.html. [news release]

Smith, A. *An Inquiry into the Nature and Causes of the Wealth of Nations.* Ed. J. Manis. Electronic Classics Series. Hazleton, PA: Pennsylvania State University, n.d. www2.hn.psu.edu/faculty/jmanis/adam-smith/Wealth-Nations.pdf.

Smith, P., et al. Children Who Have Received No Vaccines: Who Are They and Where Do They Live? *Pediatrics* 2004;114(1):187–195. http://pediatrics.aappublications.org/cgi/content/full/114/1/187.

Snow, C. P. *The Two Cultures.* Cambridge, UK: Cambridge University Press, 1960.

Snyder, J., & Chipman, K. Global Warming Skeptics Ascend in Congress. *Businessweek,* November 24, 2010. www.businessweek.com/magazine/content/10_49/b4206033143446.htm.

Sokal, A. Transgressing the Boundaries: Towards a Transformative Hermeneutics of Quantum Gravity. *Social Text* 1996;(46/47):217–252. www.physics.nyu.edu/sokal/transgress_v2/transgress_v2_singlefile.html.

Sokal, A. A Physicist Experiments with Cultural Studies. *Lingua Franca,* May/June 1996. http://linguafranca.mirror.theinfo.org/9605/sokal.html.

Soon, W., et al. Reconstructing Climate and Environmental Changes of the Past 1000 Years: A Reappraisal. *Energy and Environment* 2003;14(2–3):233–296. www.marshall.org/pdf/materials/132.pdf.

Soon, W., & Balliunas, S. *Lessons and Limits of Climate History: Was the 20th Century Climate Unusual?* Washington, DC: George C. Marshall Institute, 2003. www.marshall.org/pdf/materials/136.pdf.

South Dakota Legislative Assembly, 85th Session. *House Concurrent Resolution No. 1009.* March 2, 2009. http://legis.state.sd.us/sessions/2010/Bill.aspx?File=HCR1009P.htm.

Spiegel, A. N., et al. Museum Visitors' Understanding of Evolution. *Museums and Social Issues* 2006;1(1):69–86. www-personal.umich.edu/~evansem/SpiegelEvansGramDiamond.pdf.

State of Minnesota, House of Representatives, 87th Legislative Session. House File 1010, May 18, 2011.

State of Virginia. Virginia Fraud Against Taxpayers Act: Chapter 842, Article 19.1. April 17, 2002. www.taf.org/virginiafca.htm.

Stirling, I. Polar Bears and Seals in the Eastern Beaufort Sea and Amundsen Gulf: A Synthesis of Population Trends and Ecological Relationships over Three Decades. *Arctic* 2002;55(Suppl 1):59–76. http://arctic.synergiesprairies.ca/arctic/index.php/arctic/article/download/735/761.

Stirling, I., et al. Polar Bear Population Status in the Northern Beaufort Sea. Reston, VA: US Geological Survey, 2007. www.usgs.gov/newsroom/special/polar_bears.

Straussmann, M. Countdown to 300 Million. *CBS News,* October 12, 2006. www.cbsnews.com/video/watch/?id=2086917n&tag=mncol;lst;7. [audiovisual footage]

Stuart, C. Oh, Mann: Cuccinelli Targets UVA papers in Climategate Salvo. *The Hook,* April 29, 2010. www.readthehook.com/67811/oh-mann-cuccinelli-targets-uva-papers-climategate-salvo.

Sutton, M. *Aimee Semple McPherson and the Resurrection of Christian America.* Cambridge, MA: Harvard University Press, 2009.

Tackett, M. Laying Claim to the Nation. *Chicago Tribune,* November 7, 2004. www.chicagotribune.com/news/opinion/chi-0411070166nov07,0,595361.story.

Taylor, M. Last Stand of Our Wild Polar Bears. *Toronto Star,* May 1, 2006. http://ff.org/centers/csspp/library/co2weekly/20060505/20060505_17.html. [opinion]

Tennessee State Legislature. *Tennessee Anti-Evolution Statutes. Public Acts of the State of Tennessee Passed by the 64th General Assembly 1925. Chapter No. 27. House Bill No. 185.* March 21, 1925. http://law2.umkc.edu/faculty/projects/ftrials/scopes/tennstat.htm.

The Watergate Story: Key Players: Charles Colson. WashingtonPost.com, n.d. www.washingtonpost.com/wp-srv/onpolitics/watergate/charles.html.

Thompson, J. Polar Bear Die-Off Unlikely: GN Official. *Nunatsiaq News,* September 14, 2007. www.nunatsiaqonline.ca/archives/2007/709/70914/news/nunavut/70914_498.html.

Tocqueville, A. *Democracy in America.* Ed. F. Bowen. Trans. H. Reeve. Cambridge, MA: Sever and Francis, 1864.

Tremblay, F. The Eeeevil Evolutionists and the Bumbling Christians. Goosing the Antithesis, August 2, 2008. http://goosetheantithesis.blogspot.com/2008/08/eeeevil-evolutionists-and-bumbling.html. [blog]

Tucholsky, K. "French Joke," in *Learn to Laugh without Crying.* Berlin: Ernst Rowohlt Verlag, 1932. p. 148.

Union of Concerned Scientists. Agencies Control Scientists' Contacts with the Media. n.d. www.ucsusa.org/scientific_integrity/abuses_of_science/agencies-control-scientists.html.

Union of Concerned Scientists. Scientific Integrity in Policy Making (March 2004). n.d. www .ucsusa.org/scientific_integrity/abuses_of_science/reports-scientific-integrity.html.

United Nations. *Convention on the Prohibition of Military or Any Hostile Use of Environmental Modification Techniques.* December 10, 1976. www.icrc.org/ihl.nsf/FULL/460?OpenDocument.

US Department of Agriculture Economic Research Service. Farming's Role in the Rural Economy. *Agricultural Outlook,* June-July, 2000. pp. 19–22.

US Department of Commerce Bureau of Economic Analysis. Table 1.1.5: Gross Domestic Product. 2011. www.bea.gov/national/txt/dpga.txt.

US Department of Defense. *Quadrennial Defense Review Report,* February 2010. www.defense .gov/qdr/images/QDR_as_of_12Feb10_1000.pdf.

US Environmental Protection Agency. Cap and Trade: Acid Rain Program Results. 2003. www .epa.gov/capandtrade/documents/ctresults.pdf.

US Environmental Protection Agency. Full Cost Accounting. September 30, 2008. www.epa.gov/ osw/conserve/tools/fca/index.htm.

US Food and Drug Administration Center for Drug Evaluation and Research. Transcript of Nonprescription Drugs Advisory Committee in Joint Session with the Advisory Committee for Reproductive Health Drugs. December 16, 2003. www.fda.gov/ohrms/dockets/ac/03/ transcripts/4015T1.DOC.

US Food and Drug Administration. Thimerosal in Vaccines. March 31, 2010. www.fda.gov/ biologicsbloodvaccines/safetyavailability/vaccinesafety/ucm096228.htm.

US House of Representatives Committee on Oversight and Government Reform. Committee Holds Hearings on Political Influence on Government Climate Change Scientists. n.d. http://democrats .oversight.house.gov/index.php?option=com_content&task=view&id=2607&Itemid=2.

US House of Representatives, 110th Congress. *On Thin Ice: The Future of the Polar Bear.* Hearing Before the Select Committee on Energy Independence and Global Warming, House of Representatives, January 17, 2008. Washington, DC: US Government Printing Office, 2010. http://globalwarming.house.gov/files/HRG/FullTranscripts/110-22_2008-01-17.pdf.

US Senate Committee on Environment and Public Works Minority Staff. *"Consensus" Exposed: The CRU Controversy.* Washington, DC: US Senate, 2010.

Valvo, J., & Oberg, C. Exposing the Special Interests Behind Waxman-Markey. Arlington, VA: Americans for Prosperity, September 2009. http://americansforprosperity.org/files/Policy_ Paper_0909_0.pdf.

Van Dongen, J. On Einstein's Opponents, and Other Crackpots. *Studies in History and Philosophy of Science Part B: Studies in History and Philosophy of Modern Physics* 2010;41(1):78–80.

Vaughan, R. The Life of John Milton. In *Paradise Lost,* by J. Milton. London: Cassell, 1894.

Vonnegut, K. American Notes: Vonnegut's Gospel. *Time,* June 29, 1970.

Wahl-Jorgensen, K., ed. *The Handbook of Journalism Studies.* New York: Taylor and Francis, 2009.

WAMU. Washington Goes to the Moon. WAMU.org, May 24, 2001. http://wamu.org/programs/ special/01/washington_goes_to_the_moon.php. [radio broadcast]

Wang, T. China's Future Leaders. Forbes.com, May 17, 2009. www.forbes.com/2009/05/17/china-leaders-stars-leadership-rising-stars.html.

Washington, G. First Annual Message to Congress, Washington, DC, January 8, 1790. Miller Center of Public Affairs, University of Virginia, n.d. http://millercenter.org/scripps/archive/ speeches/detail/3448.

Watts, A. Breaking News Story: CRU Has Apparently Been Hacked—Hundreds of Files Released. Watts Up with That? November 19, 2009. http://wattsupwiththat.com/2009/11/19/breaking-news-story-hadley-cru-has-apparently-been-hacked-hundreds-of-files-released. [blog]

Webber, M. E. Webber: Don't Dumb Down Texas. Statesman.com, September 15, 2009. www.statesman.com/opinion/content/editorial/stories/2009/09/16/0916webber_edit.html. [opinion]

Webber, M. Interview by S. Otto, July 30, 2010.

Weidenbaum, D., et al. *The Annual Report of the Council of Economic Advisers.* Washington, DC: Council of Economic Advisers, February 6, 1982. p. 45. http://fraser.stlouisfed.org/ publications/erp/issue/1385/ . . . /ERP_ARCEA_1982.pdf.

Weiss, D., et al. Dirty Money: Oil Companies and Special Interests Spend Millions to Oppose Climate Legislation. Center for American Progress Action Fund, September 27, 2010. www .americanprogressaction.org/issues/2010/09/dirty_money.html.

Westphal, S. P. Interview by S. Otto, April 14, 2010.

White, M. *Isaac Newton: The Last Sorcerer.* Reading, MA: Addison-Wesley, 1997.

Whitfield, E. Biography. n.d. http://whitfield.house.gov/about/bio.shtml.

Whitfield, E. Energy. n.d. http://whitfield.house.gov/issues/energy.shtml.

Wiesner, J. B. *Vannevar Bush: 1890–1974.* Washington, DC: National Academy of Sciences, 1979. www.nap.edu/html/biomems/vbush.pdf.

Wiley, D., & Wilson, K. *Just Say Don't Know: Sexuality Education in Texas Public Schools.* Texas Freedom Network Education Fund, 2009. www.tfn.org/site/DocServer/SexEdRort09_web .pdf?docID=981.

Willer, R. Interview by S. Otto, December 16, 2010.

Williams, B. An American Milestone. *NBC Nightly News.* NBC.com, October 16, 2006. [audiovisual footage]

Wilson, B. The Cultural Contexts of Science and Mathematics Education: Preparation of a Bibliographic Guide. *Studies in Science Education* 1981;8:27–44.

Wilson, E. O. *Consilience: The Unity of Knowledge.* New York: Knopf, 1998.

Wilson, R. R. *Starting Fermilab.* Batavia, IL: Fermi National Accelerator Laboratory, 1992. http:// history.fnal.gov/GoldenBooks/gb_wilson2.html.

Winkler, A. M. *Life Under a Cloud: American Anxiety about the Atom.* Urbana, IL: University of Illinois Press, 1999. p. 117.

Working Group II to the Fourth Assessment Report of the Intergovernmental Panel on Climate Change. Chapter 10.6.2: The Himalayan Glaciers. In *Climate Change 2007: Impacts, Adaptation and Vulnerability.* New York: Cambridge University Press, 2007. www.ipcc.ch/ publications_and_data/ar4/wg2/en/ch10s10-6-2.html.

Working Group II to the Fourth Assessment Report of the Intergovernmental Panel on Climate Change. Introduction to the Working Group II Fourth Assessment Report. In *Climate Change 2007: Impacts, Adaptation and Vulnerability.* New York: Cambridge University Press, 2007. www.ipcc.ch/pdf/assessment-report/ar4/wg2/ar4-wg2-intro.pdf.

World Bank. Countries and Economies. n.d. http://data.worldbank.org/country.

Xu, J., et al. Deaths: Final Data for 2007. *National Vital Statistics Reports* 58(19), May 20, 2010. www.cdc.gov/NCHS/data/nvsr/nvsr58/nvsr58_19.pdf.

Yong, E. Arsenic Bacteria—A Post-Mortem, a Review, and Some Navel-Gazing. Not Exactly Rocket Science, Discovermagazine.com, December 10th, 2010. http://blogs.discovermagazine .com/notrocketscience/2010/12/10/arsenic-bacteria-a-post-mortem-a-review-and-some- navel-gazing. [blog]

ACKNOWLEDGMENTS

Thanks to Rebecca and to Jake, without whose patience with my penchant for going down side roads this book would still be but a dream, occasionally spoken of, like a past and once great friend. You instead let him move in and take over the house like a great again but now drunken friend, demanding and full of passion, painfully incisive, constantly dragging me away and into trouble, keeping me up late, corrupting the pets and children, ignoring the housekeeping, and generally being a very poor guest. Thank you for tolerating him; he means well.

And a special thanks to Jacque Fletcher, to Joy Tutela, and to Colin Dickerman, Gena Smith, Nancy Elgin, and Sonya Maynard for your exacting standards, your high expectations, your open doors, your passion and dedication, and your encouraging support—and to the Loft Literary Center, which I have had the honor of serving and through which I met one great agent.

Third, thank you to all the bighearted and brilliant scientists, engineers, policy makers, economists, administrators, artists, theologians, ethicists, philosophers, and writers who allowed me to draw on their profound insights for this book, and to the tens of thousands more who signed on to support the idea of science debates because they believe they are critical to the future of the nation and the planet. To the extent that the book resonates, it is due to their brilliance and insight; to the extent that it errs, its defects are mine. In either case, their caring, concern, and passion for science, art, democracy, and the United States are deeply inspirational, and I wish every reader could be as graced with their acquaintance and friendship as I have been.

Finally, it must be said that, like science itself, this book is political. Some people will doubtless disagree with some of its statements and positions, perhaps even some of those people who so graciously allowed me to interview them. But despite its political nature, this book is not

365

intended to be partisan. My family founded the Minnesota Republican Party and my wife is an elected Democratic constitutional officer in that state. Like science itself, I find value in both conservative and progressive perspectives, and also disagreements. My criticisms of today's GOP for its antiscience positions, on the one hand, and of the left's overembrace of postmodernism on the other, for example, are rooted in the fundamental American value of governance based on knowledge and facts versus authoritarianism and opinion. This is an important distinction that has been lost for many Americans. Facts may be political, but they supersede partisanship.

David Hume defined freedom as the power to choose, and by expanding that power it is knowledge that makes us free. The mission of Science Debate, the organization and movement that sent me down this road, is to elevate science in the public dialogue and restore it to its rightful place by bringing candidates for elected office together with media and the public in a safe and nonpartisan environment to debate the top science-related policy challenges. Listed below are some of the incredibly generous, brilliant, and talented individuals from across the political spectrum who have helped advance that critical effort in ways large and small, and/or who have offered me kindness, support, and assistance in making this book a reality. None of the positions I take should be construed as being those of Science Debate or of any of these individuals, but their generosity of spirit and their ongoing contributions to the greater advancement of humanity should be often and loudly noted.

Natty Adams, Gillian Adler, Peter Agre, Rick Anthes, Paula Apsell, Derek Araujo, Chuck Atkins, Randy Atkins, Norm Augustine, Jennifer Ayers, David Baltimore, Craig Barrett, Bill Bates, Erik Beeler, Rosina Bierbaum, Larry Bock, Ben Bova, Bob Breck, Douglas Bremner, David Brin, Deborah Byrd, Art Caplan, Arne Carlson, Darlene Cavalier, Bill Chameides, Matthew Chapman, Peg Chemberlin, Steven Chu, Pat Churchland, Ralph Cicerone, Rita Colwell, George Crabtree, Austin Dacey, Ronald DePinho, Keith Devlin, Calvin DeWitt, Ann Druyan,

Vern Ehlers, Harold Evans, Dick Feely, Kevin Finneran, Andrew Fire, Ira Flatow, Al Franken, Gwen Freed, Peter Frumhoff, Richard Gallagher, Jim Gentile, Jack Gibbons, Newt Gingrich, Linda Glenn, Wolfgang Goede, Bart Gordon, Kurt Gottfried, Francesca Grifo, David Guston, Jocey Hale, Michael Halpern, Philip Hammer, Bruce Hendry, Sharon Hendry, Kathryn Hinsch, Roald Hoffman, John Holdren, Rush Holt, Doug Holtz-Eakin, Al Hurd, Shirley Ann Jackson, Thomas Campbell Jackson, Mariela Jaskelioff, James Jensen, Eric Jolly, Dean Kamen, Steve Kelley, Don Kennedy, Alex King, Sheril Kirshenbaum, Barbara Kline Pope, Sara Kloek, Kevin Knobloch, Kei Koizumi, Lawrence Krauss, Paul Kurtz, Eric Lander, Neal Lane, Phoebe Leboy, Leon Lederman, Russ Lefevre, Alan Leshner, Simon Levin, Jane Lubchenco, Michael Mann, Elizabeth Marincola, Thom Mason, John Mather, Bob May, Angie McAllister, Jim McCarthy, Robert McKee, Marcia McNutt, Ann Merchant, Ken Miller, Chris Mooney, Jonathan Moreno, Jan Morrison, Elizabeth Muhlenfeld, P.Z. Myers, Hajo Neubert, Matt Nisbet, Peter Norvig, Bill Nye, Rebecca Otto, Kevin Padian, Bob Park, Ray Pierrehumbert, Stuart Pimm, Phil Plait, Steve Pinker, John Podesta, Alan Polsky, Gabe Polsky, John Porter, Stacie Propst, John Rennie, Alan Robock, Brian Rosenberg, Eric Rothschild, Allan Sandage, David Sanders, Genie Scott, John Siceloff, Kassie Siegel, Maxine Singer, Carl Johan Sundberg, Joel Surnow, Jill Tarter, Jim Tate, Al Teich, Meg Ury, Harold Varmus, Chuck Vest, Ethan Vishniac, Cynthia Wainwright, Michael Webber, Ty West, Scott Westphal, Frank Wilczek, Leah Wilkes, Robb Willer, Deborah Wince-Smith, Dennis Wint, Mary Woolley, Susan Wood, and so many more who e-mailed me ideas about what not to forget to include in this book or lent their endorsement and support.

Thank you all.
July 11, 2011

INDEX